Emerging Treatment Technologies for Waste Management

Izharul Haq · Ajay S. Kalamdhad
Editors

Emerging Treatment Technologies for Waste Management

Springer

Editors
Izharul Haq
Department of Civil Engineering
Indian Institute of Technology Guwahati
Guwahati, Assam, India

Ajay S. Kalamdhad
Department of Civil Engineering
Indian Institute of Technology Guwahati
Guwahati, Assam, India

ISBN 978-981-16-2014-0 ISBN 978-981-16-2015-7 (eBook)
https://doi.org/10.1007/978-981-16-2015-7

This Springer imprint is published by the registered company Springer Nature Singapore Pte Ltd.
The registered company address is: 152 Beach Road, #21-01/04 Gateway East, Singapore 189721, Singapore

Preface

Over the years, rapid industrialization is a significant challenge for the sustainable environment and human health. The toxic pollutants released from various industrial and anthropogenic source and their effects on the living organism are an urgent need to be focused on, and the mitigation and removal of these pollutants are necessary. This book will provide detailed knowledge of the source, characteristics, and toxic effects of various emerging pollutants and their removal through advanced treatment technologies. This book will also present various new concepts and strategies of liquid and solid waste management, coupled with treatment technologies and recycling of waste products.

Hence, this book mainly focuses on various types of advanced treatment approaches for the degradation and detoxification of toxic pollutants from wastewater and solid waste and sludge for the safety of the environment and living organisms. The chapters of the first section of the book will provide detailed information about the occurrence, source, characteristics, toxicity, and conventional and advanced treatment process of industrial wastewaters. The second section's chapters of the book will provide detailed knowledge of municipal and industrial solid wastes disposal, management, and its application in waste to energy through different processes of composting, vermicomposting, anaerobic digestion, biochar, and hydrochar methods.

This book is intended to be of substantial help to the students and researchers, industrialists, and academicians working in the field of environmental engineering and sciences, environmental microbiology and biotechnology, and toxicology. This can also be of use to many universities/institutes worldwide which have developed new prospectuses and offer specialized courses on environmental and sustainable development. The readers of the book can also get valuable information related to various environmental problems and their solutions.

Guwahati, India

Izharul Haq
Ajay S. Kalamdhad

Acknowledgments

Writing my first book has been both exigent and exciting, and I could not have done it alone. This book is the artifact of several people's support, guidance, and talent. I would like to thank my family, fellow editor, and authors who helped me along in my writing. Last but not least, I'd want to express my gratitude to my mother, a strong and compassionate spirit who taught me to trust in God, believe in hard effort, and that a lot can be accomplished with a little. My father, for always encouraging and morally supporting me throughout my life.

Contents

Editors and Contributors

About the Editors

Izharul Haq is currently working as Post-Doctoral Fellow at the Department of Civil Engineering, Indian Institute of Technology Guwahati, India. He has completed his post-graduation in environmental microbiology from BBA (Central) University, Lucknow, and Ph.D. in Microbiology from CSIR-Indian Institute of Toxicology Research, Lucknow, India. He has 3 years of research and teaching experience. He has been honored with the prestigious award "Young Scientist Award 2018" to recognize his scientific research. He has exposure in the field of bioremediation and detoxification of emerging pollutants from industrial wastewater using microbial approaches. He has published many research papers in reputed national and international journals with high impact factors. He also has published many book chapters and magazine articles in reputed international books. Currently, he is editing the four books published by Springer Nature and Elsevier. He has presented several papers relevant to his research areas in national and international conferences. He has supervised four M.Tech students in the Environmental Engineering field. He is a life member of the Association of Microbiologists of India (AMI), Biotech Research Society (BRSI), India, Indian Science Congress Association (ISCA), Kolkata, and Vigyan Bharti, India.

Ajay S. Kalamdhad is working as a Professor at the Department of Civil Engineering, Indian Institute of Technology Guwahati, Assam, India. He obtained his bachelors, masters, and Ph.D. in civil and environmental engineering from GEC Jabalpur, VNIT Nagpur, and IIT Roorkee, respectively. Prior to joining IIT Guwahati, he was an Assistant Professor at VNIT Nagpur and worked on various projects at RRL, Bhopal (now AMPRI, Bhopal), and NEERI, Nagpur.

His research team is working on solid waste management by microbial and vermicomposting, aerobic and anaerobic digestion of kitchen waste and industrial sludge for biogas production, and composting of water hyacinth using different composting technologies. He also serves as an editorial board member/reviewer for several national/international journals of repute. Currently, he is supervising 18 Ph.D. and 11 M.Tech students, and more than 10 Ph.D. and 35 M.Tech dissertations have been completed under his supervision. In addition, he has handled nine major R&D projects funded by various agencies. He has published more than 150 research papers in national and international journals of repute. He also has published six book chapters and five international books. He is a life member of Indian Public Health Engineers, Asia-Pacific Chemical, Biological & Environmental Engineering Society, National Solid Waste Association of India, International Solid Waste Association, and Indian Water Works Association.

Contributors

Oluwatosin Joseph Aladekoyi Environmental Applied Science and Management, Ryerson University, Toronto, ON, Canada

Anfal Arshi Defence Institute of Bio-Energy Research, DRDO, Haldwani, Uttarakhand, India

Tanay Bhagwat Department of Chemical Engineering, Northeastern University, Boston, MA, USA

Anulipt Chandan Discipline of Information System, School of Management, Curtin University, Perth, WA, Australia

Ram Chandra Department of Environmental Microbiology, School of Environmental Sciences, Babasaheb Bhimrao Ambedkar (A Central) University, Lucknow, India

Mudita Chaturvedi School of Environmental Sciences, Jawaharlal Nehru University, New Delhi, India

Parul Chugh Amity Institute of Biotechnology, Amity University, Noida, Uttar Pradesh, India

Devki Department of Microbiology, JECRC University, Jaipur, Rajasthan, India

Brajesh Dubey Department of Civil Engineering, Indian Institute of Technology Kharagpur, Kharagpur, West Bengal, India

Sougata Ghosh Department of Microbiology, School of Science, RK University, Rajkot, Gujarat, India
Department of Chemical Engineering, Northeastern University, Boston, MA, USA

Izharul Haq Department of Civil Engineering, Indian Institute of Technology Guwahati, Guwahati, Assam, India

Ajay S. Kalamdhad Department of Civil Engineering, Indian Institute of Technology Guwahati, Guwahati, Assam, India

Adarsh Kumar Department of Environmental Microbiology, School of Environmental Sciences, Babasaheb Bhimrao Ambedkar (A Central) University, Lucknow, India

Tithi Mehrotra Amity Institute of Biotechnology, Amity University, Noida, Uttar Pradesh, India

Ambuj Mishra School of Environmental Sciences, Jawaharlal Nehru University, New Delhi, India

Olubusola Ayoola Odeniyi Department of Microbiology, University of Ibadan, Ibadan, Nigeria

Nagarajan Padmini Department of Microbiology, Science Campus, Alagappa University, Karaikudi, Tamil Nadu, India

Pandian Prabakaran Department of Biomedical Sciences, Department of Microbiology, Science Campus, Alagappa University, Karaikudi, Tamil Nadu, India

Nagasundaram Rashiya Department of Microbiology, Science Campus, Alagappa University, Karaikudi, Tamil Nadu, India

Sundaram Ravikumar Department of Biomedical Sciences, Department of Microbiology, Science Campus, Alagappa University, Karaikudi, Tamil Nadu, India

V. R. Sankar Cheela Department of Civil Engineering, Indian Institute of Technology Kharagpur, Kharagpur, West Bengal, India
Sustainable Engineering Group, Curtin University, Perth, WA, Australia

Gaurav Saxena EMBL - Environmental Microbiology and Biotechnology Laboratory, EERG - Ecotoxicology and Environmental Remediation Group, School of Biotechnology, Shoolini University of Biotechnology and Management Sciences, Solan, Himachal Pradesh, India

Gopal Selvakumar Department of Microbiology, Science Campus, Alagappa University, Karaikudi, Tamil Nadu, India

Geetansh Sharma EMBL - Environmental Microbiology and Biotechnology Laboratory, EERG - Ecotoxicology and Environmental Remediation Group, School of Biotechnology, Shoolini University of Biotechnology and Management Sciences, Solan, Himachal Pradesh, India

Hari Bhakta Sharma Department of Civil Engineering, Indian Institute of Technology – Kharagpur, Kharagpur, West Bengal, India

Krati Sharma School of Environmental Sciences, Jawaharlal Nehru University, New Delhi, India

Pooja Sharma Department of Environmental Microbiology, School for Environmental Sciences, Babasaheb Bhimrao Ambedkar Central University, Lucknow, Uttar Pradesh, India

Ajay Kumar Singh Department of Environmental Microbiology, School of Environmental Sciences, Babasaheb Bhimrao Ambedkar (A Central) University, Lucknow, India

Amit Kumar Singh School of Environmental Sciences, Jawaharlal Nehru University, New Delhi, India

Anshu Singh Defence Institute of Bio-Energy Research, DRDO, Haldwani, Uttarakhand, India

Rachana Singh Amity Institute of Biotechnology, Amity University, Noida, Uttar Pradesh, India

Surendra Pratap Singh Plant Molecular Biology Laboratory, Department of Botany, Dayanand Anglo-Vedic (PG) College, Chhatrapati Shahu Ji Maharaj University, Kanpur, India

Surbhi Sinha Amity Institute of Biotechnology, Amity University, Noida, Uttar Pradesh, India

John Onolame Unuofin Department of Microbial, Biochemical and Food Biotechnology, University of the Free State, Bloemfontein, South Africa

Saikrishna Venna Department of Civil Engineering, National Institute of Technology, Warangal, Andhra Pradesh, India
Present Address: Department of Civil Engineering, Vardhaman College of Engineering, Hyderabad, Telangana, India

Pradeepa Virumandi Department of Zoology, Seethalakshmi Achi College for Women, Sivaganga District, Tamil Nadu, India

Thomas J. Webster Department of Chemical Engineering, Northeastern University, Boston, MA, USA

Recent Advances in Physicochemical and Biological Approaches for Degradation and Detoxification of Industrial Wastewater

1

Adarsh Kumar, Ajay Kumar Singh, and Ram Chandra

1.1 Introduction

During the past few decades, water resources are getting scarcer due to exponential growth in population, urbanization, agriculture, and industrialization (Kumar et al. 2020b). The dwindling supplies of natural resources of water have made this a serious constraint for industrial growth and a reasonable standard of urban living. This has made water treatment more costly and is becoming a massive burden for industries to comply with the discharge quality requirement itself. Owing to the release of poor quality wastewater, the degradation of water resources poses a significant hazard to human beings and aquatic life as they depend on water for sustenance (Singh and Chandra 2019; Singh et al. 2020a, b). Most water-intensive industries are the major contributors to cause environmental pollutions. For example, the traditional paper industry needs 300 m^3 of water per ton of product and often generates toxic huge amount of wastewater (Singh and Chandra 2019). Similarly, the mining industry requires 40 m^3 of water per ton of ore, and the oil industry needs 10–300 m^3 of water per ton of output (Chaphekar 2013). The synthetic fertilizer industry also requires a considerable contribution, needing up to 270 m^3 of water per ton of product, and the iron and steel industry needs 20–60 m^3 of water per ton of product (Chaphekar 2013). However, traditional methods for one-liter alcohol production process to generate 15 L wastewater in the distillery industry (Pant and Adholeya 2007). Furthermore, the food factories usually produce 0.6–20 m^3 of wastewater/tonne of products such as fruit juice or bread/butter/milk. Thus, the list of waste-generating industries, i.e. mines, pulp paper mills, tanneries, refineries, sugar production/distillery, chemicals, textile, tannery, and pharmaceuticals. A major concern for industrial wastewater contains several physicochemical

A. Kumar · A. K. Singh · R. Chandra (✉)
Department of Environmental Microbiology, School of Environmental Sciences, Babasaheb Bhimrao Ambedkar (A Central) University, Lucknow, India

I. Haq, A. S. Kalamdhad (eds.), *Emerging Treatment Technologies for Waste Management*, https://doi.org/10.1007/978-981-16-2015-7_1

parameters beyond the permissible limits, i.e. pH, color, TDS, TSS, BOD, COD, and toxic residual pollutants along with various heavy metals directly discharged into the aquatic and terrestrial environment due to causes of toxicity on flora and fauna (Jimenez 2009a; Hubbe et al. 2016; Singh and Chandra 2019; Kumar et al. 2020b). In addition to decreasing the production of wastewater, the device improvements need to be changed to minimize the accumulation of refractory chemicals in the wastewater to make it more treatment-friendly.

In the treatment of industrial wastewater several physicochemical and biological methods are available to remove organic and inorganic pollutants as well as contaminations and pathogens (Jimenez 2009b). Following physical wastewater treatment (primary treatment) also involves significant quantities of dissolved and colloidal content that needs to be separated before discharge (Samer 2015a; Hubbe et al. 2016). Several physicochemical wastewater treatment methods have been employed, i.e. flotation, screening, comminution, flow equalization, sedimentation, adsorption with activated carbon and ozonation (Samer 2015b; Sghaier et al. 2019). Besides, biological wastewater treatment (secondary treatment) process, i.e. oxidation ponds, aeration lagoons, anaerobic digestions, anaerobic lagoons, bioreactor, activated sludge and biological nutrient removals as well as phytoremediation (constructed wetlands, rhizofiltration, rhizodegradation, phytodegradation, phytoaccumulation, phytotransformation, and hyperaccumulators) (Zhang et al. 2010; Samer 2015a). Biological treatment is referred to as the treatment of wastewater after the removal of suspended solids by microorganisms, i.e. bacteria, fungi, and algae under aerobic or anaerobic conditions during which organic matter is oxidized in wastewater or absorbed into cells that can be extracted by removal or sedimentation (Samer 2015b; Sghaier et al. 2019).

The chapter discusses the methods of characterization and control of industrial wastewater, with a focus on physicochemical and biological treatment processes. It aims to offer a summary of the forms of care succinctly and consistently. To provide a simple and coherent chapter that can be used as a brief reference for environmental engineers and scholars, and to be successfully implemented in higher education teaching for undergraduate and graduate students, as well as in extension and outreach, the key purpose of this work was to summarize the work of eminent scientists in this area.

1.2 Sources and Characteristics of Industrial Wastewater

There are various types of industrial wastewater based on the different industries and the contaminants; each sector produces its particular combination of pollutants and their production method and chemicals used for product preparations. Water ecology has played an important role in all living organisms, but because of human activities, they have to destroy the world's natural habitats as per their need. Man-made industries have directly dumped huge quantities of toxic polluted wastewater into the marine and terrestrial habitats, which are the major source of industrial water pollution. This has contributed to serious aquatic environmental issues that have a negative impact on the lives of river animals and humans. Therefore, higher amounts of wastewater are released by developing countries than by developed countries. Developing countries emit higher levels of wastewater than developed countries. Hence in the twenty-first century, this fact predicts that environmental pollution has migrated from developing countries to developed countries due to industries.

1.2.1 Pulp Paper Industry Wastewater

Environmental pollution causes due to the partially treated and untreated wastewater generated from paper industry are released into the aquatic and terrestrial environment (Ugurlu et al. 2007). The annual report Central Pulp Paper Research Institute 2016 defines up to 850 units of the paper industry in India and releases wastewater from organic and inorganic contaminants into the environment. In 40–45% of pulp processing, wastewater produces 55–60% lignocellulose and approx. In processing, one ton of white paper production was used to freshwater 190–200 m^3 (Chandra and Singh 2012). The wood digestion process is used to isolate wood fibers in the presence of high-temperature sodium hydroxide (NaOH) and sodium sulfite (Na_2SO_4), occurs during this chemical reaction and fibers separated from the rest of the wood and dewatering process begins further, depositary materials dissolved in water and released pollutants containing wastewater into the environment (Haq et al. 2020). Besides, digestion, bleaching, and pulping process were used in the manufacture of paper and releases vast amounts of contaminants, i.e. lignosulphonic acid, chlorinated resin acids, chlorinated hydrocarbons, chlorinated phenols, and biocides as well as several physicochemical parameters (TSS, TDS, COD, BOD, phenols, lignin, and heavy metals) beyond the permissible limits as shown in Table 1.1 (Pokhrel and Viraraghavan 2004). These pollutants containing wastewater released into the river and bay, causing water pollution as well as toxicity in aquatic organisms, i.e. fish, tubifex, etc., have been reported by several researchers (Gavrilescu 2008; Petra et al. 2015; Kumar et al. 2020b). They have to use industrial wastewater for crop irrigation directly in developing countries like India due to sufficient irrigation facilities. Moreover, paper mill wastewater contains toxic chemical compounds and toxic heavy metals were biomagnification through the food chain and cause adverse effects on human health. It can inhibit population development in marine environments due to slime and scum formation in the water due to the high content of organic matter in the wastewater (Chandra et al. 2008). Therefore, before being discharged into the marine environment, the management of toxic wastewater is very important (Chandra and Singh 2012).

1.2.2 Distillery Industry Wastewater

The distillery is one of the most polluting industries as 88% of its raw materials are converted into waste and directly release into the environment without proper treatment causes aquatic and terrestrial pollutions as shown in Table 1.2. One-liter

Table 1.1 Physicochemical analysis of paper mill wastewater

Characteristics	Units	Values
pH	–	7.68 ± 0.21
BOD	mg/L	225.0 ± 2.24
COD	mg/L	543.04 ± 1.22
Total solids	mg/L	3280.0 ± 1.32
Dissolved solid	mg/L	3110.0 ± 2.42
Fixed solids	mg/L	2900.0 ± 2.35
Chloride	mg/L	2350.0 ± 1.14
Phosphate as P	mg/L	2.56 ± 0.12
Sulphate as SO_4^{-2}	mg/L	713.1 ± 1.11
Nitrate	mg/L	210.08 ± 2.32
Color	mg/L	1275 ± 3.11
Total phenol	mg/L	13.195 ± 0.52
Lignin	mg/L	578 ± 0.13
Lead (Pb)	mg/L	0.2550 ± 0.12
Iron (Fe)	mg/L	1.8598 ± 0.90
Chromium (Cr)	mg/L	0.3058 ± 0.01
Copper (Cu)	mg/L	1.3814 ± 0.16
Cadmium (Cd)	mg/L	0.08632 ± 0.13
Zinc (Zn)	mg/L	0.0945 ± 1.21

Table 1.2 Physicochemical characteristics of distillery wastewater

Characteristics	Units	Values
pH	–	4.0–4.5
Temperature	°C	71–81
Color	–	Dark brown
Total solids	mg/L	59,000–82,000
Volatile solids	mg/L	38,000–66,000
Total suspended solids	mg/L	2400–5000
COD	mg/L	100,000–150,000
BOD	mg/L	35,000–50,000
Nitrogen	mg/L	1660–4200
Phosphorus	mg/L	225–308
Potassium	mg/L	9600–15,475
Iron	mg/L	1550–1800
Sulfates	mg/L	2100–2300
Calcium	mg/L	2300–2500
Magnesium	mg/L	220–250
Chromium (Cr)	mg/L	21.825
Nickel (Ni)	mg/L	13.425
Lead (Pb)	mg/L	16.33
Iron (Fe)	mg/L	2403

alcohol productions were generated 15 liters of spent wash from the distillery industry (Ravikumar et al. 2007). The production of alcohol is mostly cellulosic raw materials (barley malt, grains, sugarcane juice, grapes, and sugarcane molasses) were used. The distilleries industry consists of four major steps (i.e. feed preparation, fermentation, distillation, and packaging) for alcohol production. Alcohol is the key component of the medical sector and the demand for alcohol is increased day by day and the need can be meet by the distilleries as well as the basic need of chemical industry. In 1805, Kanpur Uttar Pradesh was the first distillery in India to produce rum for the military. At present, almost 315 distilleries have a total capacity of 3250 million liters per year of alcohol and 40.4 billion liters per year of wastewater generated (Mohana et al. 2009). Vinasse, for example, produced by fermentation of sugar cane juice, is mainly used in Brazil as a fertilizer (Satyawali and Balakrishnan 2008; Ghosh et al. 2004).

A big environmental concern can emerge from the stillage of the distillery as it contaminates the water sources in several respects. Its characteristics are depending on the feed stock and various aspects of ethanol production process. Spent wash polluted the water bodies into ways; first, the highly colored nature which can block out sunlight, thus reducing oxygenation of the water by photosynthesis and hence becomes detrimental to aquatic life (Fitzgibbon et al. 1998). Second, it has a high pollution load which would result in eutrophication of contaminated water sources. Distillery wastewater, without any treatment can result in depletion of dissolved oxygen in the receiving water streams and poses serious threat to the aquatic flora and fauna (Kharayat 2012). In fact, in certain countries, by-products from the distillery industry are used for direct soil due to their high content of nitrogen, phosphorus, and organic products (spreading on fields and plowing). Another option for disposal of by-products was deep well excavation, but this was not considered a safe alternative because of insufficient storage capacity and underground water contamination.

1.2.3 Textile Industry Wastewater

One of the biggest and most diverse industrial chains in the retail sector is the clothing industry. In India, 2500 textile weaving plants and 4135 textile finishing plants are working, according to the Ministry of Textiles. The cynical effect of the textile industry is consuming dyes particularly harmful synthetic dye, i.e. azo dyes for the dyeing and coloring process (Joshi and Santani 2012). The manufacture of a textile involves many phases of mechanical processing, such as spinning, knitting, weaving, and the manufacture of garments, which tend to be separate from wet treatment processes such as sizing, desizing, scouring, bleaching, mercerizing, dyeing, printing, and finishing, but there is a close relationship between dry processes and wet treatments in sequence (Joshi and Santani 2012). The textile industry produces a wide variety of contaminants from all wool, fabric, and clothing manufacturing methods. The principal environmental issue in the textile industry is the huge amount of wastewater that contains toxic recalcitrant pollutants

Table 1.3 Physicochemicalcharacteristics of textile wastewater

Characteristics	Units	Values
pH	–	10
Dye concentration	mg/L	700
Chloride	mg/L	15,867
Sulfate	mg/L	1400
Total nitrogen	mg/L	23
COD	mg/L	1781
BOD	mg/L	363
NH_4	mg/L	17
PO_4	mg/L	17
Ca	mg/L	43
Na	mg/L	2900
Fe	mg/L	1.2

discharged into the environments (Table 1.3). Approximately 200 L of water is used to treat 1 kg of textiles. Water is primarily used to (a) add chemicals to textiles, and (b) rinsing the manufactured textiles (Ntuli et al. 2009). Water is used for washing the raw material and for different flushing phases during the entire manufacturing process. In fact, during the bleaching process, 38% of water was found to be used 16% for dyeing, 8% for printing, 14% for boilers, and 24% for other uses (Ntuli et al. 2009).

The features of textile wastewater differ and rely on the kind of textile produced and the chemicals used. The textile wastewater produces large amounts of agents, including suspended and dissolved solids, biological oxygen demand (BOD), chemical oxygen demand (COD), chemicals, trace metals, i.e. Cr, As, Cu, and Zn, and paint, that cause harm to the environment and human health (Joshi and Santani 2012). Hazardous waste management practices can range from a regulatory ban to cost-saving recycling of chemicals. These measures may show extreme heterogeneity, depending on the type of material and process. The most widely recognized approaches are wastewater treatment plants to ensure environmental protection (Joshi and Santani 2012). However, sadly, no clear therapeutic technique is appropriate or usually acceptable for any form of wastewater therapy. Wastewater maintenance is then carried out using several methods, including, depending on the pollution load, physical, chemical, and biological treatment.

1.2.4 Tannery Industry Wastewater

One of the oldest businesses worldwide is the tanning industry. It is commonly described as a pollutant-generated industry producing a broad range of high-intensity toxic chemicals. High concentrations of pollutants, including salinity, organic load (chemical load or need for oxygen, need for biological oxygen), inorganic content, dissolved, suspended solids, ammonia, total nitrogen, particular contaminants (sulfide, chromium, chlorine, sodium, and other traces of salt) and

Table 1.4 Physicochemicalcharacteristics of tannery wastewater

Characteristics	Units	Values
pH	–	8–9
COD	mg/L	5000–5500
BOD	mg/L	3000–3500
Suspended solids	mg/L	2500–3000
Cr^{+2}	mg/L	80–100
SO_4^{2}-	mg/L	1800–2000
Cl^-	mg/L	5000–6000

heavy metals, etc., are known to pose a major environmental threat as shown in Table 1.4 (Cooman et al. 2003; Tariq et al. 2005; Apte et al. 2005; Leghouichi et al. 2009). Significant water volumes are used in the tanning process, 90% of which is drained as wastewater. Nice and gaseous waste, as part of leather processing, is often poured into the atmosphere. 40% of untreated chromium salts are usually released into the final wastewater from the chrome tanning process, posing a substantial environmental threat (Groenestijn et al. 2002; Owlad et al. 2008; Leghouichi et al. 2009). Exposure to chromium, pentachlorophenol and other toxic pollutants increase the risk of dermatitis, ulcer nasal septum perforation and lung cancer (Carlos et al., 2002). Without any exceptions there is no effluents treatment plant (ETP) in leather tanning industries in the country and moreover, the owners of tannery industries are not much concerned about human health and environmental safety. The pollutant load is usually measured at concentrations of the following parameters: COD, sulfates, chlorides, chromium, nitrogen, and sulfides of ammonia. COD will have very high levels of up to 10,000 mg/L depending on the drug used and can be decreased by up to 90% (Groenestijn et al. 2002). There is a strong trivalent chromium concentration in the one arising from chrome tanning. The average water intake per quintal ranges between 40 and 90 L of polished fresh leather (Groenestijn et al. 2002). The rest of the water used is wasted by evaporation (up to 50%). The wastewater created often includes salts (chloride) used to secure hides and block putrefaction during the transport of raw materials from the processing area to tanneries during the first production process. Variable pH values describe the wastewater resulting from various process phases: from very alkaline to very acidic with values varying from 3–4 to 11–13 currently, waste treatment is a big challenge to which the tanning industry has no simple answer.

1.2.5 Pharmaceutical Industry Wastewater

The composition of pharmaceutical wastewater, which has a high concentration of organic matter, microbial toxicity, a high amount of salt, and is difficult to biodegrade, is typically complex (Table 1.5) (Guo et al. 2017). Moreover, most pharmaceutical plants are in batch operation and several raw materials and manufacturing methods are available, resulting in a wide variety of wastewater treatments in various wastewater systems (Gadipelly et al. 2014). A description of the features of

Table 1.5 Physicochemicalcharacteristics pharmaceutical wastewater

Characteristics	Units	Values
pH	–	1–8
Temperature	°C	25–80
COD	mg/L	1000–10,000
BOD	mg/L	500–2500
Total nitrogen	mg/L	500–1500
Total phenol	mg/L	50–250
Suspended solid	mg/L	200–500
Chromaticity (times)	mg/L	500–1000

Table 1.6 Physicochemicalcharacteristics of wastewater from the end-of-pipe (Building and construction chemicals factory)

Characteristics	Units	Values
pH	–	6.1–9.5
COD	mg/L	1870–3924
BOD	mg/L	210–570
Total suspended solids	mg/L	157–519
Phosphorous	mg/L	0.8–30
Organic nitrogen	mg/L	9–25
Phenols	mg/L	0.06–0.3
Oil and grease	mg/L	149–600

pharmaceutical wastewater is given in Table 1.1. Nonetheless, environmental waste is becoming more harmful to the rise of the pharmaceutical industry. There are several types of pharmaceutical wastewater because of the number of products from the pharmaceutical industry, and the variation between the volume and procedure of manufacturing. Highly polluted antibiotic wastewater, characterized by high volume disparities, low C/N, high SS concentration, high sulfate concentration, complex structure, biological toxicity, and high chromaticity, is mainly provided by biopharmaceutical wastewater (Gadipelly et al. 2014). The chemical pharmacy composition is unusual, due to a lack of nutrients. It has a high content and concentration of salt as well. In microbiology, biodegradation and toxicity are complicated (Rui 2013). In addition to the rigid government standards, conventional methods of treatment of pharmaceutical wastewater are difficult to satisfy the industry. It is also essential to advance pharmaceutical wastewater treatment.

1.2.6 Chemical Industry Wastewater

The chemical industry has importance in terms of its environmental influence. Wastewater from this field is typically highly polluted by organic and inorganic contaminants and can contain radioactive pollutants (Table 1.6). Industrial chemical wastewater usually contains organic and inorganic matter in different proportions. In the chemical industry, certain compounds are toxic, mutagenic, carcinogenic, or virtually non-biodegradable. Surfactants, emulsifiers, and petroleum hydrocarbons used in the chemical industry reduce the efficacy of such activities of the treatment plant (USEPA 1998). In general, segregating at source (Peringer 1997) and even conducting on-site treatment in recycled factory lines is the safest option for hazardous industrial wastewater (Hu et al. 1999). The first and most critical step in eliminating waste generation during production in the chemical industry is waste minimization during manufacturing (Carini 1999; Alverez et al. 2004). Bury et al. (2002) used complex modeling to map and monitor wastewater treatment plants in the chemical sector due to variations in magnitude and flow rates.

1.3 Industrial Wastewater Treatment Technology

Water poisoning has become one of the world's most severe health issues. Water emission control began at the end of the 1800s. Wastewater treatment plants are planned and constructed or upgraded to reduce the amount and variety of pollutants. In recent years, this paper discusses the emerging technology for industrial wastewater treatment, including treatment concepts, the benefits, and drawbacks of each technique, and the subsequent implementations (Gray 2005). The two famous biological technologies, Anaerobic Ammonium Oxidation and Anaerobic Membrane Bioreactor were also explored in this paper by examining their benefits, drawbacks, and costs and by providing opportunities for further scientific study. This paper will serve as a guide for those who want to learn about advanced and emerging wastewater treatment technologies in the sector (Samer 2015b).

1.3.1 Physicochemical Treatment Approaches

Environmental contamination today is one of the world's major crises. In different industries, synthetic dyestuffs are used, such as textiles, cosmetics, leather, paper manufacturing, tanning, food coloring, etc. Azo dyes represent an important class of dyes that are extensively employed for various industrial applications mainly for dyeing in textile industries because of their wide spectrum of shades, covalent attachment to the fibers, ease of application, brilliant colors, and minimal energy consumption. Among the 10 000 dyes, with worldwide production of 7×105 tonnes annually, more than 50% are azo dyes, having characteristic chromophoric azo group (–N=N–). Complex aromatic structure of the dyes makes these resistant to the degradative environmental activities. Dye contaminated wastewater poses a serious health hazard to a variety of organisms as the dyes are toxic, mutagenic and even carcinogenic. These negative compounds are very complex and structurally varied, forming azoic bonds. Previous research indicates that these dyes and their by-products, in particular aromatic amines arising from azo bond cleavage reduction, are classified as highly poisonous, carcinogenic, mutagenic, and genotoxic to humans and aquatic biota and are also capable of causing severe skin irritation (Singh and Chandra 2019). Through obstructing light penetration, they prevent underwater photosynthesis, depleting the degree of dissolved oxygen (DO) and compromising the water bodies' beauty and ecological equilibrium. Dye-filled wastewater from factories is frequently discharged into the ecosystem, untreated or poorly handled, to avoid the complexity and expense of the plant resources (Singh et al. 2020b). Underdeveloped and developing nations are at increased risk because, to cut manufacturing prices, clothing firms are using them as profitable areas for the construction of production sites. Current approaches to color reduction include human, chemical and physicochemical methods, with biological means being the most recent addition to the field. While the approaches mentioned are very capable of achieving acceptable outcomes, each of them has a fair share of benefits and drawbacks (Singh et al. 2020b). A single systemic method, capable of degrading these dyes completely and at the same time economically viable, is way out of reach to date. Since the various physicochemical methods are industrially generated as secondary contaminants, they also create large quantities of sludge, apart from being cost-intensive and energy-intensive (Singh et al. 2020a). The focus was then transferred to exploit the potential to remedy these poisonous substances in many biological organisms. Biological means, while inexpensive and capable of fully mineralizing the colors, are always limited to the scale of the facility. Researchers are targeting and integrating existing approaches for the development of new hybrid models in the search for successful treatment processes (Singh and Chandra 2019).

1.3.1.1 Screening

One of the oldest forms of treatment, wastewater filtering separates gross contaminants from the waste stream to avoid damage to downstream infrastructure, prevent contact with plant operations, and prevent the intrusion into the primary settling tanks of unacceptable floating liquid. Screening systems can consist of parallel bars, bars or cables, gratings, wire mesh, or perforated plates to intercept large floating or suspended objects (Li et al. 2009). The openings can be of any type but are usually circular or rectangular. Residues maintained from manual or mechanical washing of bar racks and screens are referred to as "screenings" and are either disposed of or returned to waste after grinding by burial or incineration. In the rugged panel sort, manually or automatically washed bar screens and trash racks are included. Bar screens consist of vertical or inclined bars of steel spaced equally across the container from which the wastewater flows (Westerman et al. 2008). This is used before automatic facilities are built, such as raw sewage pumps, grit chambers, and main sedimentation tanks. Parallel rectangular or oval steel bars with transparent openings compose of garbage racks. Normally, standard bar screens or routers accompany them. Bar height, vertical spacing, and angle, as well as channel width and wastewater approach velocity, are the criteria used in the design of large displays. Fine plates, including perforated holes, wire mesh, woven wire cloth, and wedge-shaped wire, are made up of various types of film stock. Fine windows need to be continuously cleaned with brushes, scrapers, or jets of water, steam, or air forced into the back of the openings because of their narrow gaps. The fine panel quality is dependent on the fineness of the gaps and the level at which the sewage runs into the holes (Morris and Knowles 2011).

1.3.1.2 Comminution

Comminutors are used in the waste stream to pulverize a large floating substance. They are positioned where the scanning treatment, normally between the grit chamber and the main seating tanks, is inefficient. Their use minimizes odors, insects, and loss of visibility. A comminutor can have cutters that revolve or oscillate. Either place a separate stationary screen next to the cutter or rotate the combined screen and cutter together as rotary-cutting comminutors. For a particular type of comminutor, referred to as a barminutor, a combination of a bar screen and a revolving cutter is required (USEPA 1981).

1.3.1.3 Flow Equalization

Flow equalization is a method used to level environmental parameters such as flow, pollutant level, and temperature over time to improve the efficiency of secondary and advanced wastewater treatment facilities. Until a near-constant flow rate is attained, fluctuations are dampened, mitigating the downstream effects of these parameters. In many water treatment systems, it is useful to implement flow equalization downstream of the plant, because the downstream demand is not steady, but in general, the treatment processes work more efficiently if the flow rate through them is steady. In treating wastewater, the rate at which the waste arrives at the treatment process might vary dramatically during the day, so it is convenient to equalize the flow before feeding it to the various treatment steps. In either case, the engineering issue is deciding how large an equalization basin is required to allow a steady, the treatment processes to operate with a steady, average flow.

Sedimentation

Sedimentation, a central and widely used wastewater treatment unit, requires heavy particles found in a solvent to be gravitationally separated. This method is used to extract gravel, particulate matter in the primary sewage reservoir, biological flocs in the activated sewage reservoir, and chemical flow while the chemical coagulation process is used. Three main designs are the horizontal surge, the contact solids, and the inclined floor. It is necessary to take into account that, when constructing a sedimentation basin, the structure must contain both a cleared wastewater and liquid sludge (Sincero and Sincero 2003). Four sedimentation modes, i.e. isolated, flocculent, obstructed, and compressed, exist depending on the concentration of the particle. It is prevalent for more than one sedimentation type to occur during sedimentation.

Horizontal Flow

Rectangular, square, or circle shapes may be horizontal flow clarifiers. The flow is rectilinear and parallel to the long basin axis in rectangular basins, while the stream in the mid-feed circular basins flows from the middle to the outer margins radially. Both types of basins are designed to keep the velocity and flow volumes as constant as possible to prevent the formation of waves and eddy and to stop the suspended fluid from resting. Basins are usually made from steel or reinforced concrete in general. To allow the sludge to recover, the bottom surface slopes softly (Kucharzyk et al. 2011). The tilt is at the side of the inlet and in circular and square tanks in rectangular tanks; the bottom is conical and slopes into the center of the basin.

Solid Contact Clarifiers

Large industrial plants such as refineries and petrochemical plants use Solids Contact Clarifiers as a key step in creating process water that can be used throughout the plant as a general process or cooling water. Municipal drinking water plants use Solids Contact Clarifiers to replace conventional flocculation and sedimentation basins thus reducing footprint. For both of these operations, surface water is drawn from a lake or river and chemically treated within the clarifier to settle solids, remove metals, reduce turbidity, and reduce hardness. Depending on the final use of the water, downstream treatment processes such as sand/carbon filtration, membrane filtration, or disinfection may be used for further treatment (Sincero and Sincero 2003).

1.3.1.4 Flotation

A flotation is a processing unit that is used to isolate solid or liquid particles from the liquid phase by adding a fine gas, typically an air bubble. The gas bubbles either bind to the liquid or are embedded in the particle framework of the suspended solids, increasing the buoyant strength of the particle and gas bubbles together. As a result, it is possible to expand particles with a density larger than the solvent. Flotation is mainly used in the handling of wastewater for the separation of suspended matter and the concentration of biological sludge (Singh et al. 2020a). The biggest advantage of flotation over sedimentation is that it is possible to capture comparatively tiny or light particles more fully and in a shorter timeline. It is possible to skim them out until the particles rise to the surface. Flotation uses air primarily as a floating agent, as usually practiced in urban wastewater treatment. Moreover, to facilitate the removal process, various chemical additives may be added.

1.3.1.5 Adsorption with Activated Carbon

Adsorption is the process of separating soluble compounds at an optimal interface within a solution. In wastewater treatment, activated carbon adsorption is usually followed by normal biological treatment and the bulk of residual dissolved organic matter is required to be removed. Even, particulate matter can be dissolved in the bath. Activated carbon is produced at high temperatures by heating carbon and then activated at high temperatures by exposure to oxidizing gas (Ersahin et al. 2011). In carbon, the gas forms a porous matrix and hence creates a wide internal surface area. It is then possible to segment the activated map into different sizes with different adsorption capacities. The two most common forms of activated carbon are granular activated carbon (GAC) with a diameter greater than 0.1 mm and powder activated carbon (PAC), with a diameter lower than 200 mesh (Samer 2015a). Fixed-bed columns are often used for bringing drainage into contact with the GAC. The water is applied to the column's top and separated from the column's bottom, while the carbon is left in place.

1.3.1.6 Ozonation

Ozonation is a method of chemical water treatment based on the injection of ozone into water. One of the most active oxidants is ozone, a gas composed of three atoms of oxygen (O_3). Ozonation is a form of advanced oxidation process involving the development of highly reactive oxygen species that can target a wide variety of organic compounds and all microorganisms (Kumar et al. 2020a; Cuerda-Correa et al. 2020). Water treatment with ozone has a wide variety of uses, as it is useful both for disinfection and for oxidizing organic and inorganic pollutants. Using electricity, ozone is created by subjecting oxygen (O_2) to high electric voltages or UV radiation. At the point of usage, the requisite quantities of ozone can be produced, but processing requires a lot of energy and is therefore expensive (Cuerda-Correa et al. 2020).

1.3.2 Principles of Biological Treatment

Biological wastewater management principles have been defined by (Russell 2006). A list of the tenets is given below:

1. Biological processes are very vulnerable to extreme hydraulic load shifts. Diurnal variations of more than 250% are difficult because in the clarifiers they can trigger loss of biomass.
2. The growth rate of microorganisms is strongly temperature-dependent. The rate of biological reaction is significantly decreased to half by a 10 °C drop in the temperature of wastewater.
3. In the range of 60–500 mg L^{-1}, BOD is easily handled. Wastewater containing more than 500 mg of L^{-1} BODs was successfully treated when proper dilution was added during the treatment process or when an anaerobic process was used before treatment.
4. To eliminate up to 95% of the BOD, biological treatment is successful. To remove the whole BOD, big tanks are required, which is not feasible.
5. Methods of biological recovery are insufficient to efficiently monitor "shock loads." Equalization is needed where the disparity in drainage severity is greater than 150% or if the wastewater exceeds 1000 mg L^{-1} BOD at its highest concentration.
6. Usually, the carbon: nitrogen: phosphorus (C:N:P) ratio of wastewater is optimal. For the most favorable biological system, the C:N:P ratio of toxic wastewater can differ from 100:20:1 to 100:5:1.
7. If the C:N:P ratio of the wastewater of an element is high relative to the other elements, so there will be bad handling but, this is particularly true if the wastewater is very rich in carbon. Often, the wastewater in an element should not be very weak or very strong; it is difficult to treat, although it is necessary to be very weak.

8. In a biological treatment environment, oils and solids cannot be treated because they adversely affect the treatment process. To remove solids and oils, these wastes should be pre-treated.
9. Hazardous and biological-resistant goods need extra attention before being incorporated into a biological treatment scheme and will need pretreatment.
10. While the capacity of wastewater for oxygen usage is unlimited, the capacity of any aeration device for oxygen transmission is reduced.

1.3.3 Biological Treatment Approaches

Bioremediation is a process of management that requires the addition of microorganisms to remove toxins from the polluted area. Bioremediation can be characterized as a "treatment implementing the decomposition of hazardous materials into less toxic or non-toxic materials by natural organisms." Phytoremediation, bioaugmentation, rhizofiltration, and biostimulation are some examples of bioremediation-related technologies. The microorganisms that are used to conduct bioremediation are referred to as bioremediation (Kumar and Chandra 2020; Kumar et al. 2020b). However, bioremediation does not easily extract or decompose toxins. Heavy metals such as lead and cadmium are not enthusiastically captured by bioremediation. Examples of bioremediation: small concentrations of cadmium, copper, and zinc are bioremediated by fishbone char. Autotrophs or heterotrophs may achieve the bioremediation of wastewater. A heterotroph is an organism that is incapable of carbon fixation and for its production includes organic carbon. Based on their source of energy, heterotrophs are distinguished. It is called a photoheterotroph if the heterotroph uses light as its energy source. If the heterotroph uses organic and/or inorganic compounds as sources of energy called as chemoheterotroph. Chemoheterotrophs, such as humans and other animals, are the more classic heterotrophs. They get their energy from chemicals that they acquire by eating other forms of life exclusively. Photoautotrophs are called autotrophs, such as plants and algae, which are capable of utilizing sunlight energy. Autotrophs that use inorganic compounds to create organic compounds are referred to as lithoautotrophs, such as sugars, fats, and inorganic carbon dioxide proteins. The reduced carbon compounds can be used as energy sources by autotrophs and can provide energy to heterotrophs in the food they utilized.

1.3.3.1 Aerobic Treatment

Biological control mechanisms that exist in the presence of oxygen are aerobic processes as shown in Fig. 1.1. The aerobic condition of the reactor is accomplished by the use of diffuse or mechanical aeration, which retains a fully mixed state of the blended liquor. Aerobic digestion is an alternate method of organic sludge management created by different treatment techniques. To eliminate trace volatile organic compounds (VOCs) in water, aeration has been used. This method is called "gas adsorption" or "oxidation," i.e. to oxidize iron and/or manganese, it has also been used to shift a product, such as oxygen, from the air or a gas process into water. The

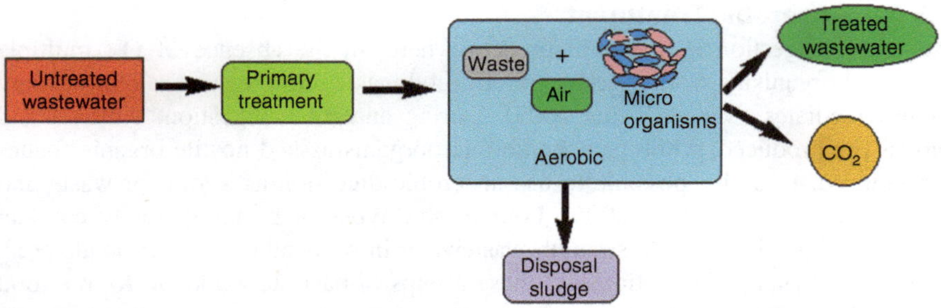

Fig. 1.1 Schematic of aerobic biological wastewater treatment process

gases CO_2 and H_2S are also given by aeration. To extract NH_3 from wastewater and to reduce toxicity and other chemicals in wastewater, air stripping has also been used successfully (Lin 2007). Biowaste aerobic treatment is also effective in reducing toxic concentrations of gaseous compounds such as greenhouse gases (CH_4 and N_2O) and ammonia (Samer et al. 2014).

Oxidation Ponds

Oxidation ponds are aerobic habitats in which the oxygen produced by heterotrophic bacteria is not only by atmospheric transport but also by photosynthetic algae (a heterotroph is an organism that does not fix carbon and uses organic carbon for growth). The algae are restricted to the euphotic zone, which is just a few centimeters deep for the most part (sunlight zone). Ponds are designed at a depth of between 1.2 and 1.8 m to ensure maximum sunlight penetration and look dark green due to thick algal growth. Samer et al. (2008, 2012) have seen the designs and buildings of aerobic treatment tanks and the construction materials used. Algae use inorganic compounds (N, P, CO_2) released by aerobic bacteria to expand using sunlight energy in oxidation ponds. They release oxygen into the water which is utilized by bacteria during the symbiotic cycle. There are two separate optional wetland areas: the upper aerobic zone where bacterial (optional) activity occurs and the lower anaerobic zone where solids settle out of suspension and create an anaerobically infected sludge.

Aeration Lagoons

Relative to oxidation ponds, aeration lagoons are deep (3–4 m), where oxygen is provided by aerators but not by the photosynthetic activity of algae, as in oxidation ponds. To provide sufficiently dissolved oxygen to allow optimum aerobic operation, the aerators retain suspended microbial biomass. Bubble aeration is typically used on the other side, where the bubbles are formed by compressed air injected into plastic tubing laid into the lagoon's base. In household wastewater, HRT of around 5 days usually achieves an 85% reduction in BOD at 20 °C. However, if the temperature drops by 10 °C, BOD elimination will decrease to 65% (Gray 2005).

1.3.3.2 Anaerobic Treatment

Anaerobic digestion is a natural process where, in the absence of O_2, multiple microbial organisms work together to turn food waste into biogas through a variety of intermediates (Mata-Alvarez 2003). During anaerobic digestion, biomass and biogas are produced, while pathogenic microorganisms and hostile organic matter are reduced. It is also possible to use anaerobic digestion as a tool for waste and energy extraction (Moletta 2005). Four trophic types of bacteria usually conduct anaerobic digestion of high-strength wastewater in sequential stages (Ranade et al. 1999). In a synergistic relationship, these groups of bacteria work and form a food chain where CH_4 and CO_2 are the final products. Several microorganisms are involved in the anaerobic digestion process, including acetic acid-forming bacteria (acetogens) and methane-forming archaea (methanogens). In anaerobic digestion, there are four primary biological and chemical processes: hydrolysis, acidogenesis, acetogenesis, and methanogenesis (Fig. 1.2).

For anaerobic treatment, different instruments were used, i.e. anaerobic lagoons, anaerobic digester, anaerobic contact reactor, anaerobic filter, UASB upflow, anaerobically fluidized, and spent bed reactor. Anaerobic digestion is the most potent alternative for high-strength organic wastewater treatment. Anaerobic treatment is a proven technique, and pilot and full-scale operations have shown numerous high-quality anaerobic reactor designs (Lata et al. 2002). The major advantages of anaerobic treatment using UASB flow are that, in addition to wastewater treatment, methane gas is also generated at this point and can be used as fuel mainly in boilers, meaning that anaerobic treatment methods not only assist in the treatment of

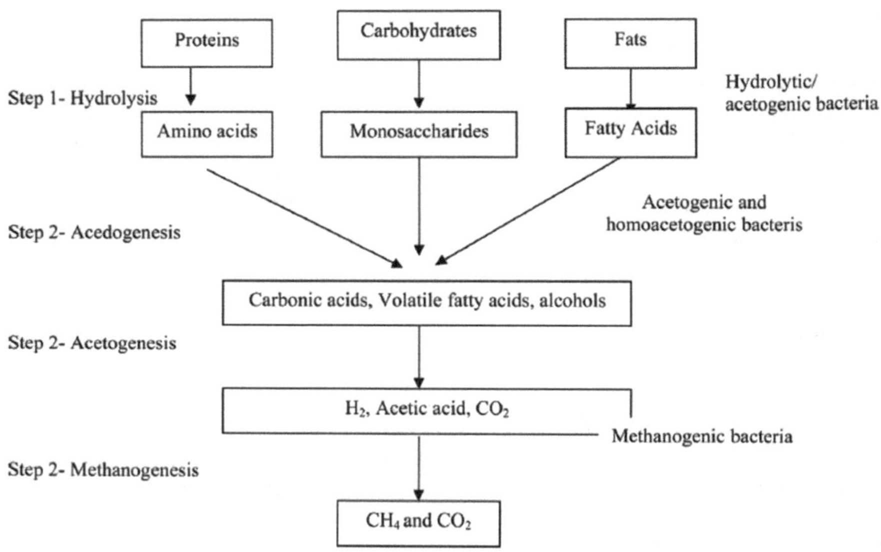

Fig. 1.2 Schematic diagram of anaerobic digestion indicating the process steps and the four bacteria groups involved in the process (Garcia-Heras 2003)

wastewater but also reduce the cost of fuel usage. Anaerobic techniques are used for the disposal of wastewater rich in biodegradable organic matter (BOD >500 mg L^{-1}) and for the further handling of sedimentary sludge. In the agriculture and food processing sectors, heavy organic waste containing significant amounts of biodegradable materials is primarily discharged. Owing to the difficulties and costs of fulfilling the high demand for oxygen to maintain aerobic environments, this wastewater is difficult to handle aerobically (Gray 2005). Furthermore, in the absence of oxygen, anaerobic oxidation takes place. There is a multitude of advantages to the treatment of solid organic wastewater, though anaerobic treatment is time-consuming. These benefits include higher purification levels, the ability to treat heavy organic loads, the storage of small volumes of comparatively solid sludge, and the production of methane as an end product (inert combustible gas). Anaerobic digestion is a complex, multiple-step mechanism in terms of chemistry and microbiology. Chemical compounds are reduced into basic constituents, finally to methane oxide, in the absence of an electron acceptor such as oxygen (Ersahin et al. 2011).

Anaerobic Digestions

Anaerobic digestion involves the degradation of organic matter biologically in an oxygen-free environment, releasing the product in the form of biogas composed mainly of methane. It is distinct from an anoxic phase, which, in contrast to an oxygen-free environment, is a reduced environment. Both systems are anoxic, but an anaerobic is an above-anoxic condition in which the oxidation-reduction potential (ORP) values are extremely negative. Nitrate is converted into ammonia and nitrogen dioxide in the anaerobic process and sulfate (SO_3^{2-}) is reduced to hydrogen sulfide (H_2S). Phosphate is also decreased as it is often transformed into the ADP-ATP chain (Russell 2006). Samer (2012) identified and explained the processes and systems of the anaerobic digester and the construction materials used. An expert system for biogas plant preparation and construction has been established by Samer (2012).

Anaerobic Lagoons

A deep lagoon is an anaerobic lagoon, effectively without dissolved oxygen, that enforces anaerobic conditions. In deep field wetlands, the anaerobic process happens, and certain basins are introduced for anaerobic pretreatment. Aerated, heated, or combined anaerobic lagoons are not used. The depth of the anaerobic lagoon, where deeper lagoons are more involved, is usually around 2.5 m deep. These depths decrease the amount of oxygen released from the soil, causing anaerobic conditions to take precedence (USEPA 2002).

1.3.3.3 Bioreactor

A bioreactor may be defined as an "engineered or constructed device or system that regulates the enclosed or enclosed bio-environment." The bioreactor is specifically a vessel in which microorganisms (e.g. bacteria, algae, fungi) or biochemical (e.g. enzymes) derived from such microorganisms are performed in the microbial method (Mallikarjuna and Dash 2020). The treatment process should be performed in aerobic or anaerobic conditions (e.g., continuous stirred-tank bioreactor). Mycoremediation is a method of bioremediation in which toxins are broken down through the use of fungi. The word "Mycoremediation" essentially refers to the use of fungal "mycelia" bioremediation (Mallikarjuna and Dash 2020). The breakdown of mycelium toxins is the primary role of fungi in the biological system. Lignin and cellulose biodegrading enzymes and acids that are essential components of vegetable fibers are excreted by mycelium, the vegetative part of the fungus. Organic lignin and cellulose molecules consist of long chains of carbon and hydrogen and are thus structurally similar to other organic contaminants. The big challenge is to find the right fungus for the breakdown of a certain pollutant. In an experiment, diesel-contaminated wastewater was inoculated with oyster mushroom mycelia. One month later, more than 93% of these polycyclic aromatic hydrocarbons (PAHs) were reduced to non-toxic components in mycelial-inoculated samples (Mallikarjuna and Dash 2020). To break down toxins, the natural microbial population is involved in the virus, ultimately being CO_2 and H_2O (Mallikarjuna and Dash 2020). Wood-degrading fungi are extremely useful in breaking down aromatic contaminants and chlorinated compounds (toxic components of petroleum) (certain persistent pesticides).

1.3.3.4 Activated Sludge

The base of the activated sludge system is a mixture of thick bacterial species suspended under aerobic conditions in wastewater (Fig. 1.3). With uncontrolled nutrients and oxygen, rapid bacterial growth and respiratory rates can be accomplished, culminating in the consumption of usable organic matter for either oxidized end products (e.g. CO_2, NO_3, SO_4^{-2}, and PO_4^{-3}) or fresh biosynthesizing microorganisms (Kabir et al. 2011; Jasim 2020). The foundation for the activated sludge system consists of five interdependent components: bioreactor, activated sludge, aeration and mixing machine, sedimentation tank, and returned sludge (Gray 2005). A widely used wastewater treatment system with low operating costs is a biological method that uses activated sludge. However, a significant amount of surplus sludge is produced in wastewater treatment plants (WWTPs), which is a big

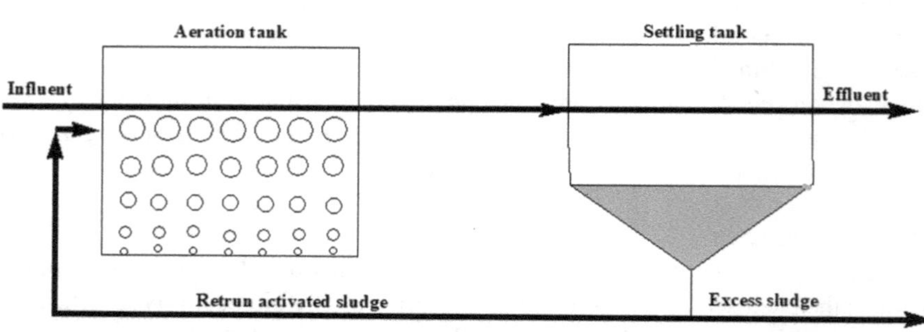

Fig. 1.3 Schematic illustration of activated sludge

economic and environmental burden. Excessive sludge holds the most moisture and cannot be washed quickly. WWTP by-products are dewatered, fried and finally burnt to ashes. Some are used as fertilizer for manure in farm fields (Kabir et al. 2011). It is recommended, however, that instead of the burning process, the dried by-products of WWTPs be fed into the pyrolysis process.

The sludge volume index (SVI) is a formula that describes the tendency to be compressed or condensed by a thickening process of aerated solids, i.e. activated sludge solids. The SVI can be calculated as follows: (a) to cause the mixed liquor sample to be sedimented within 30 minutes from the aeration tank; (b) to calculate the suspended solid content of the same mixed liquor sample; (c) to measure the SVI as the mean approximate wet volume (mL/L) of the deposited sludge at the MLSS dry weight concentration in g/L/L (Jasim 2020; Kabir et al. 2011).

During the treatment of wastewater in aeration tanks, suspended solids are found in the activated sludge phase, where the concentration of suspended solids is called mixed suspended liquor solids (MLSS), measured in milligrams per liter of solids (mg L^{-1}). In an aeration bath, mixed liquor is a mixture of raw wastewater and activated sludge (Gray 2005; Jasim 2020). MLSS consists mostly of microorganisms and non-biodegradable suspended solids. MLSS, which ensures that the quantity of organic pollutants supplied at all times requires sufficient quantities of viable biomass to degrade them, is the active and effective part of the activated sludge method. This is called the food-to-microorganism ratio (F/M ratio) or the food-to-mass ratio (Jasim 2020). If this ratio is kept at the required level, the biomass can absorb high volumes of food, thus reducing the lack of residual food during discharge. In other words, the more food consumed from biomass, the BOD would be decreased in the treated wastewater continuously. It is essential that MLSS extracts BOD to purify the wastewater for further use and hygiene (Gray 2005). With a concentration of several hundred mg L^{-1} BOD, raw sewage is sent into the wastewater treatment process. The concentration of BOD in wastewater is lowered to less than 2 mg L^{-1}, which is deemed safe to use, resulting in treatment with MLSS and other treatment techniques (Jasim 2020).

According to (Gray 2005), the key components of all activated sludge systems are:

1. The key benefit of the bioreactor is that it offers accurately aerated and mixed materials. The bioreactor is referred to as the aeration tank.
2. *Activated sludge*: bacterial biomass consists predominantly of bacteria and other flora and microfauna within the bioreactor. The sludge of these microorganisms is a flocculent suspension and is generally referred to as a mixed suspended solid liquor varying from 2000 to 5000 mg L^{-1}.
3. *Aeration and mixing system*: the activated sludge and the raw result are appropriate for aeration and mixing. Although these processes may be done independently, they are typically performed as a single unit using either surface aeration or diffuse air.

4. *Sedimentation tank*: the triggered sludge discharged from the aeration tank must be clarified or overcome. This separates bacterial biomass from the sewage water being processed.
5. *Returned sludge*: settled activated sludge is returned to the bioreactor in the sedimentation tank to retain the required concentration of the microbial population to ensure the continuation of the treatment process.

Several criteria should be addressed when operating activated sludge plants (1) Management of biomass, (2) preparation of plants, (3) settlement of sludge, and (4) activity of sludge, are the most important parameters. The other organizational component is aeration, where its key tasks are: (1) supplying the bacterial community with a sufficient and continuous supply of dissolved oxygen (DO); (2) preserving suspended bacteria and biomass; and (3) combining heavy wastewater with biomass and removing excess CO_2 from the solution resulting from organic matter oxidation (Gray 2005).

1.3.3.5 Biological Nutrient Removals

Biological Nitrogen Removal

Nitrification and denitrification processes are responsible for the production of N_2O. The processes were nitrification/denitrification for biological nitrogen removal as shown in Fig. 1.4.

Biological Phosphorus Removal

It is generally accepted that microorganisms use acetate and fatty acids to accumulate polyphosphates, such as poly-β-hydroxybutyrate and polymeric acids (Fig. 1.5). The basic approach is based on the synthesis and regeneration of adenosine diphosphate (ADP) in bacteria and includes adenosine triphosphate (ATP). Phosphate

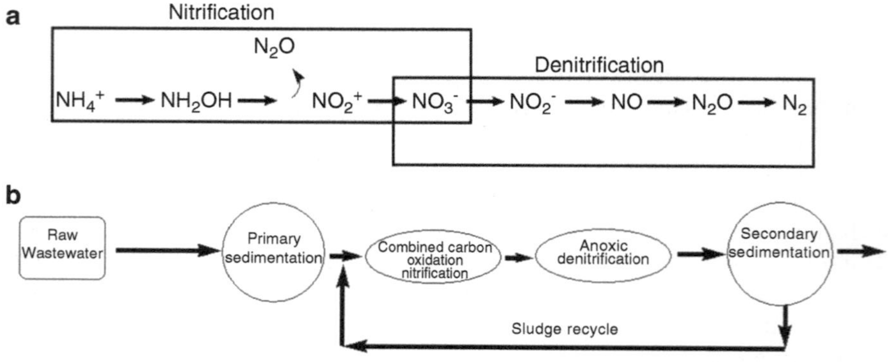

Fig. 1.4 Schematic illustration of nitrification and denitrification process (**a**) responsible for N_2O release and (**b**) biological removal of nitrogen

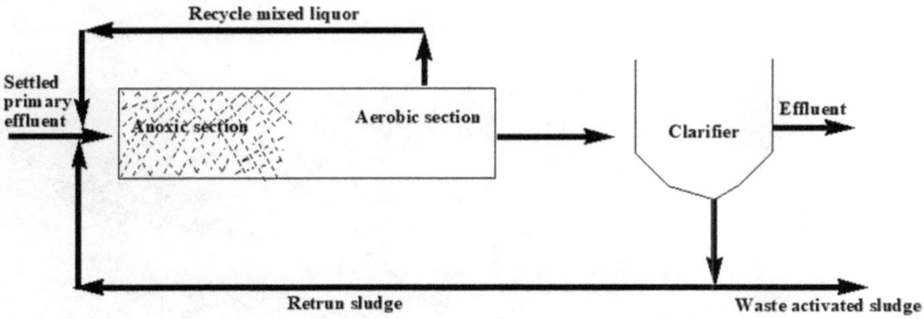

Fig. 1.5 Schematic illustration of phosphate removal process

removal needs true anaerobic conditions that only exist when no other oxygen donor is available (Russell 2006).

1.3.3.6 Phytoremediation

A phytoremediation is a plant-based approach, which involves the use of plants to extract and remove elemental pollutants or lower their bioavailability in soil (Chandra et al. 2018a, b; Berti and Cunningham 2000). Also at low concentrations, plants can absorb ionic compounds in the soil through their root system. To absorb heavy metals and modulate their bioavailability, plants expand their root system into the soil matrix and create a rhizosphere ecosystem, thereby restoring the contaminated soil and stabilizing soil fertility (Chandra and Kumar 2017). There are various advantages of using phytoremediation, i.e. (1) economically feasible phytoremediation is an autotrophic system powered by solar energy, therefore, simple to manage, and the cost of installation and maintenance is low, (2) environment and eco-friendly it can reduce exposure of the pollutants to the environment and ecosystem, (3) applicability it can be applied over a large-scale field and can easily be disposed, (4) it prevents erosion and metal leaching through stabilizing heavy metals, reducing the risk of spreading of contaminants, (5) it can also improve soil fertility by releasing various organic matters to the soil as shown in Fig. 1.6 (Yan et al. 2020).

Several phytoremediation strategies are applicable for the remediation of heavy metal-contaminated soils, including (1) phytostabilization (reduce heavy metal bioavailability in soil), (2) phytoextraction (extract and remove heavy metals from soil), (3) phytovolatilization (absorb heavy metal from soil and release into the atmosphere as volatile compounds), and (4) phytofiltration (hydroponically cultured plants to absorb or adsorb heavy metal ions from groundwater and aqueous waste) (Marques et al. 2009; Chandra and Kumar 2017; Yan et al. 2020). Moreover, the other methods of phytoremediation, i.e. phytodegradation and rhizodegradation, are used for residual organic pollutant degradation (Mustafa and Hayder 2020; Yan et al. 2020). In situations where natural plants or transgenic plants can photodegrade and

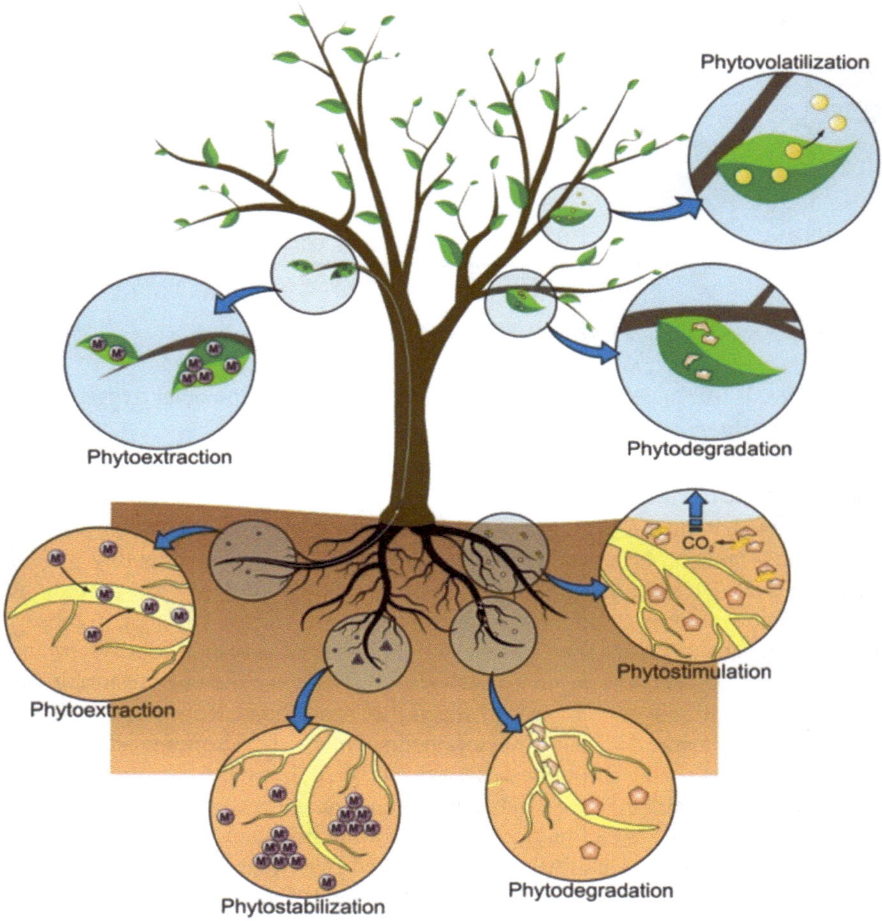

Fig. 1.6 Schematic representation of phytoremediation strategies (Paulo et al. 2014)

phytoaccumulate these toxic compounds in their substrate materials, which are then collected for extraction, phytoremediation is beneficial (Mustafa and Hayder 2020). Through incineration, heavy metals in treated wood can spread further and be used for industrial purposes. Rhizofiltration is a type of phytoremediation that involves filtering wastewater for the extraction of hazardous pollutants or excess nutrients through a root mass (Chandra and Kumar 2017; Chandra et al. 2018a, b). These contaminants are rendered non-toxic to many plants through ingestion of organic pollutants, i.e. toxic, explosives, solvents, industrial chemicals, and other xenobiotic substances, and are classified as phytotransformation (Yan et al. 2020).

1.4 Strategy and Challenges for Wastewater Management

It is important to conclude all the concerns related to the proper management and usage of our current water supplies and the healthy and effective use of water for industrial applications in general but, removing or minimizing the human risk of water shortages and the risks of environmental contamination as shown in Fig. 1.7. It is not easy to create sustainable waste management methods for the chemical and related sectors, for the same reason that most of the current wastewater treatment methodologies differ considerably, not just in terms of their principles, but also in terms of their limitations. In recent years, the increase in industrial production has greatly altered the composition of wastewater produced. Most of the non-biodegradable pollutants (e.g. plastic detergents, petrochemicals, synthetic rubber) are continuously increase as environmental waste, due to the non-biodegradable characteristics of these pollutants we have a requirement of the invention of modern non-biological technology and water waste management methods. It is only in a few cases where there are ready-to-use alternatives. If biological treatment is necessary, the use of advanced techniques should be implemented preferably and successfully. The emerging trend is to minimize needs for plant space along with costs through the use of hybrid/new technology. Process incorporation is essential for fine and specialty chemicals with strong COD and non-biodegradable chemicals. Wastewater is gradually being taken into consideration as an electricity source. A modern solution is expected to integrate innovative wastewater treatment materials, procedures, and options/technology for process integration covering a whole spectrum of chemical and related industries as shown in Fig. 1.7. At this point, there is a need to focus more on water-intensive sectors than on scattered initiatives in multiple directions.

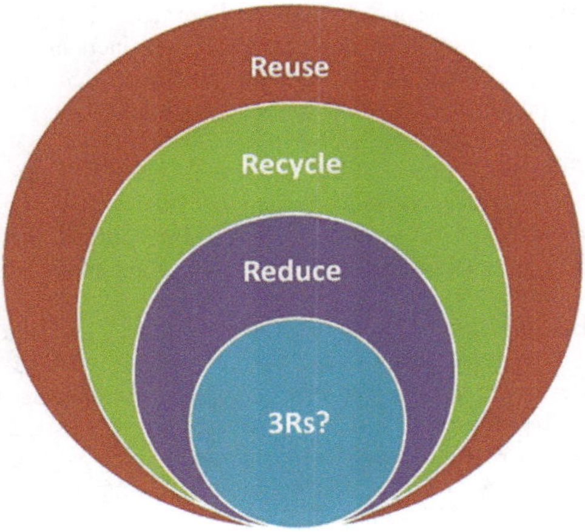

Fig. 1.7 Major key issues in industrial wastewater treatment

1.5 Conclusions

The industries that released wastewater into the environment are a great concern due to its toxicity such as color, mutagenicity, and carcinogenicity. Most industries wastewater have high concentration of toxic pollutants that is why it contains high TSS, TDS, BOD, and COD as well as recalcitrant toxic contaminants along with heavy metals, etc., causing hazardous effects on living organisms. Various physico-chemical treatment approaches are now used or suggested in the degradation and detoxification of industrial wastewater but are often not applied due to the excessive costs involved in the method and generation of secondary pollutants. Methods of biological treatment are widely used because they are the only treatment technique that produces very useful by-products that could be in high demand, and because they are also important for the sustainable development of an environmentally friendly environment. However, recent research using eco-friendly microbial techniques offers the opportunity for the future development of new biological treatment technologies. Finally, zero toxicants or color pollution through the discharge of the decolorized and detoxified industrial wastewater into the environment should be the ultimate goal.

Acknowledgments This work was supported by the Department of Biotechnology New Delhi, Letter No. BT/PR18896/BCE/8/1372/2016 dated 28-03-2018, University Grant Commission, New Delhi, to Mr. Adarsh Kumar is also highly acknowledged and NFPwD (F No.01-01/2019-Sch.) fellowship to Mr. Ajay Kumar.

References

Alvarez D, Garrido N, Sans R, Carreras I (2004) Minimization-optimization of water use in the process of cleaning reactors and containers in a chemical industry. J Clean Prod 12:781–787

Yan A, Wang Y, Ngin TS, Lokman MYM, Subhadip G, Zhong C (2020) Phytoremediation: a promising approach for revegetation of heavy metal-polluted land. Front Plant Sci 11:359. https://doi.org/10.3389/fpls.2020.00359

Apte AD, Verma S, Tare V, Bose P (2005) Oxidation of Cr(III) in tannery sludge to Cr(vi): field observations and theoretical assessment. J Hazard Mater B 121:215–222

Berti WR, Cunningham SD (2000) In: Raskin I, Ensley BD (eds) Phytostabilization of metals in phytoremediation of toxic metals: using plants to clean-up the environment. Wiley, New York, pp 71–88

Bury SJ, Groot CK, Huth C, Hardt N (2002) Dynamic simulation of chemical industry wastewater treatment plants. Water Sci Technol 45(4–5):355–363

Carini D (1999) Treatment of industrial wastewater using chemical-biological sequencing batch biofilm reactor (SBBR) processes, Ph.D. thesis, Swiss Federal Institute of Technology, Zurich

Carlos K, Katarzyna W, Kazimierg W, Juan MM, Laura E, Nava L, Gomez RG (2002) Possible adverse effects of chromium in occupational exposure of tannery workers. Ind Health 40:207–213

Chandra R, Kumar V (2017) Phytoextraction of heavy metals by potential native plants and their microscopic observation of root growing on stabilised distillery sludge as a prospective tool for in situ phytoremediation of industrial waste. Environ Sci Pollut Res 24:2605–2619

Chandra R, Kumar V, Tripathi S, Sharma P (2018a) Heavy metal phytoextraction potential of native weeds and grasses from endocrine-disrupting chemicals rich complex distillery sludge and their histological observations during in-situ phytoremediation. Ecol Eng 111:143–156

Chandra R, Raj A, Yadav S, Patel DK (2008) Reduction of pollutants in pulp paper mill effluent treated by PCP-degrading bacterial strains. Environ Monit Assess 155:1–11

Chandra R, Sharma P, Yadav S, Tripathi S (2018b) Biodegradation of endocrine-disrupting chemicals and residual organic pollutants of pulp and paper mill effluent by biostimulation. Front Microbiol 9:960. https://doi.org/10.3389/fmicb.2018.00960

Chandra R, Singh R (2012) Decolourisation and detoxification of rayon grade pulp paper mill effluent by mixed bacterial culture isolated from pulp paper mill effluent polluted site. Biochem Eng J 61:49–58

Chaphekar SB (2013) Education for water conservation. In: National conference on water sanitation and recycling. Maharashtra Chamber of Commerce. Industry & Agriculture, Mumbai

Cooman K, Gajardo M, Nieto J, Bornhardt C, Vidal G (2003) Tannery waste water characterization and toxicity effects on *Daphnia* spp. Environ Toxicol 18:45–51

Cuerda-Correa EM, Alexandre-Franco MF, Fernandez-Gonzalez C (2020) Advanced oxidation processes for the removal of antibiotics from water. An overview. Watermark 12:102. https://doi.org/10.3390/w12010102

Ersahin ME, Ozgun H, Dereli RK, Ozturk I (2011) Anaerobic treatment of industrial effluents: an overview of applications. In: Einschlag F (ed) Waste water-treatment and re-utilization. InTech., ISBN 978–978–953-307-249-4, Rijeka, pp 3–28

Fitzgibbon FJ, Singh D, Mcmullan G, Marchant R (1998) The effect of phenolic acids and molasses spent wash concentrations of distillery wastewater remediation by fungi. Process Biochem 33 (8):799–803

Gadipelly C, Perez-Gonzalez A, Yadav GD, Ortiz I, Ibanez R, Rathod VK, Marathe KV (2014) Pharmaceutical industry wastewater: review of the technologies for water treatment and reuse. J Ind Eng Chem Res 53:11571–11592

Garcia-Heras JL (2003) Reactor sizing, process kinetics and modelling of anaerobic digestion of complex wastes. In: Mata-Alvarez J (ed) Biomethanization of the organic fraction of municipal solid wastes. IWA Publishing, TJ International Ltd., Padstow, Cornwall, pp 21–62

Gavrilescu D (2008) Energy from biomass in pulp and paper mills. Environ Eng Manag J 7:537–546

Ghosh M, Verma SC, Mengoni A, Tripathi AK (2004) Enrichment and identification of bacteria capable of reducing chemical oxygen demand of anaerobically treated molasses spent wash. J Appl Microbiol 96(6):1278–1286

Gray NF (2005) Water technology: an introduction for environmental scientists and engineers, 2nd edn), ISBN 0750666331. Elsevier, Amsterdam

Groenestijn JWV, Langerwerf JS, Lucas M (2002) Reducing environmental emissions in tanneries. J Environ Sci Health 37(4):737–743

Guo Y, Qi PS, Liu YZ (2017) A review on advanced treatment of pharmaceutical wastewater. IOP Conf Ser: Earth Environ Sci 63:012025

Haq I, Mazumder P, Kalamdhad AS (2020) Recent advances in removal of lignin from paper industry wastewater and its industrial applications – A review. Bioresour Technol 312:123636

Hu HY, Goto N, Fujie K (1999) Concepts and methodologies to minimize pollutant discharge for zero-emission production. Water Sci Technol 10–11:9–16

Hubbe MA, Metts JR, Hermosilla D, Blanco MA, Yerushalmi L, Haghighat F, Lindholm-Lehto P, Khodaparast Z, Kamali M, Elliott A (2016) Wastewater treatment and reclamation: a review of pulp and paper industry practices and opportunities. Biol Res 11(3):7953–8091

Jasim NA (2020) The design for wastewater treatment plant (WWTP) with GPS X modeling. Cogent Eng 7(1):1723782. https://doi.org/10.1080/23311916.2020.1723782

Jimenez B (2009a) Coming to terms with nature: water reuse new paradigm towards integrated water resources management. In: Encyclopedia of biological, physiological and health sciences.

Water and health, vol. II: life support system. EOLSS Publishers/UNESCO, Oxford, pp 398–428
Jimenez B (2009b) Wastewater risks in the urban water cycle. In: Jimenez B, Rose J (eds) Urban water security: managing risks. UNESCO/Taylor and Francis Group, Paris, p 324
Joshi VJ, Santani DD (2012) Physicochemical characterization and heavy metal concentration in effluent of textile industry. Univers J Environ Res Technol 2(2):93–96
Kabir M, Suzuki M, Yoshimura N (2011) Excess eludge reduction in wastewater treatment plants, in: wastewater - treatment and reutilization. In: Einschlag F (ed) . InTech, Rijeka, pp 133–150. ISBN 978-978-953-307-249-4
Kharayat Y (2012) Distillery wastewater: bioremediation approaches. J Integr Environ Sci 9 (2):69–91
Kucharzyk KH, Soule T, Paszczynski AJ, Hess TF (2011) Perchlorate: status and overview of new remedial technologies. In: Einschlag F (ed) Waste water-treatment and reutilization. InTech, Rijeka, pp 171–194. ISBN 978–978–953-307-249-4
Kumar A, Chandra R (2020) Ligninolytic enzymes and its mechanisms for degradation of lignocellulosic waste in environment. Heliyon 6:e03170
Kumar A, Singh AK, Ahmad S, Chandra R (2020a) Optimization of laccase production by Bacillus sp. strain AKRC01 in presence of agro-waste as effective substrate using response surface methodology. J Pure Appl Microbiol 14(1):351–362
Kumar A, Singh AK, Chandra R (2020b) Comparative analysis of residual organic pollutants from bleached and unbleached paper mill wastewater and their toxicity on Phaseolusaureus and Tubifextubifex. Urban Water J 2020:1–11
Lata K, Kansal A, Balakrishnan M, Rajeswari KV, Kishore VVN (2002) Assessment of biomethanation potential of selected industrial organic effluents. Resour Conserv Recy 35:147–161
Leghouichi E, Laib E, Guerbet M (2009) Evaluation of chromium contamination in water, sediment and vegetation caused by the tannery of Jijel (Algeria): a case study. Environ Monit Assess 153:111–117
Li Y, Xiao Y, Qiu J, Dai Y, Robin P (2009) Continuous village sewage treatment by vermifiltration and activated sludge process. Water Sci Technol 60:3001–3010
Lin SD (2007) Water and wastewater calculations manual, 2nd edn. McGraw-Hill. ISBN 0-07-154266-3, New York
Mallikarjuna C, Dash RR (2020) A review on hydrodynamic parameters and biofilm characteristics of inverse fluidized bed bioreactors for treating industrial wastewater. J Environ Chem Eng 8:104233
Marques A, Rangel A, Castro P (2009) Remediation of heavy metal contaminated soils: phytoremediation as a potentially promising clean-up technology. Crit Rev Environ Sci Technol:39. https://doi.org/10.1080/10643380701798272
Mata-Alvarez J (2003) Fundamentals of the anaerobic digestion process. In: Biomethanization of the organic fraction on municipal solid wastes. IWA, London, pp 1–20
Mohana S, Bhavik KA, Madamwar D (2009) Distillery spent wash: treatment technologies and potential applications. J Hazard Mater 163(1):12–25
Moletta R (2005) Winery and distillery wastewater treatment by anaerobic digestion. Water Sci Technol 51(1):137–144
Morris RH, Knowles P (2011) In: Einschlag F (ed) Measurement techniques for wastewater filtration systems, in: waste water-treatment and reutilization. InTech., ISBN 978–978–953-307-249-4, Rijeka, pp 109–132
Mustafa HM, Hayder G (2020) Recent studies on applications of aquatic weed plants in phytoremediation of wastewater: a review article. Ain Shams Eng J 12:1. https://doi.org/10.1016/j.asej.2020.05.009
Ntuli F, Omoregbe I, Kuipa P, Muzenda E, Belaid M (2009) Characterization of effluent from textile wet finishing operations. WCECS 2009:10–20

Owlad M, Aroua MK, Daud WAA, Baroutian S (2008) Removal of hexavalent chromium-contaminated water and wastewater. Water Air Soil Pollut 200:59–77

Pant D, Adholeya A (2007) Biological approaches for treatment of distillery wastewater: a review. Bioresour Technol 98(12):2321–2334

Paulo JC, Pratas FJ, Varun M, D'Souza R, Paul MS (2014) Phytoremediation of soils contaminated with metals and metalloids at mining areas: potential of native flora. IntechOpen, London. https://doi.org/10.5772/57469

Peringer P (1997) Biologischer Abbau von Xenobiotica. Bio World 1:4–7

Petra C, Lindholm-Lehto J, Knuutinen S, Heidi SJ, Sirpa A, Herve H (2015) Refractory organic pollutants and toxicity in pulp and paper mill wastewaters. Environ Sci Pollut Res 22 (9):6473–6499

Pokhrel D, Viraraghavan T (2004) Treatment of pulp and paper mill wastewater-a review. Sci Total Environ 333:37–58. https://doi.org/10.1016/j.scitotenv.2004.05.017

Ranade DR, Dighe AS, Bhirangi SS, Panhalkar VS, Yeole TY (1999) Evaluation of the use of sodium molybdate to inhibit sulphate reduction during anaerobic digestion of distillery waste. Bioresour Technol 63(3):287–291

Ravikumar R, Saravanan R, Vasanthi NS, Swetha J, Akshaya N, Rajthilak M, Kannan KP (2007) Biodegradation and decolorization of biomethanated distillery spent wash. Indian J Sci Technol 1(2):1–6

Rui D (2013) Study on advanced treatment of pharmaceutical wastewater. South China University of Technology, Guangzhou, pp 1–3

Russell DL (2006) Practical wastewater treatment. Wiley, Hoboken

Samer M (2012) Biogas plant constructions. In: Kumar S (ed) Biogas. InTech, Rijeka, pp 343–368. ISBN 978-953-51-0204-5

Samer M (2015a) Biological and chemical wastewater treatment processes. Polyteknisk Forlag, Kongens Lyngby. https://doi.org/10.5772/61250

Samer M (2015b) GHG emission from livestock manure and its mitigation strategies. In: Sejian V, Gaughan J, Baumgard L, Prasad C (eds) Climate change impact on livestock: adaptation and mitigation. Springer, Berlin. ISBN 978-81-322-2264-4

Samer M, Grimm H, Hatem M, Doluschitz R, Jungbluth T (2008) Mathematical modeling and spark mapping for construction of aerobic treatment systems and their manure handling system, In: Proceedings of international conference on agricultural engineering, book of abstracts, Eur AgEng, Crete, June 23–25, p 28

Samer M, Mostafa E, Hassan AM (2014) Slurry treatment with food industry wastes for reducing methane, nitrous oxide and ammonia emissions. Misr J Agric Eng 31(4):1523–1548

Satyawali Y, Balakrishnan M (2008) Wastewater treatment in molasses based alcohol distilleries for COD and color removal: a review. J Environ Manage 86(3):481–497

Sghaier I, Guembri M, Chouchane H et al (2019) Recent advances in textile wastewater treatment using microbial consortia. J Textile Eng Fashion Technol 5(3):134–146

Sincero AP, Sincero GA (2003) Physical-chemical treatment of water and wastewater.,ISBN 1–84339–028-0. CRC Press, Boca Raton

Singh AK, Chandra R (2019) Pollutants released from the pulp paper mill: aquatic toxicity and their health hazards. Aquat Toxicol 211:202–216

Singh AK, Kumar A, Chandra R (2020a) Detection of refractory organic pollutants from pulp paper mill effluent and their toxicity on *Triticumaestivum*; *Brassicacampestris* and *Tubifex-tubifex*. JEBAS 8(5):663–675

Singh AK, Kumar A, Chandra R (2020b) Residual organic pollutants detected from pulp and paper industry wastewater and their toxicity on Triticumaestivum and Tubifex-tubifex worms. Mater Today: Proc. https://doi.org/10.1016/j.matpr.2020.10.862

Tariq SR, Shaheen N, Khalique A, Manzoor S, Jaffar M (2005) Multivariate analysis of selected metals in tannery effluents and related soil. J Hazard Mater 122:17–22

Ugurlu M, Gurses A, Dogar C, Yalcin M (2007) The removal of lignin and phenol from paper mill effluents by electrocoagulation. J Environ Manage 87:420–428

USEPA (1981) Development document for effluent limitations: guideline and standards for the metal finishing point source category, U.S. Environmental Protection Agency, EPA/440/1–83-091, Washington

USEPA (1998) Wastewater treatment technologies. In: Pollution prevention (P2) guidance manual for the pesticide formulating, packaging and repackaging industry including implementing the P2 alternative. EPA, Washington

USEPA (2002) Wastewater technology fact sheet: anaerobic lagoons. U.S. Environmental Protection Agency., EPA 832-F-02-009, Washington

Westerman PW, Shaffer KA, Rice JM (2008) Sludge survey methods for anaerobic lagoons, AG-639W, North Carolina Cooperative Extension Service, College of Agriculture & Life Sciences, North Carolina State University

Zhang BY, Zheng JS, Sharp RG (2010) Phytoremediation in engineered wetlands: mechanisms and applications. Procedia Environ Sci 2:1315–1325

Bioremediation of Hexavalent Chromium from Industrial Effluents

2

Anfal Arshi and Anshu Singh

2.1 Introduction

The heavy contamination metal that includes Chromium, Iron, Cadmium, Mercury, Cupper, Lead, Zinc, Nickel, Arsenic, Selenium, Magnesium, Manganese, etc. is a serious problem to the environment because of the anthropogenic activities from mining, processing, and applications of these metals have increased enormously during the past few decades and have become a challenge for life on earth (Haq et al. 2020). Consequently, their removal/remediation has been all the more necessary. For a few years, the disposal of solid and/or liquid waste materials containing heavy metals from production processes has acquired a lot of scrutiny and as a result, environmental regulation has eventually become more stringent. Heavy metal polluted area is typically the consequence of previous production processes, where understanding of the health and environment consequences involved with the processing, use, and storage of toxic substances has been widely understood than is actually the case (Mazumder et al. 2021). The problem is global and the total number of polluted sites is large. It is also widely accepted that heavy metal polluted land is a possible hazard to human health, and its persistent emergence in recent years has contributed to international attempts to fix many of these sites, either as a solution to the possibility of adverse health or environmental consequences caused by pollution or to enable the site to be renovated for use. The traditional methods used for remediation have been used is to unearth polluted soil to clear it from the landfill or to seal and contain the contaminated areas of the site, these methods have few pitfalls. The first approach merely transfers contaminants elsewhere which may face serious hazards in the exploration, storage, and transport of dangerous materials. It is still very complex and increasingly costly to locate new dump sites for the final disposal of the waste. The cover and containment system is just a temporary

A. Arshi (✉) · A. Singh
Defence Institute of Bio-Energy Research, DRDO, Haldwani, Uttarakhand, India

I. Haq, A. S. Kalamdhad (eds.), *Emerging Treatment Technologies for Waste Management*, https://doi.org/10.1007/978-981-16-2015-7_2

approach as pollution persists on site, requiring long-term inspection and repair of the insulation walls, with all related costs and possible liability. A simpler solution than these conventional approaches is to totally kill or at least turn toxic into harmless compounds, if possible. Some of the methods used include high-temperature incineration and different forms of chemical decomposition (e.g., base-catalyzed dechlorination, UV-oxidation). They may be very effective at reducing the levels of a number of pollutants, but have many disadvantages, in particularly their technical difficulty, the expense of small-scale usage and lack of community support especially for incineration, which can maximise the sensitivity of pollutants both on-site staff and local citizens. Bioremediation is an alternative that provides the potential of removing or making different pollutants harmless by using basic biological action (Singh et al. 2014). As such, it uses relatively low-cost, low technology methods that widely well accepted by the population and can also be done on-site. It may not always be sufficient, however, since the spectrum of pollutants against which it is effective is small, the time scale involved are relatively long and the amounts of residual contaminants that may not always be acceptable. Though its different theories are not extremely challenging, in order to develop and implement an effective bioremediation programme, substantial knowledge, and expertise may be required due to the need to comprehensively assess the suitability of a site and optimise conditions to achieve a satisfactory result. Even if bioremediation appears to be a great substitute for conventional research in the field of clean-up technology, especially in the USA, it is increasing rapidly. On a range of sites worldwide, including Europe, bioremediation has been used with differing degrees of effectiveness. As more expertise and experience is gained, methods are evolving, and there is no question that bioremediation has tremendous potential to cope with some forms of site pollution. Alas, bioremediation concepts, strategies, benefits, and drawbacks are not well recognised or understood particularly among those who would have to interact directly with plans for bioremediation, such as site owners and regulators. In, this endeavour we aimed to aid by offering a transparent, realistic view of the mechanisms involved in bioremediation, the benefits and drawbacks of the procedure, and the problems to be addressed when faced with a bioremediation plan. Some studies review the bioremediation documentation of organic and inorganic contaminants comprehensively, and other studies look at specific field application research findings.

Chromium is among the most commonly used metal pollutants, and over the last 15 years it has been rated one of the top 20 pollutants on the Superfund priority list of harmful substances. In several ways, chromium containing ores are found, but the mineral chromite is the economically recoverable type. Mineral chromite is inert and insoluble in water in its spinal phase. In 2010 and 2011, the gross world mine output was 23,700 and 24,000 × 103 metric tons, respectively. In Kazakistan (220), Southern Africa (200), India (54), and the USA (0.62), the gross world reserve is projected to be >480 × 106 metric tons of shipping quality chromite ore with ~45% Cr_2O_3 and major deposits (×106 tons). Cr(VI) species, which exhibit very different chemical compositions, alter organisms in various ways. In the maximum pH spectrum, hexavalent chromium is water soluble, whereas trivalent chromium appears to be adsorbed or precipitated on the surface of the soil as chromium hydroxide in a mildly acidic and alkaline region. Therefore, Cr(III) has a higher potential to contaminate the ecosystem, especially aquifers and surface water. In the USA, Chromium in its hexavalent form has been listed as a human carcinogen category 'A' by the Environmental Protection Agency (EPA) and is one of the major pollutants. It can cause allergic contact dermatitis in all its prevalent forms, trivalent and hexavalent chromium. Chromate, which is the most prevalent source of Cr (VI) present in solid/liquid waste caused by human activities, such as electroplating, steel and automotive production, mining, leather tanning, cement, metal refining, clothing, paint pigment and dye manufacture, wood preservation, is a harmful contaminant because it quickly spreads across initial contamination beyond the site of contamination.

The CrO_4^{2-} and $HCrO_4$ ions are the most mobile sources of chromium in soils. They can be consumed by plants and quickly leached into deeper soil layers, resulting in runoff of soil and surface water. Industrial solid waste/sludge and soil are either treated by different physical/chemical processes such as reduction, precipitation, adsorption, ion exchange, reverse osmosis, and electro-dialysis, or disposed of by landfill, in order to reduce/remove Cr(VI). The chemical, electrochemical, and biological methods for aqueous Cr(VI) reduction have recently been reviewed by Barrera-Díaz et al. (2012). Many of these methods are commercially costly and have drawbacks such as inadequate removal of metals, heavy use of reagents, and energy needs. Thus, contaminates groundwater as generation/disposal of hazardous sludge/ secondary waste. Any of the treatment methods/processes, such as their economic feasibility, have serious drawbacks only at relatively high metal concentrations and not at low concentrations (1–100 mg L^{-1}). In the other hand, bioremediation seems to have greater consequences for the issue of low heavy metal content in detoxification.

As a result, in order to extract, restore or stabilise heavy metals in soil and effluents, researchers have a strong concern in metal-microbe associations. It has been reported that bioremediation technology is more effective in removing soluble and particulate forms of metals particularly from dilute solutions, and therefore bioaccumulation and therefore, microbe-based technologies can provide an alternative to conventional metal removal/recovery techniques. Since many microorganisms have the potential to reduce Cr(VI) to comparatively less harmful Cr(III), bioremediation offers enormous possibilities for the development of Cr(VI)-contaminated soil detoxification technologies as an alternative to current physico-chemical technologies.

This study would outline some of the major attempts made to use microbes for future Cr(VI) bioremediation of soil/sludge/mine and metallurgical waste.

2.2 Chromium Industrial Applications

Chromium is used extensively in industry. In the metallurgical industry, the Cr (VI) compounds are used for the manufacture of chrome alloy and chromium oxide, and in chrome plating. They are used as oxidising agents in the pharmaceutical industry and in the processing of other chromium compounds. About 80–90% of leather is tanned with chromium chemicals, of which about 40% of the chromium used is discharged as Cr(VI) and Cr (III) into the effluent. In cloth dyeing, in the ceramics and glass industries, and in painting, Cr(III) salts are used less commonly. Chromium and its derivatives are also useful in common lives. 90% of the global supply of chrome ore is used for steel alloy and non-ferrous alloy production in the metallurgical industries. In the refractory (iron and steel, cement, glass, ceramics, and machinery) and chemical (leather tanning, plating, wood preservation, and pigment) fields, only 5% each was used. Chromium is resistant at room temperature to ordinary corrosive chemicals, which accounts for its use in protective coating electroplating. The majority of consumption is accounted for ferrous alloys, primarily stainless steel. Besides being corrosion and oxidation resistant, these steels have a broad range of mechanical properties. Cast iron can contain 0.5–30% chromium, which provides strength and durability, and resistance to corrosion and wear. Chromium is also found in conjunction with magnesium, iron-nickel, cobalt, aluminium, titanium, and copper in non-ferrous alloys. As far as chromium chemicals are concerned, the greatest quantity is consumed in order to create pigments for paint and ink use. Leather tanning, metal corrosion prevention, fracking powder, fibre dyes, catalysts, wood, and water treatment are other uses. In the refractory industry, chromite is used to manufacture bricks, mortar, and blends for ramming and gunning. Their thermal shock and slag resistance, volume stability and strength are improved by chromite. In rocks, fresh water and seawater, chromium is widely distributed.

2.3 Soil Chromium Transformations: Mobility and Bioavailability

Chromium, in the bound form composes 0.1–0.3 mg kg^{-1} of the Earth's crust. Cr (III) and Cr(VI) found in soils have contrary chemical and physical properties, as stated above. The former is immobile as it has a good affinity in soils for negative charged ions and colloids, and provides sparingly soluble compounds such as $Cr(OH)_3$. In the 4–8 pH range, such products dominate.

2.4 The Phase of Chromium

Chromium cycle indicates that Cr(VI) is the most oxidised, mobile, reactive, and toxic form of chromium, and if all chromium were to be in thermodynamic equilibrium with the atmosphere, it would be the only available form. Smaller concentrations may be the consequence of natural Cr(III) oxidation, but higher concentrations are usually the result of either Cr(VI) contamination or Cr(III) oxidation. Soils and sediments contain both oxidised manganese and reduced carbon (Eq. 2.1) at partial equilibrium with atmospheric oxygen and the reduction by soil carbon compounds from Cr(VI) to Cr(III) are both thermodynamically spontaneous reactions (Eq. 2.2). The fascinating feature of the soil chromium cycle is that there can be oxidation and reduction at the same time. Some of the Cr(III) added to an aerobic soil sample will oxidise, and decrease some of the Cr(VI) added to the same soil.

Oxidation (Apte et al. 2005) by manganese oxides of Cr(III) to Cr(VI) by surface oxidation under neutral pH conditions proceeds as follows:

$$Cr^{3+} + 1.5MnO_2 + H_2O \rightarrow HCrO^{4-} + 1.5Mn^{2+} + H^+ \tag{2.1}$$

While reduction by organic compounds, e.g. hydroquinone (with quinine formation) (James 2001) proceeds as follows:

$$\begin{aligned} C_6H_6O_2 + CrO_4{}^{2-} + 2H_2O \rightarrow 0.5Cr_2O_3 + 1.5C_6H_4O_2 + 2.5H_2O \\ + 2OH^-\_G298^{\circ} \\ = -427\ \text{kJ mol}^{-1} \end{aligned} \tag{2.2}$$

The properties of Cr(III) forms in waters and soils limit their bioavailability and mobility. In equilibrium with insoluble compounds, the concentrations of soluble Cr (III) are $<10 - 9$ M (0.05 parts per billion) in water at pH value 6 to less than 10–15 M at pH value 8. In soil, Cr(VI) is more soluble and more bio-available than Cr(III). Under most natural circumstances, Cr(VI) is a form of an anion. It is mainly present as chromate ($CrO_4{}^{2-}$) at higher pH values >6.4, while below pH 6.4 it exist mainly as bichromate ($HCrO^{4-}$) (James 2002). Compared to the oxidation of Cr(III) to Cr(VI) by oxidants in soils, the reduction of Cr(VI) to Cr(III) is simpler (Cervantes et al. 2001). The oxidation of Cr(III) to Cr(VI) was studied experimentally by Bartlett and James (1979) in several field soils containing manganese oxides. They concluded that Cr(III) to Cr(VI) was readily oxidised by soils containing oxides of manganese than by soils without manganese oxides. The quantity of Cr(III) oxidised is comparable to soil Mn oxide content and also to the decreased level of manganese oxide (Bartlett and James 1979). Oxidation of Cr(III) is however, closely linked to its soil concentration and is highly dependent on the forms of Cr(III) (Kamaludeen et al. 2003). The effect of pH on chromium's oxidation and reduction reactions in soils is a complicated mechanism, but generally high pH values increase oxidative power whereas low pH values increase reactions of reduction. Soluble and newly precipitated forms of Cr(III) such as CrC13 and Cr(OH)3 applied to soil have been observed to oxidise up to 15% under laboratory conditions (soils with near-neutral pH values, elevated levels of Mn oxides and optimum aeration conditions) (James 2002). The Potential Chromium Oxidation Score (PCOS) proposed by James et al. (1997) is based on four integrating variables such as solubility and shape of Cr(III), reactive soil manganese, soil potential for Cr(VI)-reduction, and soil pH as a modifier of the first three factors in order to design the remediation process through reduction strategies. Such parameters are being mathematically analysed and ranked; the sum of these values gives PCOS. The PCOS varies from 10 to 40, with high scores indicating a high possibility of Cr(III) oxidation and Cr(III) persistence. Thus, the measures for improving chromium-contaminated soil remediation processes should consider the probability that some forms of Cr(III) may be oxidised to Cr (IV) more favourably. In addition, precautions should be exercised with respect to the use of hydrogen peroxide for the in situ remediation of soils contaminated with plant-absorbed ionic form of chromium Cr^{3+} and Cr^{6+}; Cr^{3+} is absorbed more frequently than Cr^{6+}. Soils polluted with chromium from the refining of chromate ore or from electroplating waste release significant quantities of Cr(VI) after hydrogen peroxide treatment. CrCl3 with chemically complex waste, since Cr (VI) mobilisation may be a hazardous product of such a process (Rock et al. 2001).

The numerous (stable/unstable) oxidation states found in nature (Cr compounds) are:

0 $Cr(CO)_6$; +1 (unstable); + 2 (chromous) (unstable) [Cr(CH3COO)4, CrO, CrSO4]; + 3 (chromic) (stable)[Cr2O3, Cr2(SO4)3]; + 4 (unstable) [CrO2]; + 5 (unstable) [CrF5], and + 6 (stable) [K2Cr2O7, K2Cr2O4, CrO3] (Thacker et al. 2007).

2.5 Cr (IV) Specification

Depending on both the pH of the medium and total Cr(VI) concentration, Cr (VI) may form several species as $CrO_4{}^{2-}$, $HCrO_4{}^{2-}$or $Cr_2O_7{}^{2-}$.H_2CrO_4, a strong acid (Sperling et al. 1992) and the deprotonated form of Cr(VI) is seen at pH >1. Beyond pH 7, only $CrO_4{}^{2-}$ions remain in the solution in the entire concentration range. HCrO4-is the prevalent species at pH between 1 and 6 (Cotton and Wilkinson 1980; Nieboer and Jusys 1988)

$$H_2CrO_4 \rightarrow H^+ + HCrO_4{}^-, K1 = 10 - 0.75 \tag{2.3}$$

$$HCrO_4 \rightarrow H^+ + CrO_4{}^{2-}, K2 = 10 - 6.45 \tag{2.4}$$

$$2HCrO_4{}^- \rightarrow Cr_2O_7{}^{2-} + H_2O, K3 = 10 - 2.2 \tag{2.5}$$

Cr(VI) compounds are very soluble and hence mobile in the environment. Even so, electron donors such as organic matter or reduced inorganic species are readily reduced to trivalent forms by Cr(VI) oxyanions, which are widespread in soil, water and atmospheric systems.

2.6 Oxidation/Reduction Reactions in Soil

The existence of chromium in soil is highly dependent on the redox potential and pH. In most soils, chromium would be found primarily in the Cr(III) state. Under reduction conditions, redox reactions with aqueous inorganic species, electron transfer at mineral surfaces, reactions with non-humic organic substances such as carbohydrates and proteins or reduction of soil humic substances can reduce Cr

(VI) to Cr(III). With a drop in pH, the shift from Cr(VI) to Cr(III) increases. In aerobic soils, often at a moderately alkaline pH, the reduction of Cr(VI) to Cr(III) is possible if the soil provides a sufficient organic source of energy for the redox reaction (Bartlett and Kimble 1976). Sub-soils contain much less organic matter than soils on the top and may be less likely to restrict the movement of chromates. However, because pH is an important factor driving the rate and degree of Cr (VI) reduction in sub-soils, Cr(VI) can be significantly decreased in acidic sub-soils because acidic conditions increase the frequency of releasing of Fe (II) species from soil minerals to interact with soluble Cr(VI) species and also accelerate the amount of reduction of Cr(VI) by organic matter (Arnfalk et al. 1996; Eary and Rai 1989, 1991). Abiotic reductants such as vitamin C (Xu et al. 2005) and nano-materials (UV/TiO_2) (Xu et al. 2006) have been found to be efficient in reducing Cr(VI) across the entire pH spectrum. Cr(III) can be oxidised to Cr(VI) in some conditions, a mechanism that may lead to significant ecological consequences (Bartlett 1991; Bartlett and James 1979; Rock et al. 2001; Chuan and Liu 1996; Eary and Rai 1987; James and Bartlett 1983a, b; Milacic and Stupar 1995). Cr(III) chelated oxidation by organic compounds with low molecular weight is slower than recently precipitated Cr(III) (less crystalline) (James and Bartlett 1983a, b). The mechanism of oxidation tends to be confined to oxidation by oxygen or oxides of manganese (Bartlett and James 1979; James and Bartlett 1983a). As per Eary and Rai (1987), oxygen does not really react dramatically with Cr(III).Cr(III) oxidation can be associated with the amount of manganese reduced by hydroquinone in the soil (James 2001; Bartlett and James 1979) and does not occur in dry soil. The drying of the soil affects the surface of the manganese, reducing its ability to oxidise Cr (III). Cr(III) to Cr(VI) oxidation is also constrained by the water-soluble chromium concentration, pH, initial accessible surface area, and ionic strength (Apte et al. 2005; Eary and Rai 1987). Due to the inaccessibility of mobile Cr(III), a significant portion of chromium in the soil would not be oxidised to Cr(VI) except in the presence of manganese oxides and favourable pH conditions (Apte et al. 2005). Cr (VI) in reduction reactions is reduced to Cr(III) or Cr(III) is oxidised to Cr(VI) in oxidation reactions (Kimbrough et al. 1999). In these processes, the pH, oxygen concentration, presence and concentration of reductants are important. James and Bartlett proposed that $HCrO^{4-}$and CrO_4^{2-}mobile forms of Cr(VI) can be reduced by various inorganic reducers such as Fe(II) or S^{2-}, which is very essential in this phase called 'dechromification'. Atmospheric oxygen may be converted into chromate in the absence of such a process, which would pose a danger to life on earth (James and Bartlett 1988). Oxidation processes will oxidise Cr(III) to Cr(VI) in the vicinity of H_2O and free O_2 (James 1996) in the presence of oxidising agents such as Mn and Pb.

2.7 Toxicity of Chromium

The major anthropogenic causes of Cr(VI) pollution are: (1) chromite extraction and refining, (2) chromium plating, (3) chromium chemical production, and (4) evaporation of cooling towers (ATSDR 2000). Chromium is an important micro-nutrient in animal and human diets, since it is important for the maintaining metabolism of mammalian starch, lipids, and proteins (Mordenti and Piva 1997). Its dietary deficiency causes lipid and glucose metabolism to alter in animals and humans. Chromium is found in the glucose tolerance factor complex (GFC). In plants and microorganisms, no beneficial effects of chromium are known (Cervantes et al. 2001; Nies 1999). Raised chromium levels, though, are always hazardous, because as extent of toxicity is relevant to the chromium oxidation state. Indeed Cr (VI) extremely toxic to all types of living organisms, mutagenic to bacteria, carcinogenic, and mutagenic to humans and animals (Losi et al. 1994), although it is engaged in inducing genetic abnormalities and reducing reproductive health (Kanojia et al. 1998). In animals and humans, Cr(VI) can cause mortality if consumed in high quantities (Zayed and Terry 2003). For oral toxicity in rats, the LD50 (dose that causes 50% of the given animal population to die) is 50 to 100 mg kg^{-1} for Cr^{6+} and 1900–3000 mg kg^{-1} for Cr(III) (De Flora et al. 1990). The toxicity of Cr (VI) is due to its convenient diffusion in prokaryotic and eukaryotic species across the cell membrane and subsequent reduction of Cr(VI) in cells, which provides free radicals that can directly induce DNA alterations and detrimental effects (Arslan et al. 1987; Kadiiska et al. 1994; Liu et al. 1995). Cr(III) has indeed been measured to be 10–100 times less toxic than C(VI) (De Flora et al. 1990), since most Cr(III) complexes seem to be very impermeable to cellular membranes. Nonetheless, intracellular Cr(III), the final component of Cr(VI) reduction, forms amino acid nucleotide complexes in vivo, the mutagenic potential of which is not well known (Roundhill and Koch 2002). Bielicka et al. (2005) identified the positive and negative effects of Cr(III) and Cr(VI) on humans. Positive impacts comprise Cr (III) involvement in glucose metabolism (glucose tolerance factor), reducing levels of body fat, cholesterol, and triglycerides, activated enzyme reactions and increased muscle mass, whereas dermatitis, asthma, tumours, mutations, and teratogenic effects are adverse effects of Cr(III). Prokaryotes are considered to be more resistant to Cr(VI) than eukaryotes (Kamaludeen et al. 2003). Many authors have studied the harmful effects of Cr(VI) on bacteria, algae, and plants (Kamaludeen et al. 2003; Cervantes et al. 2001; Wong and Trevors 1988). Alternatively, there is scarce evidence on the effect of chromium on the composition and diversity of soil microbial communities (Viti and Giovannetti 2001, 2005). For total Cr in drinking water, the federal maximum concentration level (MCL) is 100 g L^{-1} (USEPA 1999), the California MCL is 50 g L^{-1}(Calder 1988). The National Occupational Health and Safety Institute (NIOSH) proposes an exposure limit of Cr(VI) of 1 g L^{-1} and an exposure limit of 500 g L^{-1} for a 10-h workday, 40-h week for Cr(0), Cr(II), and Cr (III). Cr (IV) is around 35% of the chromium that is released from all anthropogenic sources. In the atmosphere human activity further leads to chromium contribution. Total Cr concentration is limited at 0.1 mg L^{-1} in drinking water because of toxicity issues in the USA, 5 mg L^{-1} leached from solids in the toxicity trait. The standard ratio of plant chromium to soil chromium is calculated at 0.00455. A limit of 100 g Cr(III) and 50 g Cr(VI)/L of drinking water was set by the EPA. The Occupational Safety and Health Administration (OSHA) has set limits for 8-h work shifts and 40-h workweeks (Das and Singh 2011) of 500 g of water-soluble Cr(III) compounds per cubic metre of workplace air (500 g L^{-1}), 1000 g L^{-1} for metallic chromium(0) and insoluble chromium compounds, and 52 g L^{-1} for chromium(VI) compounds.

The toxicity of chromium to soil microorganisms was difficult to ascertain in several experiments, since the areas studied were mostly contaminated with organic compounds and/or various heavy metals at the same time (Turpeinen et al. 2004). Oxygenic phototrophic microorganisms and heterotrophic bacterial communities are both infected by chromium in soils severely contaminated with chromium (about 5000 mg kg^{-1} of soil) by leather tannery activity (Viti and Giovannetti 2001). Generally, chromium contamination does not influence the size of the cultivable heterotrophic bacterial population, although there is a link between the proportion of chromate resistant bacteria and the amount of chromium throughout the soil. Some studies have indicated that Gram-positive bacteria are more chromate-tolerant than Gram-negative bacteria (Viti and Giovannetti 2001, 2005; Ross et al. 1981). Shi et al. (2002) reported that chromium adversely affected the microbial behaviour of the soil and contributed to soil organic carbon accumulating. Speir et al. (1995) suggest that low exposure to Cr(VI) inhibited soil biological activities, such as phosphatase and sulphatase activity, and lowered microbial biomass (Viti and Giovannetti 2007).

2.8 Evaluation and Chemical Processing of Chromium in Various Solid Wastes

Due to manufacturing and use, contamination of chromium in solid waste/residue has been found through various sources and locations. From the mining of ore to the final use of chromium, a variety of waste is produced. In uncontaminated soils, in the form of hydroxides and oxides, the comparatively insoluble and mobile Cr(III) predominates and adsorbs on clay particles, soil organic matter, metal oxyhydroxides, and other surfaces that are negatively charged. A variety of advanced clean-up technologies, such permeable reactive barriers (Naftz et al. 2002; Flury et al. 2009; Wanner et al. 2011) and Na-dithionite liquid injections (Fruchter et al. 2000), have been introduced for the remediation of Cr(VI)-contaminated sites. Bioremediation and regulated natural attenuation are among the most cost-effective methods to the management of sub-soil contamination (Illman and Alvarez 2009). A certain major sites polluted with chromium are listed here and their methodologies for treatment are discussed.

2.9 Chromite Ore Processing Residue (COPR)

At many commercial and industrial sites in the USA, UK, China, Japan, India, and many other countries, chromite ore mining residue is found in surface soils. Hudson County in northeast New Jersey on the western side of the Hudson River opposite New York City (>100 sites) (Burke et al. 1991) and Shawfield in southeast Glasgow (>15 sites) (Bewley et al. 2001) are two large polluted COPR sites. COPR is an industrial waste material derived by the manufacture of chromate and bichromate from chromium ores. In a ball mill, the ore is pulverised to less than 100 mesh sizes (150 m), mixed with soda ash and lime, and roasted at 1100–1150 °C in rotary kilns. The combination does not fuse, but to form water-soluble sodium chromate, the molten soda ash reacts with the chromite. The overall volume of manufacturing residue produced is estimated to be in the range of 2–3 million tonnes. The reaction can be generalised as Eq. (2.6) for the chromate development method.

$$4FeCr_2O_4 + 8Na_2CO_3 + 7O_2 \rightarrow 8Na_2CrO_4 + 2Fe_2O_3 + 8CO_2 \tag{2.6}$$

Burke et al. (1991) have identified the potential for human exposure and adverse effects due to Cr(VI) mobility, and the environmental fate of chromium from chromium-contaminated sites near New Jersey City. Different treatment methods have been proposed for the remediation of chromium from COPR (both on bench and field scale), some of which are listed in brief here. Distilled water/ultrapure water (Li et al. 2011a; Sun et al. 2009, 2010; Zhang et al. 2009; Mcleod 2001), simulated rain water (Weng et al. 2001), and tap water (Higgins et al. 1998) were used to analyse the leaching and solubility of Cr(VI) from COPR. The leaching of chromium by chemical acids such as HF (Li et al. 2011b), the weak extractable acid, the reducible and oxidisable form of chromium (Sun et al. 2009) and the mobilisation of H_2SO_4 (Meegoda et al. 1999; Tinjum et al. 2008) and HNO_3 (Tinjum et al. 2008) were also examined (Geelhoed et al. 2003). Nano-scale zero-valent iron (Fe0) (Du et al. 2012), reduction with $FeSO_4$ and leaching with H_2SO_4 (Jagupilla et al. 2009), $FeSO_4$ and asphalt (Moon et al. 2009), and $FeSO_4$ (Dermatas et al. 2006) have been reported to reduce Cr(VI) in COPR. $FeSO_4$, sodium dithionate ($Na_2S_2O_4$), $FeCl_2$, citric acid, hydroxylamine sulphate, peerless iron (a commercial NaCl, NaCl + Na2SO4 in batch and column (Geelhoed et al. 2003) were investigated for the effect of mixed reductant on COPR reduction. Hydrothermal reduction of Cr(VI) in COPR was achieved using starch and H2SO4 leaching (Xu et al. 2011). Meegoda et al. (1999) investigated the extraction of chromium by high-temperature vitrification technologies to make glass from waste-containing chromium, by three immobilising mechanisms: encapsulation, silicon atom substitution, and non-bridging oxygen bonding) where 2.5 kg of soil (1/4 in. containing 5% moisture) from each site was combined with 25% (w/w) of sand (as silica source).

2.10 Leather Tannery Contaminated Soil

The chromium comprising tannery solid wastes classified as shavings are of significant concern among the various waste contaminants because of their high toxicity. The use of chromium in tanning processes enhances the rate of mobility and transfer of chromium, which greatly exceeds the rate of natural cycle processes, leading to significant complications in the tanning industry (Taylor et al. 1990). While as the major recipient of meat industry waste, the tanning industry is environmentally significant, the industry is viewed as a resource collector and a pollutant source. 200 kg of final leather commodity (3 kg Cr), 250 kg of non-tanned solid waste, 200 kg of tanned waste (3 kg Cr), and 50,000 kg of waste water (5 kg Cr) are developed by processing a single metric tonne of raw hide (Hüffer and Taeger 2004). Thus about 20% of the feedstock is processed into leather, and the solid and liquid waste comprises over 60% of the chromium. Approximately 2500 units are distributed throughout the country in India, but the major areas of concentration are Tamil Nadu, Uttar Pradesh, and West Bengal with an annual production potential of 0.7 million tonnes of hides and skins (Rajamani 1995). The use of basic chromium sulphate (BCS) has risen from 25,000 to 40,000 tonnes per year (Agrawal et al. 2006), with 80% of tanners participating in the chromium tanning process (Rajamani 1995). The probability of oxidation of Cr(III) into Cr(VI) under mild conditions by air in the broad pH range is a major concern. Oxidation can be realised mainly according to the following equations (Kolomaznik et al. 2008):

$$2Cr_2O_3 + 8OH^- + 3O_2 = 4CrO_4{}^- + 4H_2O \text{ (in alkali medium)} \quad (2.7)$$

$$2Cr_2O_3 + 3O_2 + 2H_2O = 2Cr_2O_7{}^- + 4H^+ \text{ (in acid medium)} \quad (2.8)$$

Tannery effluent and waste sludge soil contamination is fundamentally distinct from air or water pollution since the accumulation of heavy metals in soil is allegedly prolonged than in other biosphere compartments. Removal of heavy metals from contaminated soils is difficult due to versatile composition and material. Many metals, such as lead and chromium, can be practically irreversible once settled on the surface (Okeyode and Moshood 2010). Different removal and extraction methods have been investigated to remediate pollutants from infected soil from tannery waste. Chromium recovery was investigated by Erdem (2006) from tannery chrome shavings. Mineral acids were used for the extraction of Cr by Shen et al. (2001). The separation of Cr(III) with sulphuric acid and recovery by oxidation to Cr (III) was reported by Macchi et al. (1991). The effect of pH on chromium leaching actions from tannery sludge was studied by Chuan and Liu (1996). Other analysis showed that oxidation of Cr(III) to Cr(VI) and Cr(VI) was further removed using an organic solvent. Some teams used the micro-emulsion method which used the solid–liquid extraction process to treat dry, ground, and sieved tannery sludge (Dantas et al. 2009). Virotec Global Solution Company, referred to as ViroflowTM technology used ViroChromeTM reagent, for pilot scale treatment of tannery sludge where the concentration of chromium rapidly decreased below the recommended limit

(5 mg L^{-1}) in the 5d mixing period (Virotec 2011). Usually, these processes have better extraction efficiency, but due to operational problems, high chemical consumption and long processing time requirements, the implementation of chemical extraction methods is limited and costly.

2.11 Electroplating Sludge and Contaminated Soil

Large amounts of hazardous waste sludge comprising Cr, Pb, Ni, and Zn are produced by metal finishing and electroplating industries which are mainly disposed of on landfill, generating new environmental threats (Savvides et al. 2001; Wang et al. 2001). The electroplating sludge contains 7–11% Cr and 3–5% Fe together with other metals such as Ni, Co, Zn, Cd, Mo, V, Cu, etc. in various amounts based on the types of substrate used for surface treatment (Lee and Pandey 2012), especially those from the chrome plating units which settle in the bath. Due to the presence of metal ions as well as other inorganics, that pose significant disposal problems, sludge produced from the electroplating factories is extremely complicated. In solid waste such as portland cement and portland cement with fly ash, solidification stabilisation (S/S) has been used to stabilise metals. Water applied interacts chemically with cement to form hydrated silicates and aluminates in S/S procedures of cement, resulting in an aggregate mass of concrete; TCLP were smaller than the chromium value (5 mg L^{-1}) (Sophia and Swaminathan 2005). In order to generate a type of waste with high physical consistency, an optimum cement-to-waste ratio is therefore required (Landreth 1979). Rossetti et al. (2002) summarised the optimisation of the aqueous solution S/S method for Cr, Pb, Zn, and Li. The possibility for using alkali-activated pulverised fly ash (PFA) as a solidification binding material for the treatment of electroplating sludge has been explored by Asavapisit and Chotklang (2004). The prospect of slightly replacing the raw resources of cement with sludge extracted from chrome plating for cement manufacturing was investigated by Espinosa and Tenório (2000). When leaching chrome plating contaminated soils of particles ~2 mm for a significant period of time (365d) using calcium polysulfide (29% CaSx), a reduction in Cr(VI) to <10 g L^{-1} was recorded in a much shorter period of 28 or 60 days depending on the reduction concentration (Chrysochoou et al. 2010). Rock et al. (2001) investigated the oxidation and electroplating waste (0–24 mM) solubility with hydrogen peroxide, where 200-M additional Cr(VI) with leachate containing 24 mM H_2O_2 was released over 24 h. Approximately 92% of chromium was removed using 30% H_2O_2 at 60 °C from electroplating sludge (Silva et al. 2006).

2.12 Metallurgical and Construction Waste Contaminates

Metals comprising waste and by-products of the metallurgical industry are typically produced in the form of slag, sludge, ashes, tailings, and by-products, either as final or intermediary products (Lee and Pandey 2012). In addition, based on the type of metal and alloys formed, the formation of flue dust and sludge from the blast furnace, flue dust/sludge from the electric arc furnace especially that of Fe-Cr and alloying units, is abundant in many non-ferrous metals. Since slags of various sources contain a good quantity of non-ferrous metals and not just chromium, they should be removed/leached prior to land dumping. For the leaching of metals from such metallurgical waste, some literature is available. Ni and Cr leaching columns (ID 1.5 cm; height 5 cm) using EDTA and citric acid (2.5 mmol) from soil polluted with steel, super alloy, aluminium, and titanium alloy processing units have been examined. Chromium leached more than EDTA (200 g L^{-1}) with citric acid (250 g L^{-1}) and Ni leached more than citric acid with EDTA (Jean-Soro et al. 2012). Liu et al. (2010) also observed that chromium leaching is very low (in 30 days) with ultrapure water at 1:10 and 100 ratios for 15 and 30 days, respectively, 10 rpm and room temperature. Using the Cascade Leaching Test (CLT) in de-mineralised water (acidified to pH 4 with a drop of diluted HNO_3), a high Cr(VI) leaching testing was conducted how far ~2230 mg kg^{-1} of Cr(VI) was leached at neutral pH and Cr (III) was leached at acidic pH (Shtiza et al. 2008). Separate leaching methods were followed for the measurement of chromium content in the ferrochrome smelter sample (Slag, fly ash, below an acidic pH) (Shtiza et al. 2008). For the determination of chromium content in the ferrochrome smelter sample (Slag, fly ash, bottom ash) and surrounding agricultural and contaminated soil, different leaching methods were adopted (Ore˘s˘canin et al. 2006). The soil had a chromium content of 7-–1527 mg kg^{-1} in the closest proximity of the factory, which was 310 times higher than the maximum permissible concentration. As with aqua regia, the chromium leached in the presence of various lixivants was ~2230 mg kg^{-1}, acetic acid ~1.3 mg kg^{-1}, KH_2PO_4 ~ 2.9 mg kg^{-1}, and 0.9 mg kg^{-1} of chromium from steel processing waste with purified water (Facchinelli et al. 2001). Fällman (2000) mentioned leaching of chromium from low alloy steel processing scrap-based slag using ultrapure water and 0.1 mM HNO_3 in batch and column (0.19 m diameter and 0.9 m height) at L/S (liquid/solid) 5 for 24 h. The high recovery of chromium (90–96%) and iron (>96%) was accomplished with the use of Fe-Cr dust in smelting in a DC transformed plasma arc reactor at a specific energy consumption of 1250 k Wh ton^{-1} (Sokolov et al. 2005). Gomes and Pinto (2006) used the DIN 38414-S4 process (100 g L^{-1} S/L for 24 h, particle size <10 mm) to analyse the leaching of EAF chromium and ladle slags from a Portuguese steel manufacturer. They observed that even with care over 1 year, average leaching values were very low, thereby complying with the non-hazardous norm (10 mg L^{-1}).

2.13 Contamination from Waste from Mines

Depending on the properties of the ores and waste dump materials, mining can affect water and soil quality. Open cast mining and overburden dump formation will do more damage than reducing forest yields. Due to the presence of different stress factors such as low nutritional value, pH imbalance combined with toxic and heavy metal ion accumulation, and inadequate water holding capacity, overburden dumps are frequently hostile to plant growth (Dhal et al. 2011). The quantity of chromium in both marine and coastal environments has risen over the last few decades as a result of anthropogenic activities (Saha and Orvig 2010; Basu et al. 2010; Samuel et al. 2012). The mine dumps are the main cause of pollution of heavy metals like Zn, Hg, Al, and Cr, etc.

Several studies are available worldwide on the pollution of Cr(VI) from the chromite mining region. In Kemi, Finland, pollution from chromite mines and ferrochrome processing plants from dust deposited over an open pit chromite mine resulted in significantly higher levels of Cr and other heavy metals in pine tree bark (Pöykiö et al. 2001). Even worse, no studies have been performed on how these levels of emissions have impacted plant and animal health. In Zimbabwe, ferrochrome furnace dust was released from chromium-contaminated smelter soils up to 700 m away from the furnace, where 700 mg Cr kg^{-1} soil was measured (Mining Watch Canada 2012). A total of 54.6 tonnes of chromium per year was released from one stack (of six) by the furnace. A ferro-alloy factory in the Hunan Province of China dump slag directly on the field and also discharge waste water into the sewage system flowing across agricultural land (Ma and Garbers-Craig 2006). The consequence was long-term contamination of the soil and vegetables. The overall content of chromium in the soil varied from 90 to 6200 mg kg^{-1} and Cr (VI) varied from 0.1 to 252 mg kg^{-1}.The Sukinda Valley, Jajpur district, Orissa, India, occupies about 98% of India's chromite ore reserve. Approximately 7.6 million tonnes of solid waste in the form of rejected minerals, overburden materials/waste rock, and low-grade ore (Dhal et al. 2010a) are produced by open cast mining projects in this area. Owing to the leaching of these solid wastes, Cr (VI) pollutes the water sources in and around the mines (Tiwary et al. 2005). Sukinda chromite mining is among the most polluting operations primarily due to the natural oxidation of chromite from the overburden dumps, which causes soluble Cr (VI) species in the surrounding regions. According to figures, the extraction of 1 tonne of chromite produces roughly 10 tonnes of overburden, which varies in its chemical and mineralogical characteristics (Dhal et al. 2010a, 2010b). Favas et al. (2011) investigated concurrent leaching using leachants such as purified water, 1 M ammonium acetate, 0.2 M ammonium oxalate, 35% H_2O_2, $KClO_3$ + 12 M HCl + 4 M HNO_3 and HF + HNO_3 + $HClO_4$ to leach the soluble fraction of Cr(VI).

2.14 Pollution from Municipal Hazardous Waste and Polluted Soil

The involvement of hazardous heavy metals in municipal solid waste composts (MSWC) poses significant concerns about the harmful effects on the environment and unsustainable use in agricultural land (Nouri et al. 2006; Ayari et al. 2008; Mahvi 2008). Heavy metals derive primarily from a variety of origins of non-source segregated municipal solid waste: batteries, electrical equipment, newspapers, paint chips, foils, engine oils, and plastics, all of which can incorporate metal pollutants into the organic compostable fraction (Hamdi et al. 2003). Eventually high and inappropriate heavy metal concentration in soil and other media can contaminate both the food chain of humans and animals (Iwegbue et al. 2005). The leachability of metals found in MSW has recently been tested by several researchers using various accredited leaching methods. The chromium content (0.06–2.24 mg per plant) was reported by Ayari et al. (2010) in plants of municipal soil when compost/manure containing 78.87 mg kg^{-1} chromium was used. The sequential leaching process for municipal solid waste compost (653–678 mg kg^{-1}) and cow manure compost (51.8–54.5 mg kg^{-1}) was followed with purified water. Findings indicate that the chromium was insoluble as humic compounds were coupled with large molecules (Bhattacharyya et al. 2005). Esakku et al. (2003) stated digestion of chromium acid. Of the 140 mg kg^{-1} chromium content in the MSW, the water-soluble chromium was very low. Several polluted sites were evaluated for the chromium content in the waste dump produced by industrial activities. Machender et al. (2011) measured the amount of chromium (371.98 mg kg^{-1}) in samples collected from the industrial region of Balanagar, Hyderabad, India. In shoots and roots, a Switchgrass (*Panicumvirgatum* L.) plant accumulates 827.5 mg kg^{-1} of Cr when grown on soil containing 600 mg kg^{-1} chromium.

2.15 Biological Removal of Chromium in Various Industrial Waste Products

2.15.1 Chromite Trace Mining Ore

COPR was treated with reducing agents such as gastric fluid (Yu et al. 2012), earlier leaching with tap water accompanied by H_2SO_4 and reduction with cow manure (Higgins et al. 1998), acetate modified/unmodified microbes (microcosms) (Whittleston et al. 2011a), Fe(II) containing COPR with microcosms (Whittleston et al. 2011b), microbial reduction with HCl (Stewart et al. 2010), and microbial reduction with *Bacillus* sp., *Micrococcus* sp., and *Rhodococcus* sp. (Turick and Apel 1997). For the remediation of chromium from COPR, in situ physicochemical and ex situ microbiological, organic matter, sulphate reducing bacteria (SRB), and FeSO4 treatment have also been studied (Bewley et al. 2001). Many secondary biological materials such as sucrose, starch, and wheat flour were used to minimise Cr(VI) from COPR. Uptake of Cr(VI) on Eclay from the leached solution of COPR has also been studied (Mcleod 2001).

2.15.2 Leather Tannery Contained Soil

Over the last decade, the growing environmental issue and regulatory standards enforced by policymakers have driven industry to adopt an eco-friendly strategy for bioremediation of degraded soil, which has become one of the main research fields in environmental science and technology. In view of the above, it has been found that microbial bio-surfactants generated by bacteria extracellularly or as part of the cell membrane and plant-derived bio-surfactants are of specific interest in environmental applications. Because of their low toxicity, biodegradability, enhanced biodegradation characteristics, fast development and the likelihood of reuse, these bio-surfactants are also potential non-toxic cleansing agents (Mulligan 2005). In addition, even though they are spilled and released into the atmosphere, they do not cause secondary contamination. Several researchers have reported Microbial reduction of Cr(VI) from the polluted soils and sediments of the tannery. These include microbial reduction (Kathiravan et al. 2011), microbial reduction in the presence of molasses (Jayasingh and Philip 2005), microbial usage of cow manure (Sethunathan et al. 2005), and native bacterial isolates of tannery effluent sediments deposited at the site of leached Cr(VI) solution (Sundar et al. 2010). Zhou et al. (2006) and Wang et al. (2007) have used the combination of indigenous iron-and sulphur-oxidising bacteria to use bioleaching procedures for tannery sludge. The sulphur-oxidising bacteria use microbiological techniques having distinct benefit relative to the mineral acids absorbing chemicals (Skrypski-Mantele and Bridle 1995; Shen et al. 2001, 2002, 2003; Zhou et al. 2004a, b, 2005). As sulphur-oxidising bacteria play a significant role in the removal of chromium and other metals from tannery sludge and other waste, interest has emerged in the bacterial treatment of tannery sludge. However, in contrast with chemical leaching, requiring significant volumes of elemental sulphur (up to 30 g L^{-1}) (Shen et al. 2002, 2003) or sulphuric acid pre-acidification with elemental sulphur addition (Zhou et al. 2005), bacterial leaching of chromium from tannery sludge using *Acidithiobacillus thiooxidans* is time-consuming (from 8 to 25 days) (Shen et al. 2002, 2003; Zhou et al. 2004a, b, 2005). Massara et al. (2007) stated that rhamnolipids have the potential to remove 25% of the stable chromium form from polluted kaolinite and may be useful for long-term conversion Cr(VI) to Cr(III).Bio-surfactant aescin (Hong et al. 1998) and saponin extracted from plants have been used in aqueous solutions for the remediation of polluted heavy metals, oils (Hong et al. 2002), kaolin (Chen et al. 2008), and fly ash incinerators. Saponin treatment removed 24% of Cr from tannery sludge (8041 mg kg^{-1}Cr) at pH 2 at33°C in 6 h.

2.15.3 Polluted Soil and Sludge from Electroplating

The biological treatment of electroplating waste sludge and polluted soil for metal leaching and remediation of Cr (IV) is also being studied. The bioleaching of plating sludge in a completely mixed batch reactor (CMBR) of 3.0 L capacity using *A. ferrooxidans* was recently reported by Bayat and Sari (2010). The findings showed poor chromium recovery (34%) along with relatively strong leaching of other metals (97% Zn, 96% Cu, 93% Ni, 84% Pb and 67% Cd) from the dewatered metal plating sludge without sulphide or sulphate compounds in 20 days at 20 g L^{-1} pulp density, 2.0 pH and 25 °C. For the management of waste water sludge containing chromium and other metals, the hybrid method involving biochemical approach has been suggested (Drogui et al. 2005a, b). Approximately 49% Cr along with 63% Cd, 71% Cu and 80% Zn is dissolved by chemical leaching of the sludge in a 5.2 L continuous stirred tank reactor (CSTR) with bio-oxidised Fe(III) using a ferro oxidant. Fujii et al. (1988) previously reported the possible use of a bacterial isolate (*A. Ferrooxidans*) from the sludge of the steelworks' plating and pickling wastewater pit (Nippon Steel) for Fe oxidation. The oxidisability of Fe(II) from certain waters containing heavy metals and chloride ions can be used by an effective process to perform activated sludge treatment of steel works water and to ultimately recover metal. Turick et al. (1998) stated 218 mg kg^{-1} Cr(VI) leaching with mineral salt solution from the soil polluted by the active galvanising industry. Due to the development of indigenous bacterial strains with mineral salt in a glucose medium treated for 128 days, only a 64.5% decrease was achieved.

2.15.4 Metallurgical and Construction Waste Contaminates

The biological treatment of ferrochrome and steel factory slag and dust was also carried out in order to comply with environmental requirements. Banerjee (2007) showed significantly greater bio-recovery of iron and heavy metals from BF sludge and fluid dust by *A. ferrooxidants* relative to that of a fungus. The BF slag can be used by a mixed mesophilic culture at 35 °C as a neutralising agent (to pH 1.5) instead of slaked lime during pyrite oxidation (75–80%) and chromium leaching was just 16% of its value in the AOD slag when dissolving 55 mgL^{-1}Cr in 16 days of mining waste pollution. In Luria Bertani broth incubation at 35, pH 7, and 100 rpm, the microbial remediation of hexavalent chromium was also studied using indigenous bacterial species with a resistance of 2000 mgL^{-1}Cr(VI).For Cr (VI) concentrations of 10–500 mg L^{-1}, the chromate reduction rate was 3.22×10^{-2} to 6.5 × 10–3 h^{-1}. The bacterial reduction of Cr(VI) using Sukinda mine region isolates with a reduction rate ranging from 0.199 to 0.477 mg L^{-1} h^{-1} to 0.5–1.16 mg L^{-1} h^{-1} at 5–20 mg L^{-1} of actual Cr(VI) concentration, was reported by Samuel et al. (2012). In a recent analysis to decrease Cr(VI) in chromite overburden/soil comprising 500 mg Cr(VI) kg^{-1}, Bacillus sp. (4.05 × 10 cell mL^{-1}) reduced greater than 98% of Cr(VI) in the sample at 20% pulp density and 7.0 pH in 10 h (Dhal et al. 2010b).

2.16 Conclusion

In the light of the available literature relating to each and every phase of the availability, leachability and toxicity of chromium, we have concluded that there is an urgent need for low-cost biological treatment system for the remediation and toxicity reduction of chromium from different liquid and solid wastes.

References

Agency for Toxic Substances and Disease Registry (ATSDR) (2000) Toxicological profile for chromium, U.S. Department of Health and Human Services, Public Health Service, ATSDR

Agrawal A, Kumar V, Pandey BD (2006) Remediation options for the treatment of electroplating and leather tanning effluent containing chromium – a review. Miner Process Extra Metal Rev 27 (2):99–130

Apte AD, Verma S, Tare V, Bose P (2005) Oxidation of Cr(III) in tannery sludge to Cr(VI): field observations and theoretical assessment. J Hazard Mater 121(1–3):215–222

Arnfalk P, Wasay SA, Tokunaga S (1996) A comparative study of cd, Cr(III), Cr(VI), hg, and Pb uptake by minerals and soil materials. Water Air Soil Pollut 87:131–148

Arslan P, Beltrame M, Tomasi A (1987) Intracellular chromium reduction. Biochem Biophys Acta 931(1):10–15

Asavapisit S, Chotklang D (2004) Solidification of electroplating sludge using alkaliactivated pulverized fuel ash as cementitious binder. Cem Concr Res 34:349–353

Ayari F, Chairi R, Kossai R (2008) Sequential extraction of heavy metals during composting of urban waste. Chin J Geochem 27(2008):121–125

Ayari F, Hamdi H, Jedidi N, Gharbi N, Kossai R (2010) Heavy metal distribution in soil and plant in municipal solid waste compost amended plots. Int J Environ Sci Tech 7(3):465–472

Banerjee D (2007) Metal recovery from blast furnace sludge and flue dust using leaching technologies. Res J Chem Environ 11:18–21

Barrera-Díaz CE, Lugo-Lugo V, Bilyeu B (2012) A review of chemical, electrochemical and biological methods for aqueous Cr(VI) reduction. J Hazard Mater 223–224(223):1–12

Bartlett RJ (1991) Chromium cycling in soils and water: links, gaps and methods. Environ Health Perspect 92:17–24

Bartlett RJ, James B (1979) Behavior of chromium in soils: III. Oxidation. J Environ Qual 8:31–34

Bartlett RJ, Kimble JM (1976) Behavior of chromium in soils. II. Hexavalent forms. J Environ Qual 5(4):383–386

Basu A, Saha R, Mandal J, Ghosh S, Saha B (2010) Removal of hexavalent chromium by an aromatic alcohol. J Biomed Sci Eng 3:735–741

Bayat B, Sari B (2010) Comparative evaluation of microbial and chemical leaching processes for heavy metal removal from dewatered metal plating sludge. J Hazard Mater 174:763–769

Bewley RJF, Jeffries R, Watson S, Granger D (2001) An overview of chromium contamination issues in the south-east of Glasgow and the potential for remediation. Environ Geochem Health 23:267–271

Bhattacharyya P, Chakraborty A, Chakrabarti K, Tripathy S, Powell MA (2005) Chromium uptake by rice and accumulation in soil amended with municipal solid waste compost. Chemosphere 60:1481–1486

Bielicka A, Bojanowska I, Wi'sniewski A. (2005) Two faces of chromium – pollutant and bioelement. Polish J Environ Stud 14(1):5–10

Burke T, Fagliano J, Goldoft M, Hazen RE, Iglewicz R, McKeet T (1991) Chromite ore processing residue in Hudson County, New Jersey. Environ Health Perspect 92:131–137

Calder LM (1988) Chromium contamination of groundwater. In: Nriagu JO, Nieboer E (eds) Chromium in natural and human environments. Wiley, New York, pp 215–231

Cervantes C, Campos-García J, Devars S et al (2001) Interactions of chromium with microorganisms and plants. FEMS Microbiol Rev 25(3):335–347

Chen W-J, Hsia L-C, Chen KK-Y (2008) Metal desorption from copper (II) nickel (II)- spiked kaolin as a soil component using plant-derived saponin biosurfactant. Proc Biochemist 43 (5):488–498

Chrysochoou M, Dermatas D, Grubb D, Moon D, Christodoulatos C (2010) Importance of mineralogy in the geoenvironmental characterisation and treatment of chromite ore processing residue. J Geotech Geoenviron Eng 136(3):510–521

Chuan MC, Liu JC (1996) Release behavior of chromium from tannery sludge. Water Res 30 (4):932–938

Cotton FA, Wilkinson G (1980) Chromium. In: Advanced inorganic chemistry, a comprehensive text, 4th edn. Wiley, New York, pp 719–736

Dantas TNC, Oliveira KR, Dantas Neto AA, Moura MCPA (2009) The use of micro-emulsions to remove chromium from industrial sludge. Water Res 43(5):1464–1470

Das AP, Singh S (2011) Occupational health assessment of chromite toxicity among Indian miners. Indian J Occup Environ Med 15:6–13

De Flora S, Bagnasco M, Serra D, Zanacchi P (1990) Genotoxicity of chromium compounds – a review. Mutat Res 238:99–172

Dermatas D, Chrysochoou M, Moon DH, Grubb DG, Wazne M, Christodoulatos C (2006) Ettringite-induced heave in chromite ore processing residue (COPR) upon ferrous sulfate treatment. Environ Sci Technol 40:5786–5792

Dhal B, Das NN, Thatoi HN, Pandey BD (2010b) Kinetic aspects of Cr(VI) biore-mediation of water and soil contaminated from chromite mine overburden dumps. In: Singh R, Das A, Banerjee PK, Bhattacharyya KK, Goswami NG (eds) proc. XI Int. Sem. Min. Proc. Technol. NML, Jamshedpur, pp 1007–1014

Dhal B, Thatoi HN, Das NN, Pandey BD (2010a) Reduction of hexavalent chromium by bacillus sp. isolated from chromite mine soils and characterization of reduced product. J Chem Technol Biotechnol 85:1471–1479

Dhal B, Thatoi HN, Das NN, Pandey BD (2011) Environmental quality of the Boula-Nuasahi chromite mine area in India. Mine Water Environ 30:191–196

Drogui P, Mercier G, Blais JF (2005a) Hybrid process for heavy metal removal from wastewater sludge. Water Environ Res 77:372–380

Drogui P, Mercier G, Blais JF (2005b) Bio-production of ferric sulfate used during heavy metal removal from sewage sludge. J Environ Qual 34:816–824

Du J, Lu J, Wu Q, Jing C (2012) Reduction and immobilization of chromate in chromite ore processing residue with nanoscale zero-valent iron. J Hazard Mater 215-216:152–158

Eary LE, Rai D (1987) Kinetics of Cr(III) oxidation by manganese dioxide. Environ Sci Technol 21:1187–1193

Eary LE, Rai D (1989) Kinetics of chromate reduction. Am J Sci 289:180–213

Eary LE, Rai D (1991) Chromate reduction by subsurface soils under acidic conditions. Soil Sci Soc Am J 55:676–683

Erdem M (2006) Chromium recovery from chrome shaving generated in tanning process. J Hazard Mater 129(1–3):143–146

Esakku S, Palanivelu K, Joseph K (2003) Assessment of heavy metals in a municipal solid waste dumpsite. In: Workshop on sustainable landfill management, 3–5 Dec, Chennai, pp 139–145

Espinosa DCR, Tenório JAS (2000) Laboratory study of galvanic sludge's influence on the clinkerization process. Res Cons Rec 31:71–82

Facchinelli A, Sacchi E, Tararbra M, Deluca DA (2001) Assessment of chromium release from steel manufacturing waste. Environ Geochem Health 23:235–240

Fällman A-M (2000) Leaching of chromium and barium from steel slag in laboratory and field tests a solubility controlled process? Waste Manag 20(2–3):149–154

Favas PJC, Pratas J, Gomes MEP, Cala V (2011) Selective chemical extraction of heavy metals in tailings and soils contaminated by mining activity: environ-mental implications. J Geochem Explor 111:160–171

Flury B, Frommer J, Eggenberger U, Mader U, Nachtegaal M, Kretzschmar R (2009) Assessment of long-term performance and chromate reduction mechanisms in a field scale permeable reactive barrier. Environ Sci Technol 43(17):6786–6792

Fruchter JS, Cole CR, Williams MD, Vermeul VR, Amonette JE, Szescody JE, Istok JD, Humphrey MD (2000) Creation of a subsurface permeable treatment zone for aqueous chromate contamination using in situ redox manipulation. Grou Wat Mon Remed 20:66–77

Fujii M, Matsuhara S, Matayoshi Y, Fukunga K, Nomura Y, Umeda T (1988) Research on treatment by Thiobacillus ferrooxidans of pickling and plating waste waters at steelworks. Nippon Steel Tech Rep 37:59–67

Geelhoed JS, Meeussen JCL, Roe MJ, Hillier S, Thomas RP, Farmer JG, Paterson E (2003) Chromium remediation or release? Effect of iron(II) sulfate addition on chromium (VI) leaching from columns of chromite ore processing residue. Environ Sci Technol 37:3206–3213

Gomes JFP, Pinto CG (2006) Leaching of heavy metals from steelmaking slags. Rev Metal 42 (6):409–416

Hamdi H, Jedidi N, Ayari F, Yoshida M, Ghrabi A (2003) Valuation of municipal solid waste compost of Tunis (Tunisia)-agronomic aspect. In: Proc. 14th annual conf. Japan Society waste management experts, Vol. III, international session, pp 62–64

Haq I, Singh A, Kalamdhad AS (2020) Bioremediation for environmental sustainability toxicity, mechanisms of contaminants degradation, detoxification, and challenges. In: Arsenic: environmental contamination, health hazards, and bioremediation approaches for detoxification. Elsevier, pp 121–142

Higgins TE, Halloran AR, Dobbins ME, Pittignano AJ (1998) In situ reduction of hexavalent chromium in alkaline soils enriched with chromite ore processing residue. J Air Waste Manage Assoc 48(11):1100–1106

Hong K-J, Choi Y-K, Tokunaga S, Ishigami Y, Kajiuchi T (1998) Removal of cadmium and lead from soil using aescin as a biosurfactant. J Surfactant Deterg 1(2):247–250

Hong K-J, Tokunaga S, Kajiuchi T (2002) Evaluation of remediation process with plant-derived biosurfactant for recovery of heavy metals from contaminated soils. Chemosphere 49 (4):379–387

Hüffer S, Taeger T (2004) Sustainable leather manufacturing a topic with growing importance. J Am Leath Chem Assoc 99(10):423–428

Illman WA, Alvarez PJ (2009) Performance assessment of bioremediation and natural attenuation. Crit Rev Environ Sci Technol 39(4):209–270

Iwegbue CMA, Nwajei GE, Osakwe SA (2005) Recycling waste in agriculture: efficacy of composting in ameliorating trace metal availability and soil borne pathogens. Eur J Sci Res 11(4):571–577

Jagupilla SC, Moon DH, Wazne M, Christodoulatosa C, Kim MG (2009) Effects of particle size and acid addition on the remediation of chromite ore processing residue using ferrous sulfate. J Hazard Mater 168:121–128

James BR (1996) The challenge of remediating chromium contaminated soil. Environ Sci Technol 30:248–251

James BR (2001) Remediation by reduction strategies for chromate contaminated soils. Environ Geochem Health 23:175–179

James BR (2002) Chemical transformations of chromium in soils: relevance to mobility, bio-availability and remediation. In: The chromium file. International Chromium Dev. Assoc, Paris, pp 1–8

James BR, Bartlett RJ (1983a) Behavior of Cr in soils, V. fate of organically complexed Cr(III) added to soil. J Environ Qual 12:169–172

James BR, Bartlett RJ (1983b) Behavior of Cr in soils. J Environ Qual 12:173–176

James BR, Bartlett RJ (1988) Mobility and bioavailability of chromium in soil. In: Nriagu JO, Nieboer E (eds) Chromium in natural and human environments. Wiley Interscience, New York, pp 265–305

James BR, Petura JC, Vitale RJ, Mussoline GR (1997) Oxidation-reduction chemistry of chromium: relevance to the regulation and remediation of chromate contaminated soils. J Soil Contam 6:569–580

Jayasingh J, Philip L (2005) Bioremediation of chromium contaminated soil: opti-mization of operating parameters under laboratory condition. J Hazard Mater B118:113–120

Jean-Soro L, Bordas F, Bollinger J-C (2012) Column leaching of chromium and nickel from a contaminated soil using EDTA and citric acid. Environ Pollut 164:175–181

Kadiiska MB, Xiang QH, Mason RP (1994) In vivo free radical generation by chromium (VI): an electron spin resonance spin-trapping investigation. Chem Res Toxicol 7:800–805

Kamaludeen SPB, Megharaj M, Juhasz AL, Sethunathan N, Naidu R (2003) Chromium microorganism interactions in soils: remediation implications. Rev Environ Contam Toxicol 178:93–164

Kanojia RK, Junaid M, Murthy RC (1998) Embryo and fetotoxicity of hexavalent chromium: a long-term study. Toxicol Lett 95:165–172

Kathiravan MN, Karthick R, Muthukumar K (2011) Ex situ bioremediation of Cr(VI) contaminated soil by bacillus sp.: batch and continuous studies. Chem Eng J 169:107–115

Kimbrough DE, Cohen Y, Winer AM, Creelman L, Mabuni C (1999) A critical assessment of chromium in the environment. Crit Rev Environ Sci Technol 29(1):1–46

Kolomaznik K, Adamek M, Andel I, Uhlirova M (2008) Leather waste potential threat to human health and a new technology of its treatment. J Hazard Mater 160:514–520

Landreth RE (1979) Survey of solidification/stabilization technology for hazardous industrial wastes, EPA-600/2–79-056

Lee J-C, Pandey BD (2012) Bio-processing of solid wastes and secondary resources for metal extraction – a review. Waste Manag 32:3–18

Li C, Wang Q-H, Xiao B, Li Y-F (2011a) Phytoremediation potential of switchgrass (Panicumvirgatum L.) for Cr-polluted soil. Wat Res Environ Prot (ISWREP) 3:1731–1734. https://doi.org/10.1109/ISWREP.2011.5893582

Li X, Xu W, Zhou Q, Peng Z, Liu G (2011b) Leaching kinetics of acid-soluble Cr(VI) from chromite ore processing residue with hydrofluoric acid. J Cent South Univ Technol 18:399–405

Liu KJ, Jiang JJ, Shi XL, Gabrys H, Walczak T, Swartz HM (1995) Low frequency EPR study of chromium (V) formation from chromium (VI) in living plants. Biochem Biophys Res Commun 206:829–834

Liu S-Y, Gao J, Yang Y-J, Yang Y-C, Ye Z-X (2010) Adsorption intrinsic kinetics and isotherms of lead ions on steel slag. J Hazard Mater 173:558–562

Losi ME, Amrhein C, Frankenberger WT (1994) Environmental biochemistry of chromium. Rev Environ Contam Toxicol 136:91–131

Ma G, Garbers-Craig AM (2006) Cr(VI) containing electric furnace dusts and filter cake from a stainless steel waste treatment plant part 1 – characteristics and microstructure. Ironmak Steelmak 33(3):229–237

Macchi G, Pagano M, Pettine M, Santori M, Tiravanti G (1991) A bench study on chromium recovery from tannery sludge. Water Res 25(8):1019–1026

Machender G, Dhakate R, Prasanna L, Govil PK (2011) Assessment of heavy metal contamination in soils around Balanagar industrial area, Hyderabad India. Environ Earth Sci 63:945–953

Mahvi AH (2008) Application of agricultural fibers in pollution removal from aqueous solution. Int J Environ Sci Tech 5(2):275–285

Massara H, Mulligan CN, Hadjinicolaou J (2007) Effect of rhamnolipids on chromium-contaminated kaolinite. Soil Sediment Cont 16(1):1–14

Mazumder P, Haq I, Das A, Kalamdhad AS (2021) Microbial ecology of wastewater treatment plants. In: Microbial remediation of soil and water metal contaminants. Elsevier, pp 523–536

Mcleod N (2001) Chemical immobilisation of chromium wastes using modified smectite clays (E-clays). Environ Geochem Health 23:273–279

Meegoda JN, Kamolpornwijit W, Vaccari DA, Ezeldin AS, Noval BA, Mueller RT, Santora S (1999) Remediation of chromium contaminated soils bench scale investigation. Pract Period Hazard Toxic, Radio Waste Manage 3(3):24–131

Milacic R, Stupar J (1995) Fractionation and oxidation of chromium in tannery waste and sewage sludge amended soils. Environ Sci Technol 29(2):506–514

Mining Watch Canada (2012) Living near a chromite mine or ferrochrome facility, an overview of chromium, chromite and toxicity, Mining Watch Canada, Chromite Series Fact Sheet # 03

Moon DH, Wazne M, Koutsospyros A, Christodoulatos C, Gevgilili H, Malik M, Kalyon DM (2009) Evaluation of the treatment of chromite ore processing residue by ferrous sulfate and asphalt. J Hazard Mater 166:27–32

Mordenti A, Piva G (1997) Chromium in animal nutrition and possible effects on human health. In: Canali S, Tittarelli F, Sequi P (eds) Chromium environmental issues. Franco Angelis, Milan, pp 131–151

Mulligan CN (2005) Environmental applications for biosurfactants. Environ Pollut 133:183–198

Naftz DL, Morrison SJ, Davis JA, Fuller CC (2002) Handbook of groundwater remediation using permeable reactive barriers: applications to radionuclides, trace metals, and nutrients. Academic Press, San Diego

Nieboer E, Jusys AA (1988) Biologic chemistry of chromium. In: Nriagu JO, Nieboer E (eds) Chromium in the natural and human environments. Wiley, New York, pp 31–33

Nies DH (1999) Microbial heavy-metal resistance. Appl Microbiol Biotechnol 51:730–750

Nouri J, Mahvi AH, Babaei A, Ahmadpour E (2006) Regional pattern distribution of groundwater fluoride in the shush aquifer of Khuzestan County Iran. Fluoride 39(4):321–325

Okeyode IC, Moshood IA (2010) Elemental compositions of the soil of Ogun-Osun river basins development authority headquarters, Ogun state, Nigeria. Indian J Sci Technol 3(3):335–336

Oreˇsˇcanin V, Mikelic L, Lovrenˇcic I, Bariˇsic D, Mikulic N, Lulic S (2006) Environmental contamination assessment of the surroundings of the ex-ferrochromium smelter Dugi rat, Croatia. J Environ Sci Health 41(11):2547–2555

Pöykiö R, Tervaniemi O-M, Torvela H, Perämäki P (2001) Heavy metal accumulation in woodland moss (Pleurozium Schreberi) in the area around a chromium opencast mine at Kemi, and in the area around the ferrochrome and stainless steel works at Tornio, northern Finland. Int J Environ Anal Chem 81(2):137–151

Rajamani S (1995) Environment management in tanneries, feasible chromium recovery and reuse system. In: Proc. 3rd international conference on appropriate waste management technologies for developing countries, pp 965–973

Rock ML, James B, Helz GR (2001) Hydrogen peroxide effects on chromium oxidation state and solubility in four diverse, chromium-enriched soils. Environ Sci Technol 35(20):4054–4059

Ross DS, Sjogren RE, Bartlett RJ (1981) Behavior of chromium in soils: IV., toxicity to microorganisms. J Environ Qual 2:145–168

Rossetti AV, Palma LD, Medici F (2002) Assessment of the leaching of metallic elements in the technology of solidification in aqueous solution. Waste Manag 22:605–610

Roundhill DM, Koch HF (2002) Methods and techniques for the selective extraction and recovery of oxoanions. Chem Soc Rev 31:60–67

Saha, Orvig C (2010) Biosorbents for hexavalent chromium elimination from industrial and municipal effluents. Coord Chem Rev 254:2959–2972

Samuel J, Paul ML, Pulimi M, Nirmala MJ, Chandrasekaran N, Mukherjee A (2012) Hexavalent chromium bioremoval through adaptation and consortia development from Sukinda chromite mine isolates. Ind Eng Chem Res 51:3740–3749

Savvides C, Papadopoulos A, Haralambous K-J, Loizidou M (2001) Cement-based stabilization/ solidification of metal plating industry sludge. J Environ Sci Health A 36:1129–1134

Sethunathan N, Megharaj M, Smith L, Kamaludeen SPB, Avudainayagam SR, Naidu R (2005) Microbial role in the failure of natural attenuation of chromium (VI) in long-term tannery waste contaminated soil. Agric Ecosyst Environ 105:657–661

Shen SB, Tyagi RD, Blais JF (2001) Extraction of Cr(III) and other metals from tannery sludge by mineral acids. Environ Technol 22(9):1007–1014

Shen SB, Tyagi RD, Blais JF, Surampalli RY (2003) Bacterial leaching of metals from tannery sludge by indigenous sulphur-oxidizing bacteria-effect of sludge solids concentration. J Environ Eng 129:513–519

Shen SB, Vidyarthi AS, Tyagi RD, Blais JF, Surampalli RY (2002) Effect of Sulphur concentration on bioleaching of Cr(III) and other metals from tannery sludge by indigenous sulphur-oxidizing bacteria. Prac Period Hazard Toxic Radio Waste Manage 6:244–249

Shi W, Bischoff M, Turco R, Konopka A (2002) Long-term effects of chromium and lead upon the activity of soil microbial communities. Appl Soil Ecol 21:169–177

Shtiza A, Swennen R, Tashko A (2008) Chromium speciation and existing natural attenuation conditions in lagoonal and pond sediments in the former chemical plant of Porto-Romano (Albania). Environ Geol 53:1107–1128

Silva PTS, Mello NT, Duarte MMM (2006) Extraction and recovery of chromium from electroplating sludge. J Hazard Mater 128:39–43

Singh A, Haq I, Kumar A (2014) Isolation, screening and characterization of arsenic resistant bacteria from arsenic resistant contaminated soils of Hoogly, West Bengal. Ann Plant Soil Res 3:268–270

Skrypski-Mantele S, Bridle TR (1995) Environmentally sound disposal of tannery sludge. Water Res 29:1033–1039

Sokolov V, Imriˇs I, Kulczycka J (2005) The possibility of recycling flue dust of the ferrochromium productions in Kazakhstan. Manage Environ Qual An Inter J 16(6):647–655

Sophia AC, Swaminathan K (2005) Assessment of the mechanical stability and chemical leachability of immobilized electroplating waste. Chemosphere 58:75–82

Speir TW, Kettles HA, Parshotam A, Searle PL, Vlaar LNC (1995) A simple kinetic approach to derive the ecological dose value, ED50, for the assessment of Cr(VI) toxicity to soil biological properties. Soil Biol Biochem 27:801–810

Sperling M, Xu SK, Welz B (1992) Determination of chromium (III) and chromium (VI) in water using flow injection on-line preconcentration with selective adsorption on activated alumina and flame atomic absorption spectrometric detection. Anal Chem 64:3101–3108

Stewart DI, Burke IT, Hughes-Berry DV, Whittleston RA (2010) Microbially mediated chromate reduction in soil contaminated by highly alkaline leachate from chromium containing waste. Ecol Eng 36:211–221

Sun Y, Song J, Feng J, Wang G (2010) Study on process conditions of chromite ore processing residue contaminated soil washing with water. Environ Sci Inform Appl Technol 2:538–542. https://doi.org/10.1109/ESIAT.2010.5567393

Sun Y, Song J, Zhou R (2009) Morphological distribution of hexavalent chromium in the soils contaminated by chromite ore processing residue from Qing-daoHongxing chemical plant. Environ Sci Inform Appl Technol 3:594–597. https://doi.org/10.1109/ESIAT.2009.336

Sundar K, Vidya R, Mukherjee A, Chandrasekaran N (2010) High chromium toler-ant bacterial strains from Palar River basin: impact of tannery pollution. J Environ Earth Sci 2(2):112–117

Taylor MM, Diefendorf EJ, Na GC (1990) Enzymic treatment of chrome shavings. J Am Leath Chem Assoc 85(9):261–282

Thacker U, Parikh R, Shouche Y, Madamwar D (2007) Reduction of chromate by cell-free extract of Brucella sp. isolated from Cr(VI) contaminated sites. Bioresour Technol 98:1541–1547

Tinjum JM, Benson CH, Edil TB (2008) Mobilization of Cr(VI) from chromite ore processing residue through acid treatment. Sci Total Environ 391:13–25

Tiwary RK, Dhakate R, Ananda Rao V, Singh VS (2005) Assessment and predic-tion of contaminant migration in ground water from chromite waste dump. Environ Geol 48:420–429

Turick CE, Apel WA (1997) A bio-processing strategy that allows for the selection of Cr (VI) reducing bacteria from soil. J Ind Microbiol Biotechnol 18:247–250

Turick CE, Graves C, Apel WA (1998) Bioremediation potential of Cr(VI)- contaminated soil using indigenous microorganisms. Biorem J 2(1):1–6

Turpeinen R, Kairesalo T, Häggblom MM (2004) Microbial community structure and activity in arsenic- chromium- and copper-contaminated soils. FEMS Microbiol Ecol 47:39–50

U.S. Environmental Protection Agency (USEPA) (1999) Code of Federal Regulations, 40 CFR 141.32.1999

Virotec (2011) Case study treatment of chromium contaminated solid waste at the Tasman Sheepskin Tannery. http://virotec.com/download/email/webfiles/industrial-wastewater-remediation-viroflow/Case%20Study%20Tasman%20Solids.pdf

Viti C, Giovannetti L (2001) The impact of chromium contamination on soil het-erotrophic and photosynthetic microorganisms. Ann Microbiol 51:201–213

Viti C, Giovannetti L (2005) Characterization of cultivable heterotrophic bacterial communities in Cr-polluted and unpolluted soils using biolog and ARDRA approaches. App Soil Ecol 28 (20):101–112

Viti C, Giovannetti L (2007) Bioremediation of soils polluted with hexavalent chromium using bacteria: a challenge. In: Singh SN, Tripathi RD (eds) Environmental bioremediation technologies. Springer, Berlin, pp 57–76

Wang YM, Chen TC, Yeh KJ, Shue MF (2001) Stabilization of an elevated heavy metal contaminated site. J Hazard Mater 88:63–74

Wang Y-S, Pan Z-Y, Lang J-M, Xu J-M, Zheng Y-G (2007) Bioleaching of chromium from tannery sludge by indigenous Acidithiobacillus thiooxidans. J Hazard Mater 147:319–324

Wanner C, Eggenberger U, Mäder U (2011) Reactive transport modeling of Cr(VI) treatment under fast flow conditions. Appl Geochem 26:1513–1523

Weng C-H, Huang CP, Sanders PF (2001) Effect of pH on Cr(VI) leaching from soil enriched in chromite ore processing residue. Environ Geochem Health 23:207–211

Whittleston RA, Stewart DI, Mortimer RJG, Ashley DJ, Burke IT (2011a) Effect of microbially induced anoxia on Cr(VI) mobility at a site contaminated with hyper alkaline residue from chromite ore processing. Geomicrobiol J 28:68–82

Whittleston RA, Stewart DI, Mortimer RJG, Tilt ZC, Brown AP, Gerakid K, Burke IT (2011b) Chromate reduction in Fe(II)-containing soil affected by hyper alkaline leachate from chromite ore processing residue. J Hazard Mater 194:15–23

Wong PTS, Trevors JT (1988) Chromium toxicity to algae and bacteria. In: Nriagu JO, Nieboer E (eds) Chromium in natural and human environments. Wiley, New York, pp 305–315

Xu W, Li X, Zhou Q, Peng Z, Liu G, Qi T (2011) Remediation of chromite ore processing residue by hydrothermal process with starch. Proc Safety Environ Prot 89:179–185

Xu XR, Li HB, Gu J-D (2005) Kinetics of the reduction of chromium (VI) by vitamin C. Environ Toxicol Chem 24:1310–1314

Xu XR, Li HB, Gu J-D (2006) Simultaneous decontamination of hexavalent chromium and methyl tert-butyl ether by UV/TiO2 process. Chemosphere 63:254–260

Yu S, Du J, Luo T, Huang Y, Jing C (2012) Evaluation of chromium bio-accessibility in chromite ore processing residue using in vitro gastrointestinal method. J Hazard Mater 209-210:250–255

Zayed AM, Terry N (2003) Chromium in the environment: factors affecting biological remediation. Plant and Soil 249:139–156

Zhang H, Kong D, Wu S, He Z, Hu X, Hu. (2009) Remediation of chromite ore processing residue by pyrolysis process with sewage sludge. Bioresour Technol 100:2874–2877

Zhou L, Zhou S, Wang S, Fang D, Wang DZ (2004b) Cr removal and improving the settling and dehydrating capability from tannery sludge simultaneously through bioleaching approach. J Environ Sci 24:1014–1020

Zhou LX, Fang D, Wang SM, Wong JWC, Wang DZ (2005) Bioleaching of Cr from tannery sludge: the effects of initial acid addition and recycling of acidified bioleached sludge. Environ Technol 26:277–284

Zhou LX, Fang D, Zhou SG, Wang DZ, Wang SM (2004a) Removal of Cr from tannery sludge by acidophilic Thiobacilli. Environ Sci 25:62–66

Zhou S, Zhou L, Wang S, Fang D (2006) Removal of Cr from tannery sludge by bioleaching method. J Environ Sci 18(5):885–890

Integration of Nanotechnologies for Sustainable Remediation of Environmental Pollutants

3

Surbhi Sinha, Tithi Mehrotra, Parul Chugh, and Rachana Singh

3.1 Introduction

Human exercises in the last few years have caused extreme concern related to the environment and its preservation. Water shortage, water contamination, air contamination, soil debasement, poor administration of waste, loss of biodiversity are some of the ecological concerns that have caused permanent health impacts on humans as well as on animals and plants. Also, the advancement in industrialization, just as science and innovation, has prompted the enhancement of waste and lethal materials in the environment. Thus, the degradation and diminution of natural resources must be circumvented to achieve a sustainable environment. The conventional physico-chemical strategies utilized for the reclamation of the common habitat were seen as improper because of cost, lower productivity and nonspecificity. Consequently, to overcome these constraints, biological methods were amalgamated with the nanotechnology-based physiochemical techniques for the removal of pollutants from the environment (Guerra et al. 2018). The present chapter reviews the existing physical, chemical and biological methods for the treatment of pollutants along with their merits, demerits and the application of nanotechnology in the bioremediation of contaminants. Furthermore, the chapter will likewise concentrate on the biological synthesis of the nanoparticles using microbes which will provide insight into nanobioremediation for removing contaminants from the environment. This nanobioremediation approach for the expulsion of toxicants from nature will be the most dependable and suitable technology as for the cost and effectiveness relative to the financial status of the developing countries.

S. Sinha · T. Mehrotra · P. Chugh · R. Singh (✉)
Amity Institute of Biotechnology, Amity University, Noida, Uttar Pradesh, India
e-mail: rsingh2@amity.edu

I. Haq, A. S. Kalamdhad (eds.), *Emerging Treatment Technologies for Waste Management*, https://doi.org/10.1007/978-981-16-2015-7_3

3.2 Present Day Treatment Methods for the Ouster of Pollutants

Since the contaminants are lethal in nature, they have been contemplated dangerous to the environment. The treatment of these toxins in an environmentally safe way is obligatory before they are being released into the environment. The physical, chemical and biological techniques are the current treatment methods used for the expulsion of contaminants from the environment. Physical techniques incorporate methods like adsorption, reverse osmosis, electrodialysis, etc. Countless toxins are being discharged into the environment, out of which some are exceptionally hard to be treated by regular physical techniques. To solve the limitations of physical techniques, some of the chemical methods like precipitation, ion exchange, electroflotation, coagulation, flocculation, reduction and so forth were utilized for the expulsion of contaminants from the environment. In spite of the fact that the chemical methods used are productive, quick and can remove a wide range of toxins present in nature, their utilization is constrained by the significant price and sludge disposal issues. Furthermore, plenty of chemicals and high level of energy are required by these chemical methods. Considering all the above constraints, biological methods including the utilization of microorganisms (bioremediation) were utilized for the expulsion of lethal contaminants present in the environment (Ojuederie et al. 2017; Sinha et al. 2016; Behl et al. 2019). The process of bioremediation is economically attractive as well as environmentally friendly. Also, there is an advantage of minimum sludge generation, regeneration of biosorbent and possibility of metal recovery. However, the processes are slow, additional nutrition and maintenance are required. Moreover, the pollutants sometimes become toxic to the microorganisms involved in the process. Thus, every above mentioned techniques have their own benefits and disadvantages which make them insufficient to manage the issue of contaminant expulsion from nature.

3.3 Nanotechnology

The remediation of the toxicants by the existing traditional physicochemical methods and biological methods was not very efficient and effective in cleaning up the environment. Therefore, a new technology named 'nanotechnology' can be applied for the bioremediation of contaminants from the environment. Nanotechnology is derived from the Greek word 'dwarf' (El Saliby et al. 2008) and can be defined as the science of micro-engineering. Micro-engineering is the technique that deals with particles smaller than 100 nm. Nanotechnology was first proposed by Richard Feynman (1960), which now has become one of the fastest developing areas of research and development all around (Yadav et al. 2017). Presently, the field of nanotechnology is regarded as the 'Next Industrial Revolution' as in the future it will lessen the industrial costs by diminishing the consumption of energy, environmental pollution and enhancing the production efficiencies in developed countries (Roco 2005). Moreover, nanotechnology may also prove helpful in handling particular

Table 3.1 Various nanomaterials used in remediation process

Process exploited	Target compounds	Nanomaterials	Properties of nanomaterials	References
Photocatalysis	Organic pollutants, NOX, VOCs, Azo dye, Congo red dye, 4-chlorophenol and Orange II, PAHs	TiO2, ZnO, species of iron oxides (Fe III, Fe2O3, Fe3O4)	Photocatalytic activity in solar spectrum, low human toxicity, high stability and selectivity, low cost	Khedr et al. (2009)
Redox reactions	Halogenated organic compounds, metals, nitrate, arsenate, oil, PAH, PCB	Nanoscale zero-valent iron (nZVI), nanoscale calcium peroxide	Electron transfers such as photosynthesis, respiration, metabolism and molecular signalling	Zhang et al. (2003)
Adsorption	Heavy metals, organic compounds, arsenic, phosphate, Cr (IV), mercury, PAHs, DDT, dioxin	Iron oxides, carbon-based nanomaterials such as dendrimers and polymers, carbon nanotubes (CNTs)	High specific surface area and assessable adsorption sites, selective and more adsorption sites, short intra-particle diffusion distance, tunable surface chemistry, easy reuse	Bhaumik et al. (2012)
Disinfection	Diamines, phenols, formaldehyde, hydrogen peroxide, silver ions, halogens, glutaraldehyde, acridines	Nanosilver/ titanium dioxide (Ag/TiO2) and CNTs	Strong antimicrobial activity, low toxicity and cost, high chemical stability, ease of use	Amin et al. (2014)

social issues of developing nations like the necessity of clean water and treatment of epidemic diseases (Fleischer and Grunwald 2008; Schmidt 2007). Nanotechnology offers a large amount of environmental benefits in remediation, pollution prevention and contributes a lot to developing smaller, more accurate sensing and monitoring devices (Savage et al. 2008). The ability of nanotechnology to abridge contamination is in progress that can result in extensive and profound changes in pollution control (Watlington 2005). Table 3.1 lists some of the common nanomaterials utilized in the remediation process.

3.3.1 Properties of Nanoparticles

The essential part of nanotechnology is the very small particles called nanoparticles or ultrafine particles. Nanoparticles are particles somewhere in the range of

1–100 nm in size that can intensely change their physicochemical properties when contrasted with the bulk material. These particles are comprised of carbon, metal, metal oxides or organic matter and their function relies upon the type of synthesis, size and shape of the particles. They can be round, tubular, cylindrical and so on. Their surface can be uniform or irregular, while some are crystalline to amorphous with single or multi-crystal solids either free or agglomerated.

The nanoparticles are classified into organic, inorganic and carbon based. Organic nanoparticles incorporate dendrimers, liposomes, ferritin and so forth that are non-toxic, biodegradable and are likewise sensitive to thermal and electromagnetic radiation like heat and light, making them ideal for drug delivery (Tiwari et al. 2008). Inorganic nanoparticles are not comprised of carbon. They largely involve metal and metal oxide nanoparticles such as aluminium, copper, gold, iron, iron oxide, aluminium oxide, magnetite, etc. (Dreaden 2012). Carbon-based nanoparticles are totally comprised of carbon like graphene, fullerenes, carbon nanotubes (Saeed and Khan 2016).

The unusual chemical, physical, optical, thermal and electrical properties (Panigrahi et al. 2004) of nanoparticles can be used in various fields like drug delivery (Horcajada et al. 2008), medical imaging (Lee et al. 2008), optical receptors (Dahan et al. 2003), biolabelling (Liang et al. 2006), antimicrobial agents (Sanpui et al. 2008). There are other remarkable properties of nanoparticles like its small size which can cause increase in the surface area per unit mass that makes them profoundly helpful in bioremediation. Because of the small size, a lot of nanoparticles can come into contact with the surrounding medium, consequently influencing its reactivity. Nanoparticles show a remarkable property of surface plasmon resonance which helps in the detection of contaminants present in nature. Furthermore unique properties of nanoparticles likewise make them appropriate for the advancement of electrochemical sensors as well as biosensor (Peng and Miller 2011; Selid et al. 2009). Moreover, scientists have created nanosensors for the recognition of auxin and oxygen dissemination in plants (Koren et al. 2015). Because of the outstanding properties of nanoparticles, they have been proposed as a proficient, economical and environment friendly substitute to the present treatment advancements, in resource preservation as well as in ecological remediation (Friedrich et al. 1998; Dastjerdi and Montazer 2010). The ability of nanomaterials to abate pollution production is in progress and could potentially catalyse the most revolutionary changes in the environmental field in the coming decades (Fig. 3.1).

3.4 Synthesis of Nanoparticles

Nanoparticles can be synthesized by various strategies and approaches that incorporate physical, chemical and biological methods (Fig. 3.2) (Luechinger et al. 2010; Mohanpuria et al. 2008). Conventionally, the nanoparticles were produced by physicochemical strategies that enable them to be synthesized in enormous amounts with definite shape and size in a constrained timeframe; howbeit, these methods are

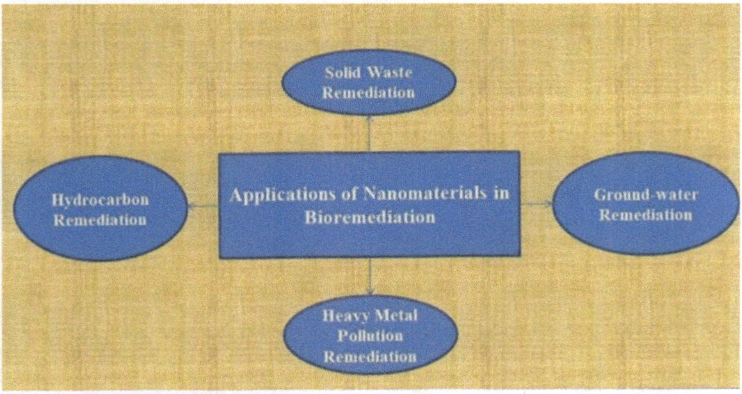

Fig. 3.1 Applications of nanomaterials in bioremediation

expensive, wasteful, complicated, utilize hazardous chemicals, require high energy and produce toxic by-products that are hazardous to the environment (Li et al. 2011a, b; Rodriguez-Sanchez et al. 2000).

Lately, the interest has been focussed on the production of economical and eco-friendly nanoparticles that do not give rise to dangerous and toxic by-products during the manufacturing procedure (Chauhan et al. 2012; Li et al. 2011a, b). Thus, recently, nanoparticles are being produced by biological methods that include microorganisms, plants and their by-products with the assistance of some biological tools. Biologically synthesized nanoparticles have striking and outstanding benefits over physical and chemical strategies like the production approaches are economical, quick and eco-friendly. In addition, the nanoparticles produced by biological path does not require any further stabilizing agents, as microorganisms and plants themselves act as stabilizing agents (Makarov et al. 2014). The biological synthesis of nanoparticles is a bottom-up approach where reducing and stabilizing agents help in synthesizing the nanoparticles (Fig. 3.2). Bio-fabrication of nanoparticles is in general achieved either through reduction or oxidation process. The biomolecules present in microbes or botanical species were found to be responsible for reduction-cum-stabilization of metal ions into their respective nanostructures (Singh et al. 2011). Biosynthesis of various nanoparticles using plants and microorganisms like bacteria, algae, fungi yeast and microbial polysaccharides is compiled below.

3.4.1 Synthesis of Nanoparticles Utilizing Plants

Biological synthesis of nanoparticles by plants is getting a lot of attention these days because of its simple, stable, rapid, cheap and eco-friendly method (Mittal et al. 2013). Additionally, plants are abundantly available, safe to handle and have a wide variability of metabolites that help in reduction. Plant extracts containing bioactive

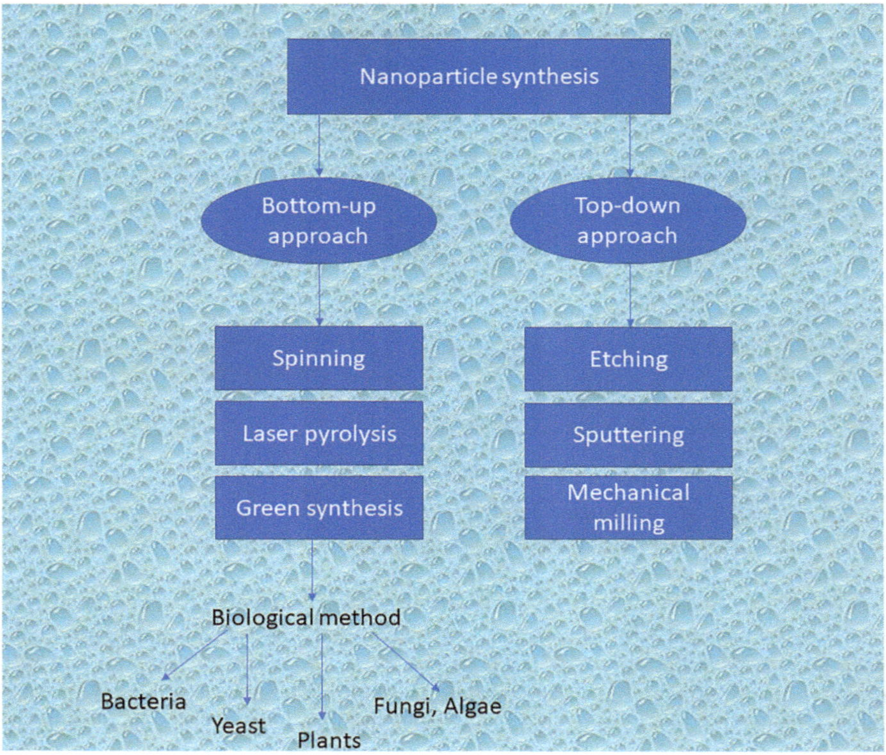

Fig. 3.2 Different methods and approaches for synthesizing nanoparticles. Source: (Siavash 2011)

alkaloids, proteins, sugars, phenolic acids, polyphenols play an important role in first reducing the metallic ions and then stabilizing them (Castro et al. 2011). Table 3.2 compiles the information on a large number of plants being utilized for the synthesis of various nanoparticles and it is clear from the information that the synthesis of nanoparticles, their size, application all vary from plant to plant.

3.4.2 Synthesis of Nanoparticles Utilizing Bacteria

Biosynthesis of nanoparticles utilizing bacteria has gained a lot of attention in the area of green nanotechnology over the globe because of their abundance in the environment and their capacity to adjust to extraordinary conditions. Additionally, these are fast-growing, inexpensive to cultivate and simple to control (Mehrotra et al. 2019a, b; Kumar et al. 2019). Moreover, the nanoparticles synthesized from bacteria have higher catalytic reactivity, more specific surface area and are of uniform size (Mehrotra et al. 2019a, b). Various species of bacteria till now have been effectively

Table 3.2 List of various plants used for the synthesis of nanoparticles

Plant species	Plant material	Type of nanoparticles	Mechanism/ causative agents	Size (in nm)	References
Azadirachta indica	Kernel	Silver, gold	Azadirachtin	50–100	Shukla et al. (2012)
Jatropha curcas L.	Latex	Lead	Curcacycline A and Curcacycline B	10–12.5	Joglekar et al. (2011)
Camellia sinensis	Leaves	Platinum	Pure tea polyphenol	30–60	Alshatwi et al. (2015)
Nephelium lappaceum L	Peels	Nickel oxide	Nickel–ellagate complex formation	50	Yuvakkumar et al. (2014)
Eucalyptus	Leaves	Iron oxide	Epicatechin and quercetin–glucuronide	20–80	Wang et al. (2014)
Syzygium aromaticum	Flower buds	Iron oxide	Eugenol	5–40	Subhankari et al. (2013)
Aloe barbadensis miller	Leaves	Zinc oxide	Phenolic compounds, terpenoids or proteins	25–40	Sangeetha et al. (2011)
Alfa sprouts	Living plant	Silver	In situ synthesis	2–20	GardeaTorresdey et al. (2003)
Asparagus racemosus	Tuber cortex	Palladium	Tuber cortex	1–6	Raut et al. (2013)

used for the synthesis of different nanoparticles like gold, silver, zinc, cadmium sulphide, palladium, etc. (Table 3.3).

3.4.3 Synthesis of Nanoparticles Utilizing Fungi and Yeast

The utilization of fungi in the synthesis of nanoparticles has gained fast interest because of their toleration and metal bioaccumulation capability (Sastry et al. 2003). A large amount of enzymes can be produced by utilizing fungi since they are magnificent secretors of extracellular proteins, which eventually can regulate the synthesis of nanoparticles (Castro-Longoria et al. 2012). Fungi is viewed as better than bacteria in the production of nanoparticles as these secrete huge volume of proteins which directly gets converted to nanoparticles, causing higher productivity (Mohanpuria et al. 2008). Furthermore, various fungal species grow very fast, making their maintenance in the research lab simple. In a similar way, easy maintenance of yeast production in the laboratory, its rapid growth and the use of simple nutrients are some of the remarkable advantages of yeast over bacteria for the mass production of nanoparticles (Skalickova et al. 2017). Fungi and yeast have supremacy over other biological systems because of their wide diversity, simple culture

Table 3.3 Some of the nanoparticles synthesized by bacteria

Nanoparticles	Bacteria	Size	Shape	Applications	References
Gold	*Bacillus subtilis* 168	5–25 nm	Octahedral	Metal uptake	Beveridge and Murray (1980)
	Marinobacter Pelagius sp.	20 nm	Spherical	Optical	Joerger et al. (2000)
	Lactobacillus sp.	20–50 nm	Hexagonal	Metal uptake	Nair et al. (2002)
	R. capsulata	10–20 nm	Spherical	–	Shiying et al. (2008)
	Stenotrophomonas maltophilia	40 nm	Spherical	Reduction in metallic toxicity	Sharma et al. (2012)
	E. coli K12	50 nm	Circular	Removal of nitroaromatic pollutants from water	Srivastava et al. (2013)
Silver	*P. stutzeri* AG259	<200 nm	Nanocrystal	Optical	Joerger et al. (2000)
	Pseudomonas aeruginosa	13 nm	Spherical	–	Kumar and Mamidyala (2011)
	Bacillus cereus	20–40 nm	Spherical	Antimicrobial	Silambarasan and Abraham (2012)
Cadmium sulphide	*Escherichia coli*	2–5 nm	Spherical, elliptical	Optical and electronic properties	Sweeney et al. (2004)
	Klebsiella aerogenes	20–200 nm	Spherical	–	Holmes et al. (1995)
Manganese sulphate and zinc sulphate	*Streptomyces sp.* HBUM171191	10–20 nm	Polymorphic	–	Waghmare et al. (2011)
Palladium	*Desulfovibrio desulfuricans* NCIMB 8307	5-10 nm	Icosahedral	Bioprocessing applications of precious metals	Naiz et al. (2002)
Magnetic	*Magnetosirillium magneticum*	50 nm	Spherical	Biomineralization	Mohanpuria et al. (2008)

methods, less time and low cost which successively lead to an eco-friendly approach for the synthesis of nanoparticles. Some of the fungal and yeast species successfully utilized for the production of the nanoparticles are documented in Table 3.4.

3.4.4 Synthesis of Nanoparticles Utilizing Algae

From the past few years, the utilization of algae for the biosynthesis of nanoparticles has increased tremendously because of their simple access and efficiency (Ogi et al. 2010; Singaravelu et al. 2007). At present, they are also called as 'biofactories' for the synthesis of nanoparticles since they are an excellent source of biomolecules (Manivasagan and Kim 2015). These biomolecules like proteins, pigments, starch, nucleic acids, fats and secondary metabolites such as alkaloids present in the algal cell wall act as reducing agents which eventually prompts the reduction and synthesis of metal and metal oxide nanoparticles at ambient conditions (Siddiqi and Husen 2016). Also, seaweeds are advantageous over different reductants because of their high metal accumulating capability, minimal effort, plainly visible structure and anti-biological fouling properties (Davis et al. 2003). In addition, seaweeds have both anti-inflammatory and inhibitory properties that can be utilized to treat diverse ailments and stifle a few types of malignant growth (Fawcett et al. 2017). The biogenic manufacturing of different nanoparticles utilizing diverse algal species is presented in Table 3.5.

3.4.5 Remediation Using Biogenic Polysaccharide

Polysaccharides are natural biopolymers of biological systems that have been extracted and put to extensive use. These biopolymers are renewable materials, environment friendly, non-toxic, biodegradable and have excellent functional properties. In recent years, polysaccharide nanomaterial composites have attracted attention of researchers in nanobioremediation due to improved processability, surface area, stability, tunable properties and cost-effectiveness. Table 3.6 provides an overview of biogenic polysaccharides that have been used in the preparation of bionanocomposites.

3.5 Nanobioremediation

Utilization of nanomaterials, synthesized from plants, algae, bacteria, fungi and yeast, to clean up the environmental pollutants such as organic or inorganic waste and heavy metals from the affected sites is termed as nanobioremediation (Yadav et al. 2017). The concept of green technology has gained immense interest in the area of nanomaterials for application in bioremediation and also due to its cost-effectiveness in large-scale use, enhanced efficiency and shortened time for the remediation process. Several other reasons contribute towards the use of

Table 3.4 Some of the nanoparticles synthesized by fungi and yeast

Nanoparticles	Fungi and yeast	Size	Shape	Applications	References
Gold	Thermophilic fungi	6–40 nm	Spherical	Biomedical	Molnar et al. (2018)
	Penicillium sp.	45 nm	Spherical	Antimicrobial	Sandhya and Suvarnalatha (2017)
	Mariannaea sp. HJ	37.4 nm	Spherical, hexagon, irregular	Photocatalytic	Pei et al. (2017)
	Pycnoporus sanguineus	29.30 nm	Spherical, pseudo-spherical, triangular	Catalytic	Shi et al. (2015)
	Magnusiomyces ingens LH-F1	10–80 nm	Triangle, hexagon, pentagon	Catalytic	Zhang et al. (2016)
	Aspergillus sp. WL-Au	4–29 nm	Spherical	Catalytic	Shen et al. (2017)
Silver	*Ganoderma enigmaticum*	15–25 nm	Spherical	Antibacterial	Gudikandula et al. (2017)
	Rhizopus stolonifer	9.47 nm	Spherical	Biomedical	Rahim et al. (2017)
	Alternaria sp	4–30 nm	Spherical	Antibacterial	Singh et al. (2017)
	Saccharomyces cerevisiae	1–10 nm	Spherical	Antimicrobial	Marquez et al. (2018)
	Saccharomyces cerevisiae, Rhodotorula glutinis and *Geotrichum candidum*	2.5–20 nm	Spherical	Biomedical	Zahran et al. (2013)
	Arthroderma fulvum	15.5 ± 2.5 nm	Spherical	Antifungal	Xue et al. (2016)
Copper carbonate	*Neurospora crassa*	10–20 nm	Spherical	–	Li and Gadd (2017)
Cadmium sulphide	*Candida glabrata* and *Schizosaccharomyces pombe*	2 nm	Spherical	–	Dameron et al. (1989)
Bioactive	*Lichen fungi* (*Usnea longissima*)	9.40–11.23 nm	Spherical	Medicinal	Shahi et al. (2003)

Table 3.5 List of some nanoparticles synthesized by algae

Nanoparticles	Algae	Size	Shape	Applications	References
Gold	*Sargassum muticum*	5.42 nm	Spherical	Pharmaceutical and biomedical	Namvar et al. (2015)
	Spirulina platensis	5 nm	Spherical	Antibacterial	Suganya et al. (2015)
	Ecklonia cava	30 nm	Spherical and triangular	Antimicrobial	Venkatesan et al. (2014)
	Chlorella vulgaris	2–10 nm	Spherical	Anti-pathogenic	Annamalai and Nallamuthu (2015)
	Turbinaria conoides	2–19 nm	Spherical and triangular	Antimicrobial	Vijayan et al. (2014)
	Sargassum myriocystum	60 nm	Spherical	Antibacterial	Ismail et al. (2018)
Silver	*Spatoglossum asperum*	20–46 nm	Spherical to oval	Antibacterial	Ravichandrana et al. (2018)
	Gelidium amansii	27–54	Spherical	Antibacterial	Pugazhendhi et al. (2018)
	Amphiroa anceps	10–80 nm	Spherical	Antibacterial	Roy and Anantharaman (2018)
	Sargassum ilicifolium	10–80 nm	Spherical, cubical and hexagonal shaped	Antimicrobial	Roy and Anantharaman (2018)
	Laminaria japonica	20 nm	Spherical- to oval-shaped	In seedling growth	Kim et al. (2017)
	Ulva compressa (L.) Kütz and Cladophora glomerata (L.) Kütz	200 to 300 nm	Irregular shaped	Antimicrobial	Minhas et al. (2018)
Aluminium oxide	*Sargassum ilicifolium*	20 nm	Spherical	–	Koopi and Buazar (2018)
Iron oxide	*Ulva flexuosa*	12.3 nm	Spherical	Antimicrobial	Mashjoor et al. (2018)
Copper oxide	*Bifurcaria bifurcata*	96–110	Spherical	Antimicrobial	Abboud et al. (2014)

Table 3.6 List of microbial polysaccharides utilized as bionanocomposites Source: Manikandan et al. (2017)

Polysaccharide	Source	Active functional group
Gellan	*Sphingomonas elodea*	OH
Dextran	*Lactobacillus sps, Streptococcus mutans*	OH
Cellulose	*Aerobacter, Acetobacter, Agrobacterium, Azotobacter, Pseudomonas*	OH
Alginate	*Azotobacter* and *Pseudomonas*	OH, COO^{-}
Chitosan	*Fungal cell walls, Cunninghamella elegans*	OH, COO
Hyaluronic acid	*Streptococcal sps* and *Bacillus subtilis*	OH
Zooglan	*Zoogloea ramigera*	OH
Pullulan	*Aureobasidium pullulans*	OH
Xanthan	*Xanthomonas campestris*	OH

nanotechnology in bioremediation. Firstly, the size in the range of nanoscale helps to increase the surface area per unit mass of a material, allowing enhanced reactivity rate. Secondly, nanomaterials exhibit quantum effect, thereby requires less activation energy to attain chemical reactions. Lastly, another feature shown by the nanomaterials is surface plasmon resonance (SPR) which can be used to detect toxic materials. There are a diverse range of multiple nanomaterials used for bioremediation, with high level of remedial versatility such as in removing wastes including hydrocarbons, heavy metals and radioactive materials like uranium, in remediation of soil, groundwater and wastewater.

The potential of nanomaterials to alleviate the pollution load is ongoing and could potentially bring about the most profound changes in the field of bioremediation sector in the upcoming years (Rizwan et al. 2014) (Table 3.7).

3.6 Conclusion

Nanotechnology has the potential to metamorphose all the existing technologies that include the techniques involving pollution control as well. This technology is gaining recognition globally for successfully removing the contaminants from the environment. The extraordinary properties of nanoparticles and their concurrence with the present day technologies offer a great opportunity to revolutionize environmental clean-up. It is clear from the reviewed literature that while much attention has been focused on the development and potential benefits of nanomaterials in water treatment processes, concerns have also been raised regarding their potential human and environmental toxicity. Biogenic synthesis of nanoparticles can solve the problem of toxicity to a great extent. Thus, the utilization of biologically synthesized nanoparticles for the process of bioremediation can go a long way in attaining a sustainable environment. Biosynthesis of nanoparticles using microbes helps to reduce the toxicity, is cheap, eco-friendly and saves time. Due to the remarkable

Table 3.7 Some of the nanoparticles used for removal of contaminants. Source: (Yadav et al. 2017; Yang et al. 2019; Vittal and Jamuna 2011)

Contaminant to be removed	Nanomaterials/nanoparticles
Lead	Ca-alginate iron oxide magnetic nanoparticles; polyacrylic acid-stabilized zero-valent iron nanoparticles (PAA-ZVIN)
Mercury	Carboxy-methylated chitosan ferromagnetic nanoparticles; thiol-functionalized silica ferromagnetic nanoparticles
Heavy metals	Thiol-functionalized super-paramagnetic nanoparticles
Arsenic	Zinc oxide nanoparticles
Cobalt and iron	Iron nanoparticles
Metal ions	Carbon nanoparticles
Lead, mercury, manganese, copper, cadmium, arsenic, chromium	Graphene based nanocomposites
Arsenic and copper metal	Iron nanoparticles
Methylene blue	Goethite nanoparticles
Tri-chloroethane (TCE)	Metallic gold nanoparticles coated with palladium
Chlorinated ethane	Metallic gold nanoparticles coated with palladium
Chlorinated methane	Metallic gold nanoparticles coated with palladium
Inorganic-mercury	Gold nanoparticles supported on alumina
Trihalomethanes (THM)	α-Fe_2O_3 sintered in zeolite form
Chlorpyrifos and malathion	Silver nanoparticles; gold nanoparticles
Escherichia coli and *Staphylococcus aureus*	Gold nanoparticles; silver nanoparticles
Pathogenic bacteria	Silver nanoparticles
Escherichia coli	Cerium oxide nanoparticles
Escherichia coli, Bacillus megaterium, Bacillus subtilis	Magnesium oxide nanoparticles; copper oxide nanoparticles
Escherichia coli	Aluminium nanoparticles; titanium dioxide nanoparticles
Escherichia coli, Pseudomonas fluorescens, Listeria monocytogenes, Salmonella enteritidis	Zinc oxide nanoparticles
Toluene, NO_2	Nanocrystalline zeolites
Heavy metal ions Benzene, toluene, ethylbenzene, xylene p-nitrophenol benzene, toluene, dimethylbenzene Heavy metal ions Trihalomethanes (THMs) Heavy metal ions THMs chlorophenols Herbicides Microcystin toxins	Carbonaceous nanomaterials CeO_2-carbon nanotubes (CNTs) Activated carbon fibres (ACFs) CNTs functionalized with polymers CNTs functionalized with Fe Single-walled carbon nanotubes Multi-walled carbon nanotubes

(continued)

Table 3.7 (continued)

Contaminant to be removed	Nanomaterials/nanoparticles
Inorganic ions Heavy metal ions Actinides and lanthanides	Self-assembled monolayer on mesoporous Supports (SAMMS) Anion-SAMMS Thiol-SAMMS HOPO-SAMMS
Heavy metal ions	Biopolymers
Polychlorinated biphenyls (PCBs) Inorganic ions Chlorinated organic compounds Heavy metal ions	Zero-valent iron nanoparticles (nZVI)
PCBs Chlorinated ethane Chlorinated methanes	Bimetallic nanoparticles Pd/Fe nanoparticles
TCE and PCBs Dichlorophenol Trichlorobenzene Chlorinated ethane Brominated organic compounds (BOCs)	Ni/Fe nanoparticles Pd/Au nanoparticles
Heavy metal ions Azo dyes Phenol Aromatic pollutants	Nanocrystalline TiO_2 Nitrogen (N)-doped TiO_2 Fe (III)-doped TiO_2 Supported TiO_2 nanoparticles

and significant capability of nanobioremediation, it is assumed that their application will enhance at a great leap in the near future and will perform a very important and indispensable part in achieving a green and renewable environment.

Acknowledgements The authors are grateful to Amity Institute of Biotechnology and Amity Institute of Marine Science and Technology, Amity University, Uttar Pradesh, for providing the research and laboratory facilities. We did not receive any specific grant from funding agencies in the public, commercial or not-for-profit sectors.

References

Abboud Y, Saffaj T, Chagraoui A (2014) Biosynthesis, characterization and antimicrobial activity of copper oxide nanoparticles (CONPs) produced using brown alga extract (Bifurcaria bifurcata). Appl Nanosci 4(5):571–576

Alshatwi AA, Athinarayanan J, Subbarayan PV (2015) Green synthesis of platinum nanoparticles that induce cell death and G2/M-phase cell cycle arrest in human cervical cancer cells. J Mater Sci Mater Med 26(1):1–9

Amin RS, Hameed RA, El–Khatib KM (2014) Microwave heated synthesis of carbon supported Pd, Ni and Pd–Ni nanoparticles for methanol oxidation in KOH solution. Appl Catal B-Environ 148:557–567

Annamalai J, Nallamuthu T (2015) Characterization of biosynthesized gold nanoparticles from aqueous extract of Chlorella vulgaris and their antipathogenic properties. Appl Nanosci 5:603–607

Behl K, Sinha S, Sharma M, Singh R, Joshi M, Bhatnagar A, Nigam S (2019) One-time cultivation of Chlorella pyrenoidosa in aqueous dye solution supplemented with biochar for microalgal growth, dye decolorization and lipid production. Chem Eng J 364:552–561

Beveridge TJ, Murray RG (1980) Sites of metal deposition in the cell wall of Bacillus subtilis. J Bacteriol 141(2):876–887

Bhaumik M, Maity A, Srinivasu VV, Onyango MS (2012) Removal of hexavalent chromium from aqueous solution using polypyrrole-polyaniline nanofibers. Chem Eng J 181:323–333

Castro-Longoria E, Moreno-Velásquez SD, Vilchis-Nestor AR, Arenas-Berumen E, Avalos-Borja M (2012) Production of platinum nanoparticles and nanoaggregates using Neurospora crassa. J Microbiol Biotechnol 22:1000–1004

Castro-Longoria E, Vilchis-Nestor AR, Avalos-Borja M (2011) Biosynthesis of silver, gold and bimetallic nanoparticles using the filamentous fungus Neurospora crassa. Colloids Surf B Biointerfaces 83:42–48

Chauhan RPS, Gupta C, Prakash D (2012) Methodological advancements in green nanotechnology and their applications in biological synthesis of herbal nanoparticles. Int J Bioassays 1(7):6–10

Dahan M, Levi S, Luccardini C, Rostaing P, Riveau B, Triller A (2003) Diffusion dynamics of glycine receptors revealed by single-quantum dot tracking. Science 302(5644):442–445

Dameron CT, Reese RN, Mehra RK, Kortan AR, Carroll PJ, Steigerwald ML, Brus LE, Winge DR (1989) Biosynthesis of cadmium sulphide quantum semiconductor crystallites. Nature 338:596–597

Dastjerdi R, Montazer M (2010) A review on the application of inorganic nano-structured materials in the modification of textiles: focus on anti- microbial properties. Colloid Surf B 79(1):5–18

Davis TA, Volesky B, Mucci A (2003) A review of the biochemistry of heavy metal biosorption by brown algae. Water Res 37(18):4311–4330

Dreaden EC, Alkilany AM, Huang X, Murphy CJ, El-Sayed MA (2012) The golden age: gold nanoparticles for biomedicine. Chem Soc Rev 41:2740–2779. https://doi.org/10.1039/C1CS15237

El Saliby IJ, Shon HK, Kandasamy J, Vigneswaran S (2008) Water and wastewater treatment technologies - nanotechnology for wastewater treatment: in brief. EOLSS 2008:7

Fawcett D, Verduin JJ, Shah M, Sharma SB, GEJ P (2017) A review of current research into the biogenic synthesis of metal and metal oxide nanoparticles via marine algae and seagrasses. J Nanosci 2017:8073850. https://doi.org/10.1155/2017/8013850

Feynman R (1960) There's plenty of room at the bottom, talk given on December 29th 1959. Sci Eng 23:22–36

Fleischer T, Grunwald A (2008) Making nanotechnology developments sustainable. A role for technology assessment. J Clean Prod 16:889–898

Friedrich KA, Henglein F, Stimming U, Unkauf W (1998) Investigation of Pt particles on gold substrates by IR spectroscopy particle structure and catalytic activity. Colloids Surf 134 (1–2):193–206

Gardea-Torresdey JL et al (2003) Alfalfa sprouts: a natural source for the synthesis of silver nanoparticles. Langmuir 19(4):1357–1361

Gudikandula K, Vadapally P, Charya MAS (2017) Biogenic synthesis of silver nanoparticles from white rot fungi: their characterization and antibacterial studies. Open Nano 2:64–78

Guerra FD, Attia MF, Whitehead DC, Alexis F (2018) Nanotechnology for environmental remediation: materials and applications. Molecules (Basel, Switzerland) 23(7):1760

Holmes JD, Smith PR, Evans-Gowing R, Richardson DJ, Russell DA, Sodeau JR (1995) Energy-dispersive X-ray analysis of the extracellular cadmium sulfide crystallites of Klebsiella aerogenes. Arch Microbiol 163(2):143–147

Horcajada P, Serre C, Maurin G, Ramsahye NA, Balas F, Vallet-Regi M, Férey G (2008) Flexible porous metal-organic frameworks for a controlled drug delivery. J Am Chem Soc 130 (21):6774–6780

Ismail AM, Ali HSJ, Parthasarathy M (2018) Biosynthesis of gold nano particles using Sargassum myriocystum and evaluation of their antibacterial activity. Int J Pure App Biosci 6 (1):1340–1350
Joerger R, Klaus T, Granqvist CG (2000) Biologically produced silver–carbon composite materials for optically functional thin-film coatings. Adv Mater 12(6):407–409
Joglekar S et al (2011) Novel route for rapid biosynthesis of lead nanoparticles using aqueous extract of Jatropha curcas L. latex. Mater Lett 65(19):3170–3172
Khedr MH, Halim KSA, Soliman NK (2009) Synthesis and photocatalytic activity of nano-sized iron oxides. J Mater Lett 63:598–601
Kim DY, Saratale RG, Shinde S, Syed A, Ameen F, Ghodake G (2017) Green synthesis of silver nanoparticles using Laminaria japonica extract: characterization and seedling growth assessment. J Clean Prod 172:2910–2918. https://doi.org/10.1016/j.jclepro.2017.11.123
Koopi H, Buazar F (2018) A novel one-pot biosynthesis of pure alpha aluminum oxide nanoparticles using the macroalgae Sargassum ilicifolium: a green marine approach. Ceram Int 44:8
Koren K, Brodersen KE, Jakobsen SL, Kuhl M (2015) Optical sensor nanoparticles in artificial sediments–a new tool to visualize O2 dynamics around the rhizome and roots of seagrasses. Environ Sci Technol 49(4):2286–2292
Kumar CG, Mamidyala SK (2011) Extracellular synthesis of silver nanoparticles using culture supernatant of Pseudomonas aeruginosa. Colloids Surf B Biointerfaces 84(2):462–466
Kumar N, Sinha S, Mehrotra T, Singh R, Tandon S, Thakur IS (2019) Biodecolorization of azo dye acid black 24 by Bacillus pseudomycoides: process optimization using box Behnken design model and toxicity assessment. Bioresour Technol Rep 8:100311
Lee HY, Li Z, Chen K, Hsu AR, Xu C, Xie J, Chen X (2008) PET/MRI dual-modality tumor imaging using arginine-glycine-aspartic (RGD)-conjugated radiolabeled iron oxide nanoparticles. J Nucl Med 49(8):1371–1379
Li Q, Gadd GM (2017) Biosynthesis of copper carbonate nanoparticles by ureolytic fungi. Appl Microbiol Biotechnol 101(19):7397–7407. https://doi.org/10.1007/s00253-017-8451-x
Li X, Xu H, Chen ZS, Chen G (2011a) Biosynthesis of nanoparticles by microorganisms and their applications. J Nanomat 2011:1–16
Li X, Xu H, Chen ZS, Chen G (2011b) Biosynthesis of nanoparticles by microorganisms and their applications. J Nanomat 2011:5
Liang SC, Tan XY, Luxenberg DP, Karim R, Dunussi-Joannopoulos K, Collins M, Fouser LA (2006) Interleukin (IL)-22 and IL-17 are coexpressed by Th17 cells and cooperatively enhance expression of antimicrobial peptides. J Exp Med 203(10):2271–2279
Luechinger NA, Grass RN, Athanassiou EK, Stark WJ (2010) Bottom-up fabrication of metal/metal nanocomposites from nanoparticles of immiscible metals. Chem Mater 22(1):155–160
Makarov VV, Loeve AJ, Sinitsyna OV, Makarova SS, Yaminsky IV, Taliansky ME, Kalinina NO (2014) Green nanotechnologies: synthesis of metal nanoparticles using plants. Acta Nat 6 (1):35–44
Manikandan M, Hui-Fen W, Judy G, Iyyakkannu S, ISechul C (2017) Exploiting microbial polysaccharides for biosorption of trace elements in aqueous environments- scope for expansion via nanomaterial intervention. Polymer 9:721
Manivasagan P, Kim SK (2015) Biosynthesis of nanoparticles using marine algae: a review. In: Kim S-K, Chojnacka K (eds) Marine algae extracts: processes, products, and applications, 1st edn. Wiley, New York
Márquez GI, Ghiyasvand M, Massarsky A, Babu M, Samanfar B, Omidi K (2018) Zinc oxide and silver nanoparticles toxicity in the baker's yeast, Saccharomyces cerevisiae. PLoS One 13:3. https://doi.org/10.1371/journal.pone.0193111
Mashjoor S, Yousefzadi M, Zolgharnain H, Kamrani E, Alishahi M (2018) Organic and inorganic nano-Fe3O4: alga Ulva flexuosa-based synthesis, antimicrobial effects and acute toxicity to briny water rotifer Brachionus rotundiformis. Environ Pollut 237:50–64

Mehrotra T, Nagabooshanam S, Singh R (2019a) Electrochemical evaluation of bacillus species for rapid biosynthesis of silver nanoparticles: application in domestic wastewater treatment. In: 6th international conference on signal processing and integrated networks (SPIN). IEEE, pp 456–460

Mehrotra T, Shukla A, Singh R (2019b) In vitro toxicological evaluation of domestic effluent treated by formulated synthetic autochthonous bacterial consortium. World J Microbiol Biotechnol 35(12):184

Minhas FT, Arslan G, Gubbuk IH, Akkoz C, Ozturk BY, Asıkkutlu B, Arslan U, Ersoz M (2018) Evaluation of antibacterial properties on polysulfone composite membranes using synthesized biogenic silver nanoparticles with Ulvacompressa (L.) Kütz. and Cladophora glomerata (L.) Kütz. extracts. Int J Bio Macromol 107(2018):157–165

Mittal AK, Chisti Y, Banerjee UC (2013) Synthesis of metallic nanoparticles using plant extracts. Biotechnol Adv 31(2):346–356. https://doi.org/10.1016/j.biotechadv.2013.01.003

Mohanpuria P, Rana NK, Yadav SK (2008) Biosynthesis of nanoparticles: technological concepts and future applications. J Nanopart Res 10(3):507–517

Molnár Z, Bódai V, Szakacs G, Erdélyi B, Fogarassy Z, Sáfrán G, Varga T, Kónya Z, Tóth-Szeles E, Szűcs R, Lagzi I (2018) Green synthesis of gold nanoparticles by thermophilic filamentous fungi. Sci Rep 8:3943. https://doi.org/10.1038/s41598-018-22112

Nair B, Pradeep T (2002) Coalescence of nanoclusters and formation of submicron crystallites assisted by Lactobacillus strains. Cryst Growth Des 2(4):293–298

Namvar F, Azizi S, Ahmad M, Shameli K, Mohamad R, Mahdavi M, Tahir P (2015) Green synthesis and characterization of gold nanoparticles using the marine macroalgae Sargassum muticum. Res Chem Intermed 41:5723–5730

Ogi T, Saitoh N, Nomura T, Konishi Y (2010) Room temperature synthesis of gold nanoparticles and nanoplates using Shewanella algae cell extract. J Nanopart Res 12:2531–2539. https://doi.org/10.1007/s11051-009-9822-8

Ojuederie OB, Babalola OO (2017) Microbial and plant-assisted bioremediation of heavy metal polluted environments: a review. Int J Environ Res Public Health 14(12):1504

Panigrahi S, Kundu S, Ghosh S, Nath S, Pal T (2004) General method of synthesis for metal nanoparticles. J Nanopart Res 6(4):411–414

Pei X, Qu Y, Shen W, Li H, Zhang X, Li S, Zhang Z, Li X (2017) Green synthesis of gold nanoparticles using fungus Mariannaea sp. HJ and their catalysis in reduction of 4-nitrophenol. Environ Sci Pollut Res Int 24(27):21649–21659. https://doi.org/10.1007/s11356-017-9684-z

Peng HI, Miller BL (2011) Recent advancements in optical DNA biosensors: exploiting the plasmonic effects of metal nanoparticles. Analyst 136:436–447

Pugazhendhi A, Prabakar D, Jacob JM, Karuppusamy I, Saratale RG (2018) Synthesis and characterization of silver nanoparticles using Gelidium amansii and its antimicrobial property against various pathogenic bacteria. Micro Pathogen 114:41–45

Rahim KA, Mahmoud SY, Ali AM (2017) Extracellular biosynthesis of silver nanoparticles using Rhizopus stonililifer. Saudi J Bio Sci 24(1):208–216

Raut RW, Haroon AS, Malghe YS, Nikam BT, Kashid SB (2013) Rapid biosynthesis of platinum and palladium metal nanoparticles using root extract of asparagus Racemosus Linn. Adv Mater Let 4(8):650–654

Ravichandrana A, Subramaniana P, Manoharana V, Muthua T, Periyannana R, Thangapandib M, Ponnuchamyc K, Pandid B, Marimuthua PN (2018) Phyto-mediated synthesis of silver nanoparticles using fucoidan isolated from Spatoglossum asperum and assessment of antibacterial activities. J Photochem Photobio 185:117–125

Rizwan M, Singh M, Mitra CK, Morve RK (2014) Ecofriendly application of nanomaterials: nanobioremediation. J Nanoparticles 2014:431787

Roco MC (2005) The emergence and policy implications of converging new technologies integrated from the nanoscale. J Nanopart Res 7:129–143

Rodriguez-Sanchez L, Blanco MC, Lopez-Quintela MA (2000) Electrochemical synthesis of silver nanoparticles. J Phys Chem 104(41):9683–9688

Roy S, Anantharaman P (2018) Biosynthesis of silver nanoparticles by Sargassum Ilicifolium (turner) C. Agardh with their antimicrobial activity and potential for seed germination. J Appl Phys Nanotechnol 1(1):002

Saeed K, Khan I (2016) Preparation and characterization of singlewalled carbon nanotube/nylon 6,6 nanocomposites. Instrum Sci Technol 44:435–444

Sandhya A, Suvarnalatha Devi P (2017) Extracellular biosynthesis and characterization of gold Nanoparticles using penicillium sps and studies on antimicrobial activity. Int J Pharma Bio Sci 8 (3):1026–1032

Sangeetha G, Rajeshwari S, Venckatesh R (2011) Green synthesis of zinc oxide nanoparticles by aloe barbadensis miller leaf extract: structure and optical properties. Mater Res Bull 46 (12):2560–2566

Sanpui P, Murugadoss A, Prasad PD, Ghosh SS, Chattopadhyay A (2008) The antibacterial properties of a novel chitosan–ag-nanoparticle composite. Int J Food Microbiol 124(2):142–146

Sastry M, Ahmad A, Khan MI, Kumar R (2003) Biosynthesis of metal nanoparticles using fungi and actinomycete. Curr Sci 85(2):162–170

Savage N, Wentsel R, Conlon M, McKinney D, Davis JM, Moris J, Hubal E (2008) Draft nanomaterial research strategy (NRS). Environmental Protection Agency, Washington, pp 1–2

Schmidt KF (2007) Nanofrontiers, visions for the future of nanotechnology, project on emerging technologies. Woodrow Wilson International Centre for Scholars, National Institutes of Health, Washington

Selid PD, Xu H, Collins EM, Striped Face-Collins M, Zhao JX (2009) Sensing mercury for biomedical and environmental monitoring. Sensors 9(7):5446–5459

Shahi SK, Patra M (2003) Microbially synthesized bioactive nanoparticles and their formulation active against human pathogenic fungi. Rev Adv Mater Sci 5:501–509

Sharma N, Pinnaka AK, Raje M, Fnu A, Bhattacharyya MS, Choudhury AR (2012) Exploitation of marine bacteria for production of gold nanoparticles. Microb Cell Fact 11(1):86

Shen W, Qu Y, Pei X, Li S, You S, Wang J, Zhang Z, Zhou J (2017) catalytic reduction of 4-nitrophenol using gold nanoparticles biosynthesized by cell-free extracts of Aspergillus sp. WL-Au J Haz Mat 321:299–306

Shi C, Zhu N, Cao Y, Wu P (2015) Biosynthesis of gold nanoparticles assisted by the intracellular protein extract of Pycnoporus sanguineus and its catalysis in degradation of 4-nitroaniline. Nanoscale Res Lett 10(1):147

Shiying H, Zhang Y, Guo Z, Gu N (2008) Biological synthesis of gold nanowires using extract of Rhodopseudomonas capsulata. Biotechnol Prog 24(2):476–480

Shukla R, Chanda N, Zambre A, Upendran A, Katti K, Kulkarni RR et al (2012) Laminin receptor specific therapeutic gold nanoparticles (198AuNP-EGCg) show efficacy in treating prostate cancer. Proc Natl Acad Sci U S A 109:12426–12431

Siavash I (2011) Green synthesis of metal nanoparticles using plants. Green Chem 13:2638

Siddiqi KS, Husen A (2016) Fabrication of metal and metal oxide nanoparticles by algae and their toxic effects. Nanoscale Res Lett 11:363. https://doi.org/10.1186/s11671-016-1580-9

Silambarasan S, Abraham J (2012) Biosynthesis of silver nanoparticles using the bacteria Bacillus cereus and their antimicrobial property. Int J Pharm Pharm Sci 4:536–540

Singaravelu G, Arockiamary JS, Kumar VG, Govindaraju K (2007) A novel extracellular synthesis of monodisperse gold nanoparticles using marine alga, Sargassum wightii Greville. Colloids Surf B Biointerfaces 57:97–101. https://doi.org/10.1016/j.colsurfb.2007.01.010

Singh M, Manikandan S, Kumaraguru AK (2011) Nanoparticles: a new technology with wide applications. Res J Nanosci Nanotechnol 1(1):1–11

Singh T, Jyoti K, Patnaik A, Sing A, Chauhan R, Chandel SS (2017) Biosynthesis, characterization and antibacterial activity of silver nanoparticles using an endophytic fungal supernatant of Raphanus sativus. J Genet Eng Biotechnol 15(1):31–39

Sinha S, Singh R, Chaurasia AK, Nigam S (2016) Self-sustainable Chlorella pyrenoidosa strain NCIM 2738 based photobioreactor for removal of direct Red-31 dye along with other industrial pollutants to improve the water-quality. J Hazard Mater 306:386–394

Skalickova S, Baron M, Sochor J (2017) Nanoparticles biosynthesized by yeast: a review of their application. Kvasny Prum 63(6):290–292
Srivastava SK, Yamada R, Ogino C, Kondo A (2013) Biogenic synthesis and characterization of gold nanoparticles by Escherichia coli K12 and its heterogeneous catalysis in degradation of 4-nitrophenol. Nanoscale Res Lett 8(1):70
Subhankari I, Nayak P (2013) Synthesis of copper nanoparticles using Syzygium aromaticum (cloves) aqueous extract by using green chemistry. World J Nano Sci Technol 2(1):14–17
Suganya UKS, Govindaraju K, Kumar GV, Dhas ST, Karthick V, Singaravelu G, Elanchezhiyan M (2015) Blue green alga mediated synthesis of gold nanoparticles and its antibacterial efficacy against gram positive organisms. Mater Sci Eng C 47:351–356
Sweeney RY, Mao C, Gao X, Burt JL, Belcher AM, Georgiou G, Iverson BL (2004) Bacterial biosynthesis of cadmium sulfide nanocrystals. Chem Biodivers 11(11):1553–1559
Tiwari DK, Behari J, Sen P (2008) Application of nanoparticles in waste water treatment. World Appl Sci J 3(3):417–433
Venkatesan J, Manivasagan P, Kim SK, Kirthi AV, Marimuthu S, Rahuman AA (2014) Marine algae-mediated synthesis of gold nanoparticles using a novel Ecklonia cava. Bioprocess Biosyst Eng 37(8):1591–1597
Vijayan SR, Santhiyagu P, Singamuthu M, Ahila NK, Jayaraman R, Ethiraj K (2014) Synthesis and characterization of silver and gold nanoparticles using aqueous extract of seaweed, Turbinaria conoides, and their antimicrofouling activity. Sci World J 938272:10
Vittal RR, Jamuna BA (2011) Nanoparticles and their potential application as antimicrobials. In: Mendez-Vilas A (ed) Science against microbial pathogens: communicating current research and technological advances, Formatex 3(1):197–209
Waghmare SS, Deshmukh AM, Kulkarni SW, Oswaldo LA (2011) Biosynthesis and characterization of manganese and zinc nanoparticles. Univ J Environ Res Technol 1:1
Wang T, Jin X, Chen Z, Megharaj M, Naidu R (2014) Green synthesis of Fe nanoparticles using eucalyptus leaf extracts for treatment of eutrophic wastewater. Sci Total Environ 466:210–213
Watlington K (2005) Emerging nanotechnologies for site remediation and wastewater treatment. US Environmental Protection Agency, Washington, DC
Xue B, He D, Gao S, Wang D, Yokoyama K, Wang L (2016) Biosynthesis of silver nanoparticles by the fungus Arthroderma fulvum and its antifungal activity against genera of Candida, Aspergillus and fusarium. IntJ Nanomed 11:1899–1906
Yadav KK, Singh JK, Gupta N, Kumar V (2017) A review of nano bioremediation technologies for environmental clean-up: a novel biological approach. JMES 8(2):740–757
Yang Y, He J, Li Q, Gao L, Hu J, Zeng R et al (2019) Self-healing of electrical damage in polymers using superparamagnetic nanoparticles. Nat Nanotechnol 14(2):151–155
Yuvakkumar R et al (2014) Rambutan (Nephelium lappaceum L.)peel extract assisted biomimetic synthesis of nickel oxide nanocrystals. Mater Lett 128:170–174
Zahran MK, Mohamed AA, Mohamed FM, El-Rafie MH (2013) Optimization of biological synthesis of silver nanoparticles by some yeast fungi. Egypt J Chem 56(1):91–110
Zhang H, Gilbert B, Huang F, Banfield JF (2003) Water-driven structure transformation in nanoparticles at room temperature. Nature 424:1025–1029
Zhang X, Qu Y, Shen W, Wang J, Li H, Zhang Z, Li S, Zhou J (2016) Biogenic synthesis of gold nanoparticles by yeast Magnusiomyces ingens LH-F1 for catalytic reduction of nitrophenols. Colloids Surf 497:280–285

Arsenic Removal Using Nanotechnology 4

Sougata Ghosh, Tanay Bhagwat, and Thomas J. Webster

4.1 Introduction

Arsenic (As) is one of the most hazardous pollutants readily distributed in the groundwater either naturally or from anthropogenic sources. As is abundant in the earth's crust, especially in soil, minerals, surface, and groundwater which facilitates easy contamination (Antman 2001). As exposure in the USA is monitored by three panel reviews: The National Academy of Sciences, The National Drinking Water Advisory Council and the U.S. Environmental Protection Agency's (EPA's) Science Advisory Board. Currently the maximum As contaminant level in drinking water was lowered to 10 μg/L, indicating the continual research going into As and the increasing alarm As possess to health (USEPA 2001, 2002).

As (III/V) exposure via drinking water has shown to cause respiratory symptoms, such as cough and bronchitis, skin lesions and hyperpigmentation. The urinary, cardiovascular and gastrointestinal systems are also affected due to a prolonged exposure to As. Further, bladder and lung cancers have also been reported apart from an increased risk of high blood pressure, diabetes and reproductive abnormalities (Haq et al. 2020). It is expected that lowering of the As guidelines from 50 to 10 μg/L could prevent death from bladder, lung and skin cancers and from heart disease (USEPA 2002). High As concentrations in groundwater and its consumption may also lead to hyperkeratosis, keratosis, melanosis, loss of appetite, anaemia, gangrene, and skin cancer (Karim 2000; Bagla and Kaiser 1996). Due to the utilization of As contaminated groundwater for irrigation of agricultural soils, even staple crops and

S. Ghosh (✉)
Department of Microbiology, School of Science, RK University, Rajkot, Gujarat, India

Department of Chemical Engineering, Northeastern University, Boston, MA, USA

T. Bhagwat · T. J. Webster
Department of Chemical Engineering, Northeastern University, Boston, MA, USA

I. Haq, A. S. Kalamdhad (eds.), *Emerging Treatment Technologies for Waste Management*, https://doi.org/10.1007/978-981-16-2015-7_4

vegetables are becoming sources of As entering the local market. Numerous factors like redox potential (Eh), adsorption/desorption, precipitation/dissolution, As speciation, pH, presence and concentration of competing ions, biological transformation, etc. control As concentration and transport in groundwater.

Various technological options that have been used to tackle As menace in groundwater for ensuring a supply of As free water can be classified as follows: (1) In situ remediation of As from an aquifer system, (2) Ex-situ remediation of As from tapped groundwater by As removal technologies, (3) As an alternative to the contaminated groundwater source, use of surface water source and (4) Tapping alternate safe aquifers for the supply of As free groundwater. In situ remediation of As from aquifer systems or decontamination of aquifers is considered as the best technological option. Ex-situ remediation of As from tapped groundwater using appropriate effective removal technologies may be a short-term alternative to provide potable As free groundwater for domestic purposes. However, this method is expensive and unsustainable for the supply of irrigational water and only limited to tapped groundwater but not from the aquifer system. Thus, the use of a surface water source is a logical proposition, although availability and supply of surface water flow and organized water supply system should be ensured for the supply of both drinking and irrigation water. Additionally, tapping alternate safe aquifers for the supply of As free groundwater seems to be helpful although a major drawback is the requirement of extensive studies and analyses for the mapping of groundwater availability, freshwater reserves and to examine mobilization of As in the aquifer, both on the spatial and temporal scale, due to forcing perturbation.

In view of the background novel strategies being explored to remove As effectively involve microorganisms like bacteria, fungi, yeasts and algae (Singh et al. 2014). In this chapter, we present an overview of As poisoning and their conventional treatment methods. Emerging microbial technology for addressing As toxicity, its removal and bioconversion is discussed in detail.

4.2 As Toxicity

For centuries, As has been used in cosmetics and pesticides, protecting crops from pests, and paints (copper acetoarsenite or 'Paris green'). The industrial use of As includes in the manufacturing of pesticides, fungicides, insecticides, pesticides, herbicides, paints, wood preservatives and cotton desiccants. Crystals of aluminium gallium arsenide or gallium arsenide are components of light emitting diodes, lasers, semiconductors and a variety of transistors. Hence, there is a high risk of occupational As exposure primarily by the consumption of contaminated drinking water and inhalation, absorption through the skin. The unbound As generates reactive oxygen species (ROS) during redox cycling and metabolic activation leading to DNA damage and lipid peroxidation (Cobo and Castineira 1997). Thiol or sulphydryl groups specifically form bonds with As (III) in tissue proteins of the lungs, liver, spleen, kidney, gastrointestinal mucosa and keratin-rich tissues (skin, hair and nails). Methyltransferase CYT19 catalyse the transfer of a methyl group from S-adenosyl-L-methionine to As as the As triglutathione (ATG) complex to form monomethylarsonicdigulathione (MADG), which acts as substrate of Cyt19 for further methylation to dimethyl As glutathione (DMAG). Hydrolysis of MADG and DMAG and subsequent oxidation lead to formation of dimethylarsinic acid (DMA) and monomethylarsonic acid (MMA) which can be found in hair. However, dimethylarsinous acid is reported in fingernails at concentrations exceeding MMA and DMA levels (Katz 2019). Fish, algae and seafood in general are the major sources of As in the body. Further, due to As contaminated irrigational water, As compounds, organic and inorganic, may enter the food chain via crops (Tamaki and Frankenberger 1992).

4.3 Acute Poisoning

Acute As poisoning mostly occurs due to the ingestion of insecticides or pesticides, accidentally, which shows symptoms like vomiting and diarrhoea. The lethal dose for acute As ranges from 100 to 300 mg (Schoolmeester and White 1980). Other associated clinical features include gastrointestinal disorders, nausea and colicky abdominal pain. Excessive salivation, seizures, diffusive skin rash, acute psychosis and cardiomyopathy due to toxicity may occur intermittently (Campbell and Alvarez 1989). Sometimes it is lethal and death occurs due to massive fluid loss from the gastrointestinal tract, severe dehydration, reduced circulating blood volume and consequent circulatory collapse apart from oesophagitis, gastritis and hepatic steatosis (Ghariani et al. 1991). Haematological abnormalities include bone marrow depression, haemoglobinuria, severe pancytopenia, intravascular coagulation, and normocytic normochromic anaemia and basophilic stippling. Renal and respiratory failure accompanied by peripheral neuropathy may lead to rapid, severe ascending weakness. Encephalopathy, hypoglycaemia and hypocalcaemia are also common (Ratnaike 2003).

4.4 Chronic Poisoning

Persons exposed to As for longer time periods show various clinical features specific to chronic poisoning which include abdominal pain, diarrhoea, sore throat and dermatological changes at the onset. Chronic As poisoning is often diagnosed on the basis of hyperpigmentation, palmar and solar keratosis. Unlike diarrhoea in case of acute As poisoning, chronic toxicity is characterized by diarrhoea in recurrent bouts and is occasionally associated with vomiting. As causes direct myocardial injury, cardiac arrhythmias and cardiomyopathy. Peripheral neuropathy resembling Guillain-Barré syndrome with similar electromyographic findings is also common. Increased mortality from prostate cancer and nephritis apart from transitional cell carcinomas of the bladder, kidney, ureter and all urethral cancers in both males and females can also occur due to the prolong consumption of As contaminated water. As may become deposited in lungs leading to the increased risk of bronchitis.

Malignancies of the skin, lung, nasal cavity, bone, liver, larynx, colon and stomach as well as lymphoma can also occur due to As poisoning. High toxicity related to As poisoning may be due to its role in the impairment of DNA repair, methylation of DNA, and increased free radical formation and activation of the protooncogenec-myc. As acts as co-carcinogen, tumour promoter, or tumour progress or under specific conditions. Significantly elevated As levels may also lead to teratogenicity (disruption of development of embryo or foetus).

4.5 Conventional Methods for As Removal

As removal from the contaminated water can be achieved by various techniques like precipitation, membrane processes, ion exchange and adsorption. However, the adsorption technique has gained more attention recently due to its simplicity of operation, potential for regeneration and sludge free operation (Tang et al. 2011). The process for As removal depends majorly on the chemical species of the As contaminating the water. Although arsenate can be effectively removed by available technologies, the trivalent form of As called arsenite cannot be subjected to precipitation, adsorption or ion exchange due to its unavailability because arsenite, below pH 9.2 is non-charged. Thus, initially arsenite is converted to arsenate which is removed thereafter using the following discussed techniques.

4.6 Oxidation

Although oxidation cannot remove As itself, it is a significant intermediate step where soluble arsenite is converted to arsenate which can be removed further using techniques like adsorption, coagulation or ion exchange. This process involves treatment with various chemical oxidative agents like oxygen, ozone, chlorine, chlorine dioxide, monochloroamine, hypochlorite, hydrogen peroxide, potassium permanganate and photocatalytic oxidation (UV/H_2O_2). Microbes like *Gallionella ferruginea* and *Leptothrix ochracea* have been reported for the effective removal of As where it is present in a concentration range of 50–200 mg/L (Katsoyiannis and Zouboulis 2004). Microbes are able to directly oxidize arsenite in water. However, interfering substances (like competing ions and organic matter) in groundwater plays a major role in determining the oxidation efficiency. The key anions like carbonate, phosphate, sulphate, silicate and chloride commonly found in groundwater can interfere in As removal. Therefore, it is very important to carefully select the appropriate oxidizing agent depending on the composition of the water to be treated and the type of contamination species (Nicomel et al. 2016) (Table 4.1).

Table 4.1 Different oxidants used to oxidize arsenite to arsenate, their operating conditions, properties and efficiencies. Reprinted from Nicomel et al. (2016)

Oxidants	Operating pH	Initial As concentration (μg/L)	Type of water	Remarks
Oxygen and ozone	7.6–8.5	46–62	Groundwater	Oxidation of As(III) by ozone is faster than by pure oxygen or air. In less than 20 min, complete oxidation was obtained using ozone, whereas 5 days were needed to oxidize 57% and 54% of As(III) using pure oxygen and air, respectively
Chlorine	8.3	300	Deionized water	As(III) was completely oxidized to As(V) by active chlorine when its initial concentration was greater than 300 μg/L. Stoichiometric rate was 0.99 mg Cl_2/mg As(III)
Chlorine dioxide	8.12	50	Groundwater	After 1 h contact time, 86% oxidation yield was achieved. This relatively high value is mainly due to the presence of some metals in water that could assist the catalysis of As (III) oxidation
Monochloroamine	8.12	50	Groundwater	Very long contact time is needed to obtain effective As(III) oxidation An oxidation yield of only 60% was achieved after 18 h
Hypochlorite	7	500	Groundwater	Given a hypochlorite concentration of 500 μg/L, there was a complete oxidation of As(III) to As (V)
Hydrogen peroxide	7.5–10.3	50	Freshwater and seawater	The efficiency of As(III) oxidation improved as pH was increased from 7.5 to 10.3
Potassium permanganate	8.12	50	Groundwater	Oxidation was completed after 1 min
Photocatalytic oxidation (UV/H_2O_2)	8	100	Groundwater	Combining hydrogen peroxide (H_2O_2) with

(continued)

Table 4.1 (continued)

Oxidants	Operating pH	Initial As concentration (μg/L)	Type of water	Remarks
				ultraviolet (UV) radiation resulted in an efficient As (III) oxidation. As UV dose increases, oxidation efficiency also increases. 85% of As(III) was oxidized to As(V) at a UV dose of 2000 mJ/cm^2
Biological oxidation (e.g., hemoautotrophic arsenite-oxidizing bacteria (CAOs))	–	–		CAOs can participate in the oxidation of arsenite to arsenate through the use of oxygen (or nitrate) as terminal electron acceptors during the fixation of inorganic carbon into cell material
In situ oxidation	–	–	Groundwater	Oxygenated water is pumped into the groundwater aquifer to reduce As concentrations to <10 μg/L

4.7 Coagulation-Flocculation

The most commonly used technique for As removal is by coagulation and flocculation. On addition of a positively charged coagulant like aluminium sulphate or ferric chloride, the negative charge of colloids is reduced which leads to collision of the particles resulting in the formation of larger aggregates. This process is called coagulation. In flocculation anionic agents called flocculant is added that leads to the formation of larger particles called flocs. Soluble As is rapidly converted to an insoluble solid by chemicals during these processes which are later precipitated.

Incorporation into metal hydroxides leads to co-precipitation by soluble As species. The efficiency of As removal varies with pH, type of coagulants used and type of As species. $Al_2(SO_4)_3$ and $FeCl_3$ show identical efficiency below pH 7.6. However, $FeCl_3$ is a better coagulant compared to $Al_2(SO_4)_3$ at a pH higher than 7.6. Moreover, arsenate is more efficiently removed compared to arsenite. Aluminium-based coagulants have been widely used as they can reduce the initial concentration of 280 μg/L to below the maximum concentration level (MCL) of 10 μg/L. The major drawback however is of coagulation-flocculation which produces high amounts of As-concentrated sludge which is difficult to manage eventually adds up to secondary pollution of the environment. Sludge treatment is costly, which is another major drawback that limits its field of application (Nicomel et al. 2016).

4.8 Membrane Technologies

Membrane filtration is one of the most effective techniques used for the generation of clean and potable water. Two categories of widely used pressure-driven membrane filtrations include: (1) low-pressure membrane processes, such as microfiltration (MF) and ultrafiltration (UF) and (2) high-pressure membrane processes, such as reverse osmosis (RO) and nanofiltration (NF). Sometimes a combination of techniques proved to be effective as using membranes with pore sizes between 0.1 and 10 μm for membrane filtration alone cannot be used to remove dissolved As species. In such cases, initially, the particle size of the As-bearing species increases by coagulation and flocculation prior to membrane filtration. However, this As flocculation prior to filtration depends on the pH of the water and the presence of other ions. Since arsenite has a neutral charge in the pH range of 4–10, it is difficult to be removed using the above strategy. On the other hand, being negatively charged at this pH range, arsenate can bind by surface complexation resulting in efficient arsenate removal and formation of arsenate by complete oxidation of arsenite is a prerequisite to exploit this technique beneficially. Similarly, since ultrafiltration alone is insufficient, a surfactant-based separation process (such as micellar-enhanced ultrafiltration (MEUF))has been developed for As removal. The addition of a cationic surfactant (like hexadecylpyridinium chloride (CPC)) at concentration above the critical micelle concentration (CMC) to contaminated drinking water of the water will lead to formation of micelles, which bind to negatively charged As species which are effectively removed (96%).

Both nanofiltration and reverse osmosis are applicable when the dissolved compound to be removed has a molecular weight above 300 g/mol. Diatomaceous earth (5–100 μm) filtration, not a membrane system technically, but similar to membrane filters, is also used for As removal occasionally. The combined effect of small pore sizes and high porosity are major advantages to remove small particles at high water filtration rates (Nicomel et al. 2016).

4.9 Adsorption and Ion Exchange

In adsorption, a solid (adsorbent) is used to remove substances (adsorbate) from gaseous or liquid solutions. It is a surface phenomenon where the target adsorbate molecules are accumulated on the surface of adsorbent due to van der Waals forces and electrostatic forces. The efficiency depends on surface properties (e.g., surface area, polarity, etc.) of the adsorbent. The diverse types of adsorbents include activated carbon, coal, red mud, fly ash, chitosan, chicken feathers, zeolites, kaolinite, zero-valent iron, montmorillonite, goethite, titanium dioxide, iron hydroxide, activated alumina and cation-exchange resins. Removal of As is accomplished by

Table 4.2 Comparison of advantages and disadvantages of different technologies for As removal. Reprinted from Ashraf (2019)

Methods	Advantages	Disadvantages
Air oxidation	Easy and low cost, oxidizes other organic and inorganic constituents in water	Slow process and increases in the rate of oxidation, eliminates As (v) mostly
Chemical oxidation	Natural process, operates rapidly, causes oxidation of other impurities and minimal waste production	Depends highly on pH and requires an oxidation step
Adsorption (activated alumina, iron oxides, titania, etc.)	Simple and easily controlled. No other reagents and additional steps required. Adsorbent can be regenerated	
Nanofiltration	High removal capacity. Well defined process	Economically not an effective process, high water rejection
Reverse osmosis	Toxic waste is not produced. High efficiency for the removal of As	Difficult to operate
Electrodialysis	Can also be used for the removal of other pollutants	Production of toxic wastewater

iron where it acts as a contaminant-immobilizing agent, co-precipitant, or by behaving as a reductant. Several advantages of adsorption include relatively high As removal efficiencies, ease of operation, cost effectiveness and no sludge production. Low pH is favourable for arsenate adsorption and maximum arsenite adsorption occurs between pH 4 and 9. Similarly, contaminating ions (like phosphate and silicate) may also compete for the adsorption sites. Moreover, the nature of the adsorbents is also a determining factor for the adsorption efficiency as irregular pore structures, low specific surface areas, lack of selectivity, relatively weak interactions with metallic ions, and regeneration difficulties as some limiting factors (Nicomel et al. 2016) (Table 4.2).

In the following section, various novel nanostructure based adsorbents for As removal are discussed along with their process of fabrication. Further, their As removal efficiency, limiting conditions and various significant process parameters are carefully accounted.

4.10 Carbon Nanotubes

The inefficiency of removal of As from water by conventional techniques has been emphasized for the exploration of a novel nanostructure based water treatment processes. Carbon nanotubes (CNTs) are nanomaterials, extensively used for the removal of not only heavy metals but also pollutants and dyes. However, significant disadvantages of CNTs are their difficulties in aggregation, solubility and lack of feasibility in manipulation. These limitations can be overcome by rational surface functionalization for tunable affinity towards the chemical species to be removed.

AlOmar et al. (2016) reported the use of a deep eutectic solvent (DES) as functionalization agents for carbon nanotubes (CNTs) in order to fabricate novel adsorbents for the removal of As ions (As^{3+}) from water. Two DESs systems used in the study were methyltriphenylphosphonium bromide (MTPB) and benzyltriphenylphosphonium chloride (BTPC) as salts, conjugated with glycerol (Gly) as a hydrogen bond donor. These novel materials were prepared by drying the pristine MWCNTs (P-CNTs) overnight at 100 °C and further oxidation by two different methods. The first method involved oxidation by potassium permanganate ($KMnO_4$) via sonication for 2 h to produce K-CNTs. In another method, acidification of P-CNTs was performed by refluxing with 65% HNO_3 for 1 h at 140 °C to produce N-CNTs. The resulting functionalized CNTs were repeatedly washed with distilled water and filtered. KM-CNTs achieved the highest removal percentage at pH 2.7 while KB-CNTs recorded the highest removal at pH 6. The maximum adsorption capacity of a MTPB-DES-functionalized CNTs adsorbent was found to be 23.4 mg/g (AlOmar et al. 2016).

In another study, ceria supported on CNTs (CeO_2-CNTs) was developed for the effective removal of arsenate from water. Initially, CNTs were prepared by catalytic pyrolysis of the propylene-hydrogen ($C_3H_6/H_2 = 2{:}1$) mixture at 750 °C in a ceramic tube furnace in the presence of Ni particles as catalysts. After washing and grinding, the CNTs were further oxidized by refluxing with nitric acid and sulphuric acid at 140 °C to introduce oxygen groups, such as hydroxyl (–OH) and carboxyl (–COOH), on the surface. Introduction of these groups resulted in superior suspensibility and adhesion of $Ce(OH)_3$ particles. Ce supported CNTs were prepared by mixing 1 g CNTs into 20 mL distilled water in which a $CeCl_3$ (0.035 g/mL) solution was added drop by drop. The resulting mixture was agitated for 1 h followed by titrating with 0.5 wt.% NaOH until the pH value reaches 9. This mixture was then dried up and heated at 450 °C for 20 min to oxidize $Ce(OH)_3$particles to CeO_2 nanoparticles. CeO_2-CNTs exhibited a pH dependent effective adsorption for arsenate. Cations (like Ca^{2+} and Mg^{2+}) significantly enhanced its adsorption capacity. Under natural pH conditions, an increase from 0 to 10 mg/L in the concentration of Ca^{2+} and Mg^{2+} resulted in an increase from 10 to 81.9 and 78.8 mg/g in the amount of As(V) adsorption, respectively. A regeneration efficiency of 94% was achieved using 0.1 mol/L NaOH, which signifies that these nanomaterials are promising candidate for generating As free drinking water (Peng et al. 2005).

Iron-manganese based amorphous binary oxide coated functionalized multiwalled CNTs(f-MWCNTs) were synthesized by a co-precipitation method for As removal. In short, multiwalled CNTs (MWCNTs) were synthesized by catalytic chemical vapour deposition, where acetylene was decomposed on the surface of a $MmNi_3$ alloy. The MWCNTs were purified, acid treated and reacted with of $KMnO_4$ and $FeSO_4$at pH between 7 and 8 using a 5 M NaOH solution. The resulting reaction mixture was continuously stirred for 1 h, aged at room temperature for 12 h, washed with deionized water, filtered and then dried at 105 °C for 4 h. These materials exhibited an adsorption capacity of As (III) (113.95 mg/g) more than that of As (V) (95.8 mg/g). This variation might have been attributed to the oxidation of As (III), which disturbs the electron cloud and hence the electric dipole moment of both the adsorbent and adsorbate, thus, improved van der Waals interactions were formed between them and adsorption capacity was enhanced. To extract the isotherm and

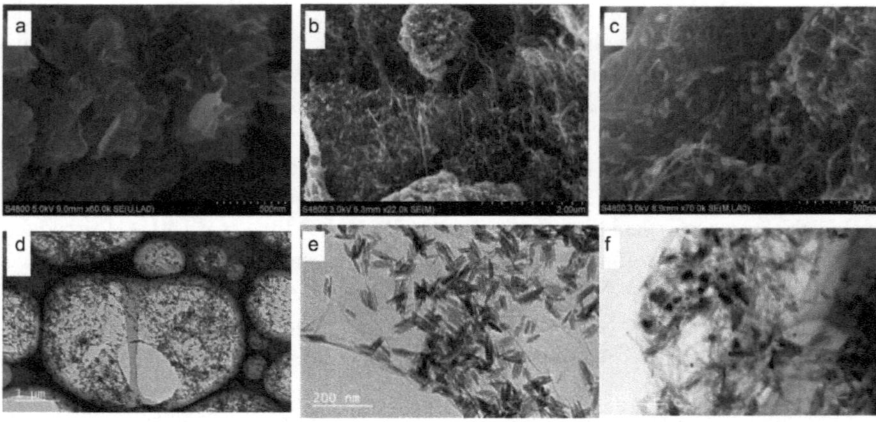

Fig. 4.1 SEM images of rGO (**a**) and α-FeOOH@GCA (**b** and **c**) and TEM images of α-FeOOH@GCA (**d** and **f**). Reprinted with permission from Fu et al. (2017). Copyright © 2017 Elsevier Inc.

kinetic constants, Langmuir and Elovich equations were used respectively. This nanocomposite shows fast adsorption kinetics and high adsorption capacity (Tamilarasan and Ramaprabhu 2012).

The goethite impregnated graphene oxide (GO)-CNTs aerogel (-α-FeOOH@GCA) was prepared via a facile self-assembly method of GO-CNTs induced by in situ Fe^{2+} reduction. Briefly, graphene oxide(GO) was prepared from a natural graphite powder by a modified Hummers method. α-FeOOH@GCA was prepared by mixing 0.07 g GO, 0.03 gCNTs (7:3) and 1.81 g $FeSO_4.7H_2O$ (0.065 M) into 100 mL deionized water. The resulting reaction mixture was sonicated for 0.5 h at room temperature to obtain a uniform mixture which was further hydrolysed in an oil bath for 6 h at 95 °C to form an α-FeOOH@GCA hydrogel. α-FeOOH@GCA showed excellent adsorption capacities of 56.43, 24.43 and 102.11 mg g^{-1} for As (V), dimethyl As sodium (DMA) and *p*-arsanilic acid (p-ASA), respectively. This nanocomposite is advantageous as it prevents aggregation of GO-CNTs and inhibits the growth of α-FeOOH nanoparticles (Fig. 4.1). Thus, both the diffusion and adsorption capacity of As species were enhanced. Other advantages of α-FeOOH@GCA lie in its wide range of working pH, high adsorption kinetics and excellent reusability. Upon binding, As species were proven to form an inner sphere complex on the surface of α-FeOOH@GCA through a different ligand exchanging mechanism greatly dependent on the molecular structures of the As species. The results showed that α-FeOOH@GCA could be promising adsorbent for As removal (Fu et al. 2017).

Three-dimensional graphene-CNTs-iron oxide nanostructures were synthesized using a microwave assisted method for the efficient removal of As from contaminated water. The synthesis of these materials was achieved by grafting organometallic ferrocene onto the graphene surfaces and subliming it under

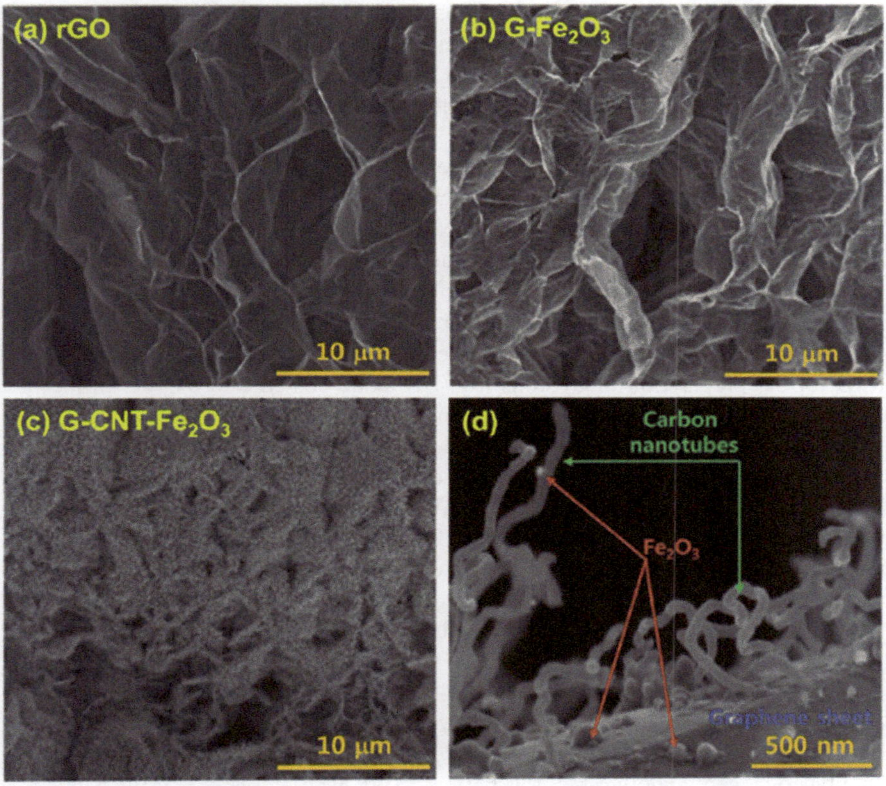

Fig. 4.2 SEM micrographs: (**a**) expanded graphene worm under microwave radiation, (**b**) 2D iron oxide decorated graphene, (**c**) graphene-carbon nanotube-iron oxide 3D nanostructures at the final stage and (**d**) magnified 3D nanostructures; carbon nanotubes are vertically grown on graphene sheets and iron oxide nanoparticles and are decorated on both graphene sheets and carbon nanotubes. Reprinted with permission from Vadahanambi et al. (2013). Copyright © 2013 American Chemical Society

microwave radiation. Initially, commercially available expandable graphite (0.5 g) was dispersed in hydrogen peroxide and subjected to ultrasonication for 30 mins to yield mild graphite oxide. This was then filtered, washed with deionized water and eventually dispersed in acetonitrile. Azodicarbonamide (ADC, 0.1 g) and ferrocene (0.5 g) were added into the above mixture and ultrasonicated for 25 min and irradiated in a microwave reactor at 700 W for 180 s. This process resulted in the fabrication of unique three-dimensional nanostructures where CNTs were vertically standing on graphene sheets and iron oxide nanoparticles were decorated on both the graphene and the CNTs (Fig. 4.2). Bamboo like CNTs showcased inner graphene compartment layers which were fully formed, elongated and had completely bridged the inner diameter of the outer tube with formation of complete knots. The diameter of each knot was found to be in the range of 100–120 nm with the length of the knots about 200–250 nm. Iron nanoparticles adhered to the outer tip of the nanotubes. Nanostructures with iron oxide nanoparticles showed excellent absorption for As removal from contaminated water, due to its high surface-to-volume ratio and open pore network of the graphene-CNTs-iron oxide three-dimensional nanostructures (Vadahanambi et al. 2013).

Another study reported the synthesis of ethylenediamine functionalized multiwalled CNTs (e-MWCNT) loaded with iron(III) oxide in the goethite form. The precipitation of adsorbed Fe^{3+} and oxidized Fe^{2+} using a base helped to obtain e-MWCNT/Fe^{3+} and e-MWCNT/Fe^{2+} adsorbents, respectively. Modification of o-MWCNT by ethylenediamine was performed by carboxylic acid residue which was pre-activated using a N-HATU coupling agent. This led to a highly efficient formation of amide linkage and amino terminal functionality. Two methods were used for the preparation of hydrous iron(III) oxide coated e-MWCNT. In the first method, 50 mg of e-MWCNT were immersed in 10 mL of a 0.1 mol L^{-1} $FeCl_3$ solution and sonicated for 2 h. Fe (III) adsorbed e-MWCNTs were isolated by vacuum filtration, processed further, and dried in a vacuum oven at 40 °C for 8 h. In the second method, 50 mg of e-MWCNTs were immersed in 10 mL of a 0.1 mol dm^{-3} $FeSO_4$ solution, sonicated for 2 h at 30 °C in an inert atmosphere of nitrogen, vacuum-filtered, and dispersed in 5 ml DI water followed by the addition of an equimolar quantity either of 0.05 mol dm^{-3} $KMnO^4$ or 30% hydrogen peroxide (H_2O_2), with respect to moles of adsorbed Fe(II) ion, to perform ferrous–ferric oxidation at pH ~2. The nanostructures were recovered by filtration and vacuum drying. Maximum adsorption capacities using Langmuir model for As(V) on e-MWCNT/Fe^{2+} and e-MWCNT/Fe^{3+} were 23.47 and 13.74 mg g^{-1} at 25 °C, respectively. Thermodynamic parameters showed that the adsorption of As(V) was spontaneous and endothermic for both e-MWCNT/Fe^{2+} and e-MWCNT/Fe^{3+}(-Velickovic et al. 2012).

Saleh et al. (2011) reported the synthesis of MWCNT/MnO_2 for the simultaneous oxidation of arsenite and sorption of arsenate. Initially oxidized-MWCNTs (o-MWCNTs) were synthesized by mixing the pristine MWCNTs (outer diameter, 30–50 nm; inside diameter, 5–10 nm; 30–50 nm; length, 10–20 μm) with a mixture of 75 mL of concentrated H_2SO_4 and 25 mL of HNO_3 (3:1). The mixture was subjected to sonication for 60 min ensuring breaking up of large aggregates and efficient dispersal of the MWCNTs. Following sonication, at 80 °C, the mixture was refluxed for 6 h, with vigorous stirring to accomplish oxidation. Extraction was achieved by repeated cycles of dilution with triple ionized water, and decanting the solutions until the resistivity of the supernatant was around 0.5 MΩ and the pH was between 5 and 6. The oxidized product was further dried for 12 h at 100 °C and milled. MWCNTs with MnO_2 nanoparticles were synthesized by a reduction oxidation method where initially o-MWCNTs were dispersed in deionized water by ultrasonic vibration for 3 h onto which a 0.5 M solution of $KMnO_4$ was added drop-wise at a rate of 1 mL/2 min (Fig. 4.3). With the initiation of heating, the development of a dark brown colour indicated the formation of MnO_2. After refluxing at 130 °C for 8 h, the nanostructures were filtered, washed and dried overnight at 110 °C. As removal by the nanostructure was found to be pH dependent

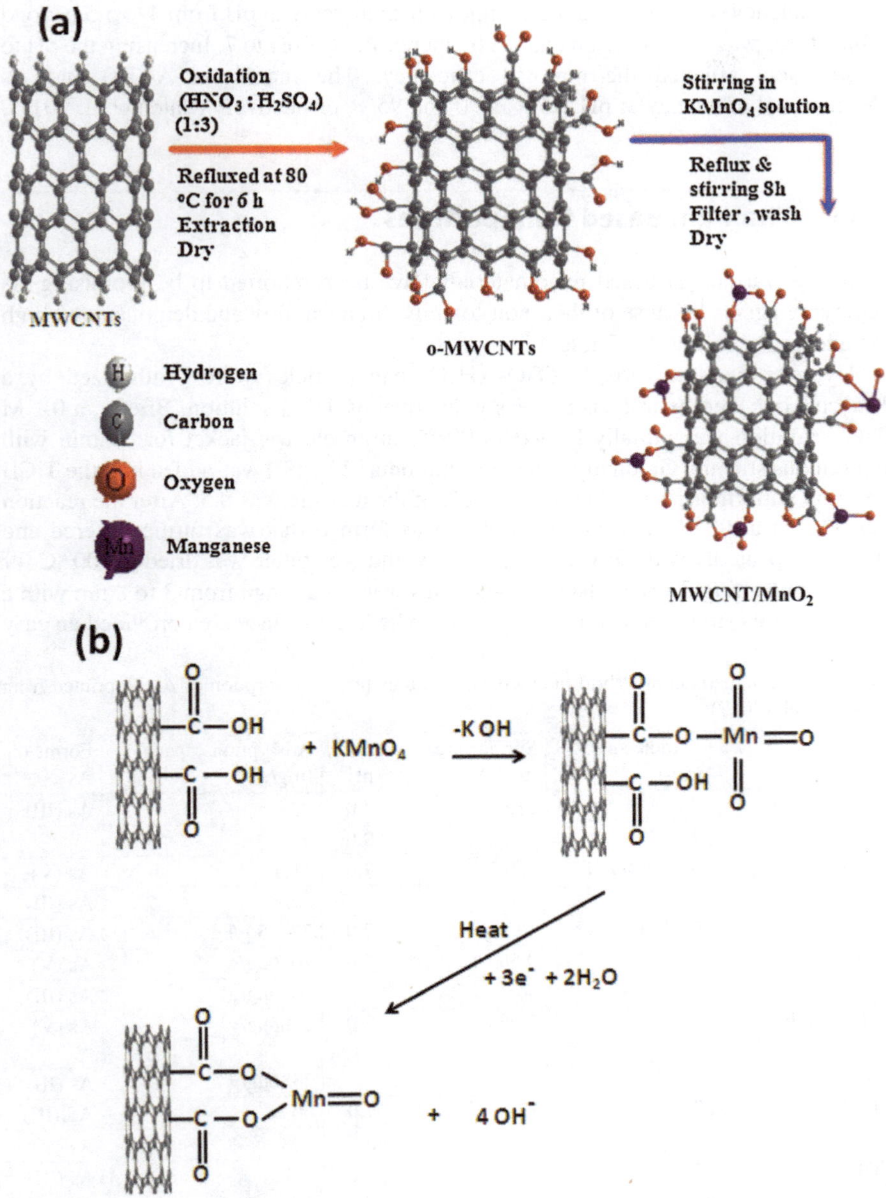

Fig. 4.3 Scheme of MWCNTs oxidation and MWCNT/MnO_2 preparation: (**a**) simplified representation of the interaction between MnO_2 and the MWCNT to form the composite MWCNT/MnO_2. (**b**) Reprinted with permission from Saleh et al. (2011). Copyright © 2011 Elsevier B.V.

with efficiency of As removal increasing with an increase in pH from 4 to 5.5 beyond which there was no significant change by increasing the pH to 7. Increasing the pH to 9 adversely affected the removal efficiency. The maximum As(III) and As (V) removal efficiency at pH 5.5 was 90 and 95%, respectively (Saleh et al. 2011).

4.11 Titanium Based Nanoparticles

Recently, various Ti based nanomaterials have been reported to be promising As removing agents because of their non-toxicity, high stability and demonstrated high As adsorbing capability (Table 4.3).

Hydrous titanium dioxide ($TiO_2.xH_2O$) nanoparticles were synthesized by a low-cost one-step hydrolysis process with aqueous $TiCl_4$ solution. Briefly, a 0.1 M $TiCl_4$ solution was initially heated to 90 °C in an electric jacket for 30 min with continuous stirring. Gradually, aqueous ammonia (25 wt%) was added into the $TiCl_4$ solution with rigorous stirring until the pH of the mixture was 8.5. After the reaction for 30 min at 90 °C, a white precipitate was formed that was further filtered and washed repeatedly with DI water. Eventually, the precipitate was dried at 100 °C for 8 h. The resulting $TiO_2 \times H_2O$ nanoparticles were in a range from 3 to 8 nm with a highly porous structure forming aggregates. The large surface area provided an easy

Table 4.3 Comparison of hybrid titania nanostructures for the adsorption of As. Reprinted from Ashraf et al. (2019)

Materials	Particle size (nm)	Surface area (m^2/g)	pH	Adsorption capacity (mg/g)	Forms of As
Hydrous TiO_2	3–8	312	7.0	83	As (III)
			9.0	96	
Granular TiO_2	0.15–0.6 mm	250.7	7.0	41.4	As (V)
				32.4	As (III)
CHTO	0.14–0.29 mm	163.6	7.0	72–75 g/kg	As (III)
TICB		0.56	7.0	2050μg/g	As (V)
				2099μg/g	As (III)
TICB under UV-light		3.06	7.0	2988μg/g	As (V)
				3536μg/g	As (III)
TNTs	2–6	312	7.0	60	AS (III)
			3.0	208	As (V)
TiO_2-CNTs	2.7	196	7.0	1.8	As (III)
				1.3	As (V)
TiO2/MMT	1.83–4	176.4	7.0	4.58	As (III)
Dark				4.86	As (V)
UV				5.16	As (V)

Abbreviations: *CHTO* crystalline hydrous titanium oxide, *TICB* titania impregnated chitosan bead, *TNTs* titania nanotubes, *CNTs* carbon nanotubes, *MMT* montmorillonite.

and effective As (III) removal capability from the aqueous environment which reached over 83 mg/g at a near neutral pH environment, and over 96 mg/g at pH 9.0. These nanostructures could effectively remove As(III) from contaminated natural lake water samples as well (Xu et al. 2010).

Manna et al. (2004) reported the synthetic crystalline hydrous titanium(IV) oxide (CHTO) mediated removal of As(III). The absolute As(III) sorption capacity was 72 to 75 g kg^{-1} at a pH of 7.0. Pre-drying above 350 °C led to the reduction of sorption density from 11.50 (±0.10) to <9.0 mg g^{-1}. Almost 70% of As(III) sorption was complete within 0.5 h. As rich CHTO could be regenerated using a 1.0 M NaOH solution, which regained 90% and 75 to 80% of its initial efficiency using batch and column methods, respectively. Efficient As recovery (~ 99.25%) from the regenerate was achieved (Manna et al. 2004).

Bang et al. (2005) reported a novel granular titanium dioxide (TiO_2) to adsorb more arsenate [As(V)] than arsenite [As(III)] in US groundwater at pH 7.0. The adsorption capacities for As(V) and As(III) were 41.4 and 32.4 mg g^{-1} of TiO_2, respectively. It is important to note that a similar adsorption capacity for As(V) and As(III) (approximately 40 mg g^{-1}) was achieved when simulated Bangladesh groundwater was used. Entities like silica (20 mg L^{-1}) and phosphate (5.8 mg L^{-1}) did not have any deleterious effect on the removal of As(V) and As (III) by TiO_2 at a neutral pH. Approximately 45,000 bed volumes of groundwater containing an average of 39 μg L^{-1} of As(V) was treated which confirmed that the granular TiO_2 adsorbent was very effective for the removal of As in groundwater (Bang et al. 2005).

Another novel sorbent for As, TiO_2-impregnated chitosan beads (TICB), was synthesized from a solution of chitosan and 0.1 M HCl (1 g/60 mL) that was stirred with a magnetic stir bar until dissolution (4 h). TiO_2-impregnatedchitosan beads (TICB) were fabricated by adding TiO_2nanopowder (<25 nm) to the same chitosan solution (0.4242 g TiO_2:1 g chitosan) and stirring until a homogenous solution was achieved. Beads were formulated by dispensing the dissolved chitosan-TiO_2 solution into 0.1 M NaOH and kept overnight before filtering through a Buchner funnel. The beads were rinsed with deionized water until the filtrate was of neutral pH. The beads were dried in the hood (Miller and Zimmerman 2010).The point of zero charge (PZC) for TICB was determined to be 7.25. Without exposure to UV-light, TICB removes 2198 μg As(III)/g TICB and 2050 μg As(V)/g TICB. UV irradiation enhanced As removal where sorption capacities reached 6400 mg As/g TICB and 4925 mg As/g TICB, for As(III) and As(V), respectively. TICB was found to be superior to TiO_2nanopowder for As removal (Miller and Zimmerman 2010).

A mesoporous material, titanate nanotubes (TNs), with different surface areas (197–312 m^2 g^{-1}) and pore size diameters (2–6 nm) were synthesized by an alkaline hydrothermal method. Nanosized titania particle (TP, 40–50 nm, 15 m^2 g^{-1}) and sodium hydroxide were used to synthesize the TNs. The surface area and pore size diameters of TNs could be effectively tuned by controlling the reaction temperature and contact time of TP and 10 M NaOH solution. The reaction was initiated by dispersing 3 g of titania into 100 mL of a 10 M NaOH solution and heated at different temperatures like 110, 130, 150 °C for 24 h, or at 180 °C for 10 h or 48 h, respectively, to obtain products coded as TN (110), TN (130), TN (150), TN

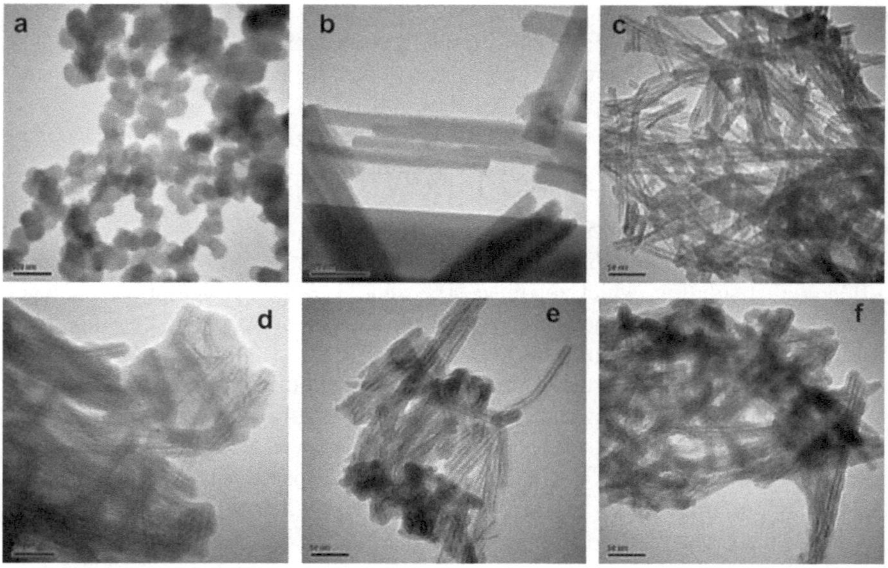

Fig. 4.4 TEM image of (**a**) TP, (**b**) TN (180–2), (**c**) TN (180–1), (**d**) TN (150), (**e**) TN (130) and (**f**) TN (110). Reprinted with permission from Niu et al. (2009). Copyright © 2009 Elsevier Inc

(180–1) and TN (180–2), respectively (Fig. 4.4). After washing with DI water to neutral, the product was soaked in a 0.5 M hydrochloric solution for 5 h which led to the exchange of interlayer sodium ions for the protons. The resulting protonated titanate nanotubes were washed to pH 7 with deionized water and dried at 100 °C. These nanotubes exhibited better adsorption of arsenate [As (V)] in an acidic solution, while the uptake of arsenite [As (III)] was preferred in an alkaline solution. The maximum uptake of As (V) and As(III) as calculated by the Langmuir equation was 208 mg g^{-1} (pH 3.0) and 60 mg g^{-1} (pH 7.0), respectively, achieved on a TN (180-1) adsorbent (312 m^2 g^{-1}, internal diameter, 5 nm), which was 33 and 10 times greater than those of nanosized titania particles (40–50 nm, 15 mg g^{-1}). More than 80% of As (III) and 95% of As (V) adsorbed on TNs could be desorbed with 1.0 M NaOH solution within 1 h which clearly indicated that TNs can be regarded as an efficient, low-cost adsorbent for the removal of As (Niu et al. 2009).

Liu et al. (2014) prepared a TiO_2-coated carbon nanotube (CNT) network filter using a simple filtration–steam hydrolysis method. Initially, 2.5 mL of titanium isopropoxide (TTIP) was dispersed in 10 mL of pure ethanol by bath sonication for 5 min. Followed to this, the CNT filter was placed on a PTFE membrane support within a vacuum filtration setup with an inner diameter of 35 mm. First, 10 mL of ethanol was passed through the filter to wet the CNTs, followed by a TTIP/ethanol mixture. Hydrolysis of TiO_2 was brought about by passing water vapour at 65 °C through the CNT filter for 15 min. An effective filtration area of 7 $cm^2$was found to be comprised of 14.7 mg of TiO_2 coated onto 16.3 mg of CNTs (31 mg total). TheTiO_2 coating (with a specific surface area of 196 m^2 g^{-1}) was 5.5 ± 2.7 nm thick which was fully covered by the CNT network. An increase in the flow rate showed a stronger effect. The first-order As sorption rate constants were found to be 4.3 and 4.4 s^{-1} for As(III) and As(V), respectively when the flow rate was adjusted to 6 mL min^{-1}. Increased mass transport due to internal convection and pore radius range, improved sorption site accessibility (due to porosity), and TiO_2 dispersion were major contributing factors for efficient As removal by these nanostructures. Additionally, a negative surface charge of reduced TiO_2 due to anodic capacitance also helped in the sorption. These nanofilters were also effective for treating groundwater contaminated with 44 ppb As (Liu et al. 2014). The As level could be significantly lowered by single-pass filtration, and 12,500 bed volumes (residence time of 4.5 s; 127 L m^{-2} h^{-1}; 5.8 mg m^{-2} h^{-1}). Further, a TiO_2 filter could be successfully regenerated by treating with 5 mM NaOH for both As(III) and As(V).

Although natural materials, such as kaolinite, montmorillonite and red mud, can remove As from water, their efficiency can further be enhanced by modifying them with nanomaterials. Various modifying polyoxy cations, such as Al^{3+}, Fe^{3+}, Mn^{4+} and Ti^{4+} can be used for pillaring. However, titania pillared clay is very attractive due to its mesoporous structure. Very small TiO_2 particles can act as pillars between silicate layers, and show high adsorption ability due to its large specific surface area. Li et al. (2012a, b) reported three main steps for the synthesis of TiO_2/MMT which can be summarized as follows:

Step 1: Initially, 100 mL of titanium tetrachloride was mixed with 200 mL of a 3 mol L^{-1} HCl solution under vigorous stirring. The resultant solution was mixed with a 0.5 mol L^{-1}HCl solution at 1:2.5 ratios. This step leads to the preparation of a TiO_2 pillared solution.

Step 2: In this phase, 10 g of Na-montmorillonite (MMT) was added to 490 mL of deionized water under continuous stirring for 2 h and then was allowed to stand for 12 h at 298 K.

Step 3: This is the final phase when the pre-formed TiO_2 pillared solution was added to the Na-MMT solution and stirred vigorously for 1 h followed by treatment in a water bath at 353 K for 6 h. Finally, TiO_2/MMT was separated from the mixture solution and washed several times with deionized water to neutral pH. After drying in a freeze dryer at 258 K under a vacuum, the TiO_2/MMT final TiO_2 content was found to be 15.1 wt%. This TiO_2/MMT adsorbent was used to remove As(III) and As(V) from the aqueous solution with adsorption capacities of 4.58 and 4.86 mg g^{-1}, respectively. Upon UV irradiation, the adsorption capacities of As(III) and As(V) could be enhanced to 5.19 and 5.16 mg g^{-1}, respectively. Hereby, TiO_2/MMT was found to be promising adsorbent for As (III) and As(V) removal from As contaminated water (Li et al. 2012a, b).

4.12 Iron Based Nanoparticles

Iron oxides are considered for As removal not only due to their easy availability and high affinity to As but also low cost. Various iron oxides like amorphous hydrous ferric oxide (FeOOH), goethite (α-FeOOH) and haematite (α-Fe_2O_3) are being explored extensively for As removal from natural water resources. Amorphous FeOOH and goethite are easily decomposable or form crystalline iron oxides as they are unstable with low surface area leading to a reduction in As removal capacity. However, α-Fe_2O_3 with high stability has great As removal capacity. Tang et al. (2011) reported the synthesis of ultrafine iron oxide (α-Fe_2O_3) nanoparticles by a solvent thermal process. Briefly, a 0.1 M $FeCl_3$solution was prepared in 80 mL ethanol in which 0.8 g of NaOH was added and reacted at room temperature under constant stirring resulting in a yellowish-brown precursor solution. This precursor solution was autoclaved at 150 °C for 2 h resulting in the formation of red precipitate which was washed with distilled water and alcohol, and dried at 80 °C for 12 h. Spherical ultrafine α-Fe_2O_3 nanoparticles with an average size of about 5 nm were obtained. These nanoparticles aggregated into a highly porous structure with a high specific surface area of ~162 m^2/g with their surface covered by high affinity hydroxyl groups. The adsorption capacities of the α-Fe_2O_3 nanoparticles for As(III) and As(V) were found to be 95 mg/g and 47 mg/g, respectively, near the neutral pH. Moreover, these α-Fe_2O_3 nanoparticles were effective in removal of both As(III) and As(V) from a contaminated natural lake water sample meeting the USEPA drinking water standard for As without any further modification treatments like preoxidation and/or pH adjustment (Tang et al. 2011).

Among various types of iron ores, magnetite is most commonly distributed in the lithosphere, pedosphere and biosphere. This iron oxide ore is often formed via bacterial activity in aquifers, soils and sediments. A study indicated that magnetite nanoparticles (20 nm) can exhibit maximum As adsorption at pH 2, with a value of approximately 3.70 mg/g for both As(III) and As(V) when the initial concentration of both As species was maintained at 2 mg/L. The efficiency of As removal by magnetite nanoparticles depends on various factors such as initial concentration of As, pH, contact time, phosphate concentration in the water and the adsorbent concentration. As adsorption involved the formation of weak As–iron oxide complexes at the magnetite surface. As removal was reported to be inversely proportional to the phosphate concentration in the water. In presence of phosphate (46 mg/L), magnetite nanoparticles could remove<50% of As from water (Chowdhury and Yanful 2011). Nanocrystalline magnetite (Fe_3O_4) was reported to enhance As removal capacity. Fe_3O_4 nanocrystals were fabricated by stirring a mixture of mixture of FeO(OH) (2.00 mmol),1-octadecene (20.00 mmol) and oleic acid (8.00 mmol)at 320 °C. Magnetic separations were achieved with a high gradient magnetic field column separator (Fig. 4.5) consisting of an S.G. Frantz Canister Separator (model L-1CN), that had a stainless-steel canister column (6.3, 25.4, and 222.3 mm in length, 35.5 cm^3). This stainless-steel column was packed with stainless-steel wool (~50μm wire diameter), with a packing volume of 5% of the canister (~15 g stainless-steel wool). The sample was passed through the column after application of a magnetic field. This helped in the retention of the nanoparticles in the column. The nanoparticles were retained in the column as the solvent passed through. Fe_3O_4 nanoparticles could be recovered by removing the magnetic field followed by passing of fresh solvent through the column. When the size of Fe_3O_4

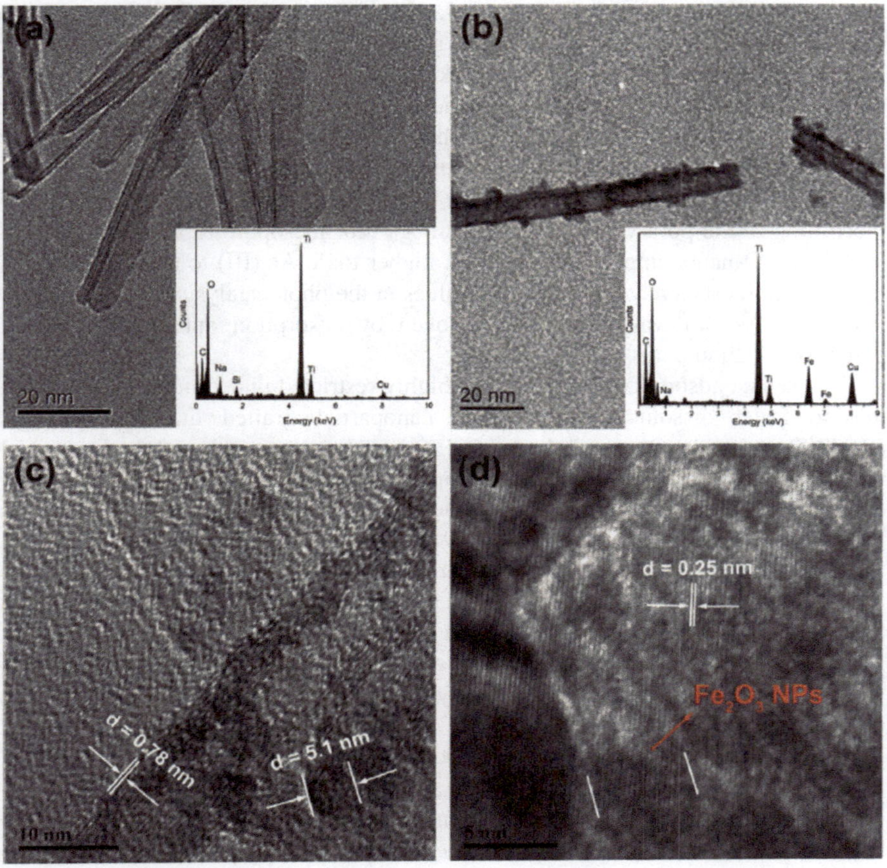

Fig. 4.5 TEM images of TNTs (**a**) and Fe-TNTs (**b**). HRTEM images of the tubular structure (**c**) and lattice distance of Fe_2O_3 (**d**). Reprinted with permission from Wang et al. (2015). Copyright © 2014 Elsevier Inc

nanoparticles decreased from 300 to 12 nm, the adsorption capacities for both As(III) and As(V) increased nearly 200 times. It is interesting to note that such a dramatic increase provided strong scientific rationale that nanoscale iron oxide materials may remove As via a mechanism altogether different than bulk systems (Mayo et al. 2007).

Since we have already seen the potential of TiO_2 nanoparticles for As removal, attempts have been made to fabricate bifunctional mesoporous TiO_2 (*meso*-TiO_2)/α-Fe_2O_3 composites by impregnation of Fe^{3+} into *meso*-TiO_2 followed by calcination at 300 °C. These 'bifunctional' composites possess a synergy of the photocatalytic ability of *meso*-TiO_2 for oxidation of As (III) to As (V) and the adsorption performance of α-Fe_2O_3 for As (V). The bifunctional *meso*-TiO_2/R-Fe_2O_3 composites were prepared by wet impregnation, drying, ethanol washing and calcination processes. Initially, 30 mL of 0.6 M (0.3 M, 1.5 M) $Fe(NO_3)_3{\cdot}9H_2O$ in ethanol (EtOH) were added to 1.5 g of a *meso*-TiO_2 powder, stirred for 30 min at room temperature and then sonicated (40 kHz, 250 W cm^{-2}) for 30 min. The ethanol was evaporated at 50 °C followed to which the sample was calcined at 300 °C for 10 min. Once again the sample was washed by ethanol thoroughly and eventually calcined at 300 °C for 6 h. With this process, the achieved loading of α-Fe_2O_3 in the composites was about 50, 30 and 70 wt. % for 0.6, 0.3 and 1.5 M $Fe(NO_3)_3$ $\cdot 9H_2O$ (according to the pore volume of 0.22 cm^3 g^{-1} for mesoporous TiO_2). The *meso*-TiO_2/R-Fe_2O_3nanocomposites oxidized a higher toxic As (III) to a lower toxic As (V) with high efficiency at various pH values in the photocatalysis reaction. Simultaneously, As (V) was effectively removed by adsorption onto the surface of composites (Zhou et al. 2008).

As arsenate adsorption onto Fe_2O_3 is highly restricted under an acidic condition due to faster dissolution, iron oxide nanoparticle-grafted titanate nanotubes (Fe-TNTs) were synthesized by a facile one-step water–ethanol hydrothermal method for effective removal of As(V) from aqueous solutions. These nanomaterials exhibited both hollow and open-ended tubes with a rougher nanotube surface. The inner and outer diameters of Fe-TNTs were found to be approximately 8 nm and 11 nm, respectively. The interlayer distance of TNTs was about 0.78 nm and the lattice distance of the attached nanoparticles was 0.25 nm, which were ascribed to the (110) plane of iron oxide. Fe_2O_3 nanoparticles (5.1 nm) were bonded to the surface of TNTs (Fig. 4.5). This novel adsorbent was not only acid-resistant, but also showed a large As(V) adsorption capacity of 90.96 mg/g. Moreover, tubular TNTs not only inhibited dissolution of Fe_2O_3 at low pH, but also maintained a good sedimentation property. The surface associated hydroxyl groups on Fe-TNTs were reported to be most significant for As(V) adsorption (Wang et al. 2015).

The combination of nanostructures in order to increase the efficiency of the removal of hazardous anions has become essential to supply safe drinking water. Hence, attempts have been made to study mesoporous Fe_2O_3/TiO_2 dispersed over the reduced graphene oxide, to expose more number of active sites present in the nanoparticles and to suppress the aggregation of the nanoparticles. Fe_2O_3/TiO_2 was synthesized by a sol–gel route using Pluronic P123 as the structure directing agent. The size of the Fe_2O_3/TiO_2 and their composites were found to be 20–50 nm. $Fe_2O_3/TiO_2$was intercalated in between the reduced graphene oxide sheets. Both As(III) and As(V) attained equilibrium adsorption within 15 min with maximum adsorption capacity for As(V) and As(III) found to be 99.5 and 77.7 mg/g at pH 6 and 7, respectively. The sorption was high in the pH range of 6–7. The low adsorption at lower and higher pH values other than 6 and 7 was due to a zero point charge of nanocomposite which is equal to 6 (Babu et al. 2016).

4.13 Ceria Nanoparticles

Ceria nanoparticles (CeNPs) are widely used in industries as UV-blocking agents, automotive catalytic converters, as well as single, nanowire-based gas sensors. They have shown As removal efficiency. In particular, CeNPs synthesized using a

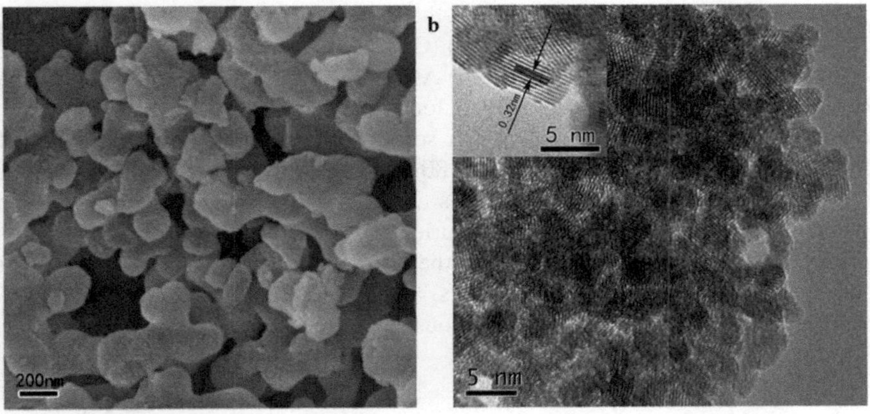

Fig. 4.6 Electron micrographs. (**a**) FESEM image and (**b**) HRTEM image of HCO nanoparticles. Reprinted with permission from Li et al. (2012b). Copyright © 2012 Published by Elsevier B.V

precipitation method showcased average particle diameter of 6.6 ± 0.9 nm and high specific surface areas about 86.85 m^2 g^{-1}. The hydrodynamic diameter of CeNPs was about 44.8 ± 1.6 nm. Adsorption of As onto CeNPs is an endothermic process favoured by high temperature. Adsorption increased with an increasing pH of the solution from pH 1 to 6. However, adsorption decreased with a further increasing pH from 6 to 13 for arsenate. In case of arsenite, adsorption increased with an increasing solution pH from 1 to 8 (the maximum adsorption was 70%) and then decreased with increasing pH from 8 to 13. The overall adsorption behaviour of As on CeNPs was reported to be strongly dependent on pH and independent of ionic strength, indicating that the electrostatic effect on the adsorption of these elements was relatively not important compared to surface chemical reactions (Feng et al. 2012).

Various other forms of ceria have also been used among in which hydrous cerium oxide and other nanohybrids are significant. Hydrous cerium oxide (HCO) nanoparticles of size 4 nm and a high specific surface area of 198 m^2/g were synthesized by a simple precipitation process (Fig. 4.6). The presence of high affinity surface hydroxyl groups on HCO nanoparticles was attributed to the extensive adsorption properties against both As(III) and As(V). The As adsorption capacity of HCO reached over 170 mg/g and 107 mg/g for As(III) and As(V), respectively, at neutral pH. Removal of As(III) was found to be very effective over a wide range of pH from 3 to 11 which might be due to strong inner sphere complexion (Li et al. 2012a, b).

In another study, ceria incorporated manganese oxide (NCMO) was prepared by sol–gel and co-precipitation calcination methods using ammonium ceric nitrate (ACN) and manganous chloride. Nanohybrid samples were agglomerates with irregular surface morphology (Ce:Mn = 1:1). The sample assigned as NCMO-1b was prepared using a metal hydroxide and calcinated at 573 K for 3.0 h. It was formed as a nano-crystal (70–90 nm). The optimum pH for the As(V) sorption is

3.0–7.0 at 303 ± 1.0 K. However, the presence of phosphate could reduce As(V) removal efficiency of NCMO-1b (Gupta et al. 2011).

A much higher sorption capacity for As(V) than both pure titanium dioxide and cerium oxide adsorbents was reported for novel Ce doped TiO_2 nano-adsorbents obtained by hydrolysis-precipitation. A solution of titanium sulphate was initially hydrolysed in the presence of PVA at 80 °C for 2 h following which 0.05 M Ce $(NO_3)_3$ was added. NaOH solution was added to adjust the pH to around 8. The precipitates were filtered, washed and dried at 80 °C for 2 h. The resulting amorphous Ce–Ti nanohybrid adsorbents in the size range of 100–200 nm aggregated to form the porous hybrid adsorbents. High sorption capacity for As(V), 9.4mgg^{-1}could be due to the amorphous compositions and small nanoparticles. (Deng et al. 2010).

4.14 Zirconium Based Nanoparticles

Mesoporous hydrous zirconium oxide (301-HZO and 302-HZO) was used for As removal from drinking water. These particles were synthesized from zirconium salt granular precursor, with a narrow particle size distribution (25–35 μm). In order to synthesize 301-HZO, an initially slurry of zirconium salts was prepared and gradually converted to its hydroxide by slow addition of an aqueous base solution until the equilibrium pH exceeded 10.302-HZO was similarly prepared with an additional stage of slurry treatment under alkali conditions at 90–100 °C for 1–3 h. The resulting zirconium hydroxides particles (20–50 μm) were washed and dried. The particles dried at 100 °C adsorbed ~150–160 mg As/g, 15–25 times more than 301-HZO. After rising in a calcination temperature of up to 350–400 °C, a gradual decrease in uptake of As was reported (up to ~1.4–2.4 times) (Bortun et al. 2010).

A novel oxide adsorbent of amorphous zirconium oxide (am-ZrO_2) nanoparticles was reported as a superior adsorption of As species. This material was synthesized employing a simple hydrothermal process using $ZrOCl_2.8H_2O$ (0.035 mol) as a precursor. am-ZrO_2 nanoparticles exhibited superior adsorption of both As(III) (arsenite) and As(V) (arsenate) due to their high specific surface area (327.1 m^2/g), large mesopore volume (0.68 cm^3/g), and the presence of high affinity surface hydroxyl groups. Faster adsorption kinetics were reported at pH ~7, with a high adsorption capacity of over 83 mg/g for As(III) and over 32.4 mg/g for As(V). Immobilizing am-ZrO_2 nanoparticles on a glass fibre cloth enhanced As removal performance indicating potential application in treating As contaminated water (Cui 2012).

The synthesis of hydrated zirconium oxide ($ZrO(OH)_2$) coated carbon nanotubes (CNTs) was achieved by filtration-steam hydrolysis method that could efficiently remove As(III) and As(V) from drinking water. The nanoparticles were prepared by mixing 7 mL of zirconium n-propoxide and 200 mg of CNTs in 100 mL of absolute ethanol followed by sonication for 30 min. The resulting suspension was filtered and subjected to water vapour at 75 °C for 90 min in order to induce hydrated zirconium oxide hydrolysis on the surface of CNTs. Eventually, the nanostructures were

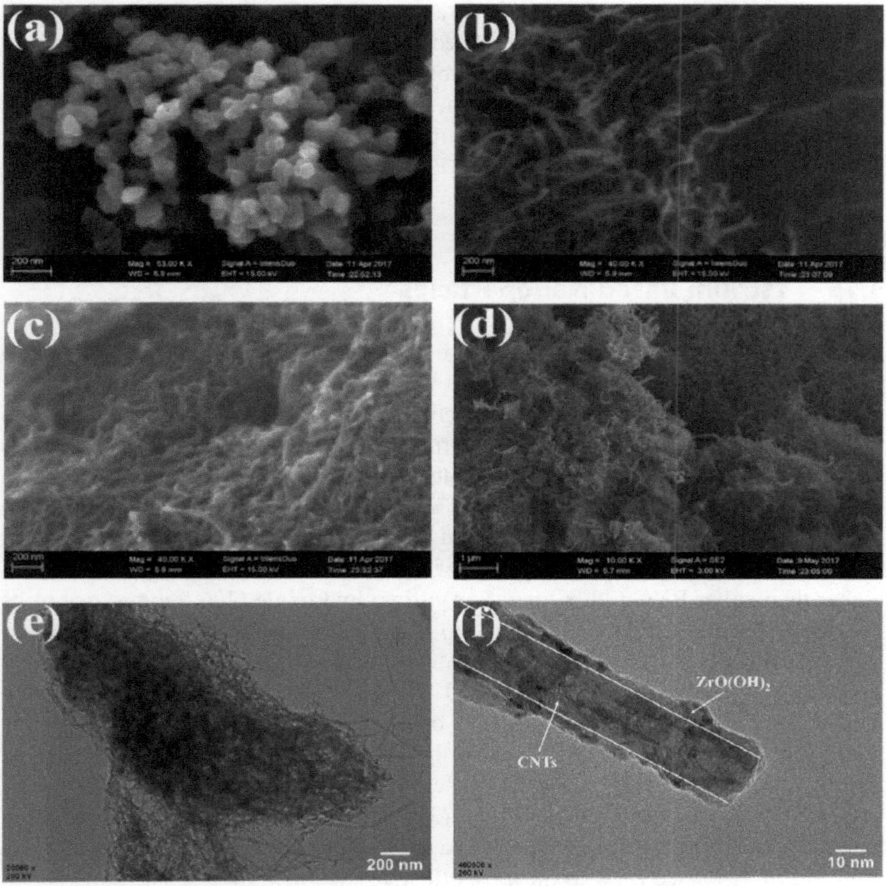

Fig. 4.7 SEM images of $ZrO(OH)_2$ (**a**), CNTs (**b**), and $ZrO(OH)_2$/CNTs (**c** and **d**), as well as TEM images of $ZrO(OH)_2$/CNTs (**e** and **f**). Reprinted with permission from Liu et al. (2018). Copyright © 2017 Elsevier Inc.

washed with an ethanol based solution and dried at 105 °C overnight. The amorphous ZrO $(OH)_2$ coated CNTs showed a coating thickness of 1–8 nm (Fig. 4.7). The $ZrO(OH)_2$/CNTs nanocomposite showed high adsorption for both As(III) and As (V) with maximum adsorption capacities of 78.2 and 124.6 mg/g, respectively. Moreover, the spent adsorbent was regenerated by re-coating $ZrO(OH)_2$ as the preparation method. However, a minor alteration in adsorbed amounts of As(III) and As(V) on the $ZrO(OH)_2$/CNTs occurred within six cycles (Liu et al. 2018).

Zirconium based ($Zr_2(OH)_6SO_4.3H_2O$) nanoparticles with particle sizes ranging from 60 to 90 nm showed excellent As removing potential. The optimal pH for As (V) adsorption was between 2.5 and 3.5. The Langmuir equation described well the adsorption isotherm, at the optimal pH, indicating the maximum adsorption capacity as 256.4 mg As/g (Ma et al. 2011). The presence of humic acid, phosphate or silicate in aqueous solution significantly reduced the uptake of As(V). A similar study reported that a multiwall carbon nanotube–zirconia nanohybrid (MWCNT–ZrO_2) with 4.85% zirconia showed an absorption capacity of 2000 μg g^{-1} and 5000 μg g^{-1} for As(III) and As(V), respectively. The major advantage of these MWCNT–ZrO_2 was that the adsorption capacity was not influenced by a variation in pH (Ntim and Mitra 2012).

4.15 Yttrium Based Nanoparticles

Recently, alongside rare-earth metals like zirconium and cerium, yttrium has also been studied to develop novel adsorbents considering its high adsorption capacity. Yttrium-based nanostructures can be successfully applied to naturally contaminated water which can possess a superior performance for removing As. Lee et al. (2015) reported various yttrium-based adsorbents, with basic yttrium carbonate (BYC), Ti-loaded basic yttrium carbonate (Ti-loaded BYC) and yttrium hydroxide, were prepared using a co-precipitation method. Using a Langmuir isotherm, confirmed that the maximum adsorption capacity of Ti-loaded BYC (348.5 mg/g) was 25% higher than either BYC (289.6 mg/g) or yttrium hydroxide (206.5 mg/g) due to its increased specific surface area (82 m^2/g) and surface charge (PZC: 8.4). Further, pseudo first- and second-order kinetic models indicated a faster rate of As removal in Ti-loaded BYC compared to BYC and yttrium hydroxide. The most probable mechanism behind As removal might be the carbonate-arsenate ion exchange process for BYC and Ti-loaded BYC. On the other hand, yttrium hydroxide removed As by a co-precipitation process. Higher adsorption affinity was displayed by Ti-loaded BYC over a wide pH range (3–11) and in the presence of coexisting anionic species (such as phosphate, silicate and bicarbonate) which rationalize the promises of Ti-loaded BYC as the most effective adsorbent for arsenate remediation in drinking water (Lee et al. 2015).

4.16 Perlite Nanocomposites

Perlite, a naturally occurring glassy volcanic siliceous rock, is found in the earth's crust that can expand up to 20 times its original volume, forming a lightweight, glassy material with a cellular structure when heated to 760–1100 °C. This expanded perlite possesses a chemically inert non-toxic chemistry with high porosity (and consequently, low density). Perlite can be used in filters and in the processing of vegetable fat, beer, wine and juice in the food industry; cleaning of dams and ponds in aquatic environments; and for obtaining of a clear liquid in filtration (Thanh et al. 2011). It can efficiently adsorb thorium, cadmium, methyl violet and can remove oil from aqueous solutions, making it a promising material for environmental remediation. However, its low specific surface area and acidic surface are major disadvantages for As. Hence, incorporation of nanoparticles, like γ-Fe_2O_3(PI) and

Fig. 4.8 TEM images of the (**a**) ball-milled expanded perlite (P); (**b**) PM composite and (**c**) PI composite. Reprinted with permission from Thanh et al. (2011). Copyright© 2011 Elsevier B.V

α-MnO_2 (PM), leads to a novel hybrid nanocomposite with superior As removal quality (Fig. 4.8). As(V) adsorption capacity was reported to be 4.64 and 7.09 mg g^{-1} for γ -Fe_2O_3 and α-MnO_2 containing adsorbents, respectively, which is significantly higher as compared to 0.0025 mg g^{-1} for perlite alone.

4.17 Biochar Nanocomposite

Biochar (charcoal or biomass-derived black carbon) is mainly prepared by pyrolysing biomass (e.g., wood, crop residues, manure, etc.) in the absence or limited supply of air. Biochar can significantly improve not only soil fertility but also enhance soil nutrients and water holding capacity. It is also responsible for increasing crop production and reducing emissions of other greenhouse gases. Recent developments suggest that nanomaterials can be grown, added or deposited on the biochar surface of nanomaterials (biochar nanocomposites) for the removal of heavy metals. Biochar/AlOOH nanocomposites have been designed by initially immersing the biomass into an $AlCl_3$ solution for 2 h. The mixture was dried at 80 °C for 24 h under air and then pyrolysed at 600 °C under N_2flow for 1 h resulting in the formation of a biochar/AlOOH nanocomposite which was then gently crushed and sieved into two size fractions: <0.5 and 0.5–1 mm. The Langmuir maximum capacity of a biochar/AlOOH nanocomposite to As(V) was evaluated to be around 17,410 mg kg^{-1} (Zhang and Gao 2013).

4.18 Polymeric Nanocomposites

Polymer based nanocomposites have emerged recently as synthetic adsorbents. The flexibility in tuning polymeric surfaces with desired functional groups has made them an attractive material for the removal of toxic metals. The coupling of functional groups (carboxy (COOH) or thiol (SH) groups) to a polymer material is one strategy to remove As. Synthetic adsorbents like aluminium nanoparticles (Alu-NPs, <50 nm) incorporated into amine rich cryogels (Alu-cryo), molecular imprinted polymers (<38 mm) in polyacrylamide cryogels (MIP-cryo) and thiol functionalized cryogels (SH-cryo) have been fabricated for the removal of As(V). Incorporation of nanoparticles into the polymeric matrix increased the mechanical stability. It was observed that polymer backbones of pure polyacrylamide (MIP-cryo) had better stability than the amine containing polymer backbone (Alu-cryo); the adsorption capacities for the composites were 20.3 ± 0.8 mg/g adsorbent (Alu-cryo) and 7.9 ± 0.7 mg/g adsorbent(MIP-cryo), respectively. These composites could remove As(V) well in the pH range of pH 2–8. The presence of co-ions (nitrate, sulphate and phosphate) hindered the As removal capacity more for Alu-cryo compared to MIP-cryo (Önnby et al. 2012).

Malana et al. (2011) synthesized aluminium substituted manganese copper ferrite (average size of 13 nm), doped in poly methyl acrylate (MA), vinyl acetate (VA) and acrylic acid (AA)(used as monomers) to make the nanocomposite. In this synthesis process, ethylene glycol dimethacrylate (EGDMA) and chloroform were used as a cross-linking agent and porogen, respectively. The monomers were used at a volumetric ratio of 2.5:1:2.5, respectively, while the porogen and the cross-linking agent were used at 50% (v/v) of the total monomer volume. Ethanol (50 mL) and benzoyl peroxide (1%, w/v of the total volume) were added to the reaction mixture as the solvent and initiator, respectively. In this, 0.05 g of aluminium doped manganese copper ferrite ($Mn_{0.5}Cu_{0.5}Fe_{1.2}Al_{0.8}O_4$) was added and mixed. Slow heating was used for the polymerization (radical) reaction where the temperature of the reaction mixture was gradually increased and maintained constant for at least 1 h for every 5 K rise in temperature. The reaction mixture was constantly stirred up to 328 K. At 333 K and 338 K, the temperature of the reaction mixture was maintained for 2 h. Precipitation of the resulting composite occurred at 343 K and this temperature was maintained for 6 h to allow completion of the polymerization reaction, following which the precipitates were filtered, washed and dried at 353 K for 2 h. The maximum adsorption capacity (qm) of As on the nanocomposite was found to be 0.053 mg g^{-1} (Malana et al. 2011).

Novel nano-alumina dispersed in chitosan-grafted polyacrylamide was reported for As removal by Saha and Sarkar (2012). Initially, the alumina nanoparticles were prepared by a reverse microemulsion technique where the microemulsion was prepared using cyclohexane, triton X-100 (surfactant) and n-butyl alcohol (co-surfactant). The aqueous solution of aluminium nitrate $Al(NO_3)_3$ was mixed as a microemulsion and ammonia was added slowly and pH adjusted to 8. After reacting for 24 h, the precipitate was washed in ethanol, ultrasonicated, centrifuged and dried at 80 °C for 5 h. The dried powder was calcined at 850 °C in a muffle furnace for 3 h. These nano-alumina particles were immobilized on chitosan-grafted polyacrylamide matrix by in situ dispersion. Briefly, chitosan (0.5 g) was dissolved in 30 mL of an acetic acid solution (1%,w/v) and purged with nitrogen. The addition of 0.1 g of ammonium persulphate (APS) led to the generation of radicals by chitosan. Further, a mixture of 1.50 g acrylamide, 0.2 g methylene-bis-acrylamide (MBA) was also added after 10 mins and later the sonicated alumina nanoparticles (0.5 g) were mixed and allowed to react at 70 °C for 12 h to complete the polymerization. After washing with RO water, sodium alginate was added, mixed and dispensed into a saturated solution of $CaCl_2$ by a peristaltic pump at a constant rate to form uniform spherical hydrogel beads of about 2.5 mm in diameter. These beads were maintained for 1 h in a $CaCl_2$ solution to allow crosslinking with Ca^{2+} and the polymeric nano-adsorbents could efficiently remove As with a high loading capacity of 6.56 mg/g. As removal was pH dependent with maximum removal obtained at pH 7.2 while the equilibrium time was 6 h. It is interesting to note that the adsorbent retained 94% adsorption capacity even in the fifth cycle indicating its efficient ability of regeneration (Saha and Sarkar 2012).

4.19 Conclusions and Future Perspectives

The elevation of As levels in surface and groundwater due to anthropogenic activity has become a global concern. As toxicity due to prolonged exposure may lead to mortality and morbidity. Hereby, there is a growing need to develop innovative and efficient technologies for As removal to provide clean and safe drinking water. Recent advances in nanotechnology and associated multidisciplinary approaches have enabled researchers to extend its application in advanced water treatment and thereby improving water quality. Due to their high adsorption capacities, large surface areas and their nanostructures metals, polymers and hybrids are being used for membrane separation or adsorption. Chemical modification of the nanomaterials by surface functionalization may help enhance As removal efficiency. Low cost highly efficient nanostructures can be fabricated from cheaper precursors in order to make the water treatment process affordable. This will enable one to translate the laboratory scale model to full-scale field applications. Nanotechnology based As removal has a huge scope of further research where biocompatible polymeric matrices like carboxymethyl cellulose (CMC) or starch can be used for the impregnation of nanoparticles for the enhancement of As removal.

Moreover, multimetallic nanohybrids like iron-manganese nanoparticles can be further doped with other metals like silver or copper for greater efficiency for As removal. Carbon nanotubes and their composites have also attracted great attention over the last few years owing to their efficient As removal capacity, high sensitivity and selectivity. Its combination with polymeric materials, zeolites and metals may further enhance its activity. Hence, need for more detailed systematic studies on immobilization mechanisms and technical improvements in preparing and utilizing exotic nanostructure based water treatment technology is necessary. Finally, future As removal studies employing nanoscale biosorbents should also consider economic affordability and scalability apart from strategic reduction in the disposal of As treatment waste. This will ensure a clean environment using a green technology.

Acknowledgements Dr.Sougata Ghosh acknowledges the Department of Science and Technology (DST), Ministry of Science and Technology, Government of India and Jawaharlal Nehru Centre for Advanced Scientific Research, India for funding under the Post-doctoral Overseas Fellowship in Nano Science and Technology (Ref. JNC/AO/A.0610.1(4) 2019-2260 dated August19,2019).

References

AlOmar MK, Alsaadi MA, Hayyan M, Akib S, Hashim MA (2016) Functionalization of CNTs surface with phosphonuim based deep eutectic solvents for arsenic removal from water. Appl Surf Sci 389:216–226

Antman KH (2001) Introduction: the history of arsenic trioxide in cancer therapy. Oncologist 6 (Suppl 2):1–2

Ashraf S, Siddiqa A, Shahida S, Qaisar S (2019) Titanium-based nanocomposite materials for arsenic removal from water: a review. Heliyon 5:e01577

Babu CM, Vinodh R, Sundaravel B, Abidov A, Peng MM, Cha WS, Jang HT (2016) Characterization of reduced graphene oxide supported mesoporous Fe_2O_3/TiO_2 nanoparticles and adsorption of as(III) and as(V) from potable water. J Taiwan Inst Chem Eng 62:199–208

Bagla P, Kaiser J (1996) India's spreading health crisis draws global as experts. Science 274:174–175

Bang S, Lippincott L, Meng X (2005) Removal of arsenic from groundwater by granular titanium dioxide adsorbent. Chemosphere 60:389–397

Bortun A, Pardini J, Khainakov SA, Garcia JR (2010) Synthesis and characterization of a mesoporous hydrous zirconium oxide used for arsenic removal from drinking water. Mater Res Bull 45(2):142–148

Campbell JP, Alvarez JA (1989) Acute arsenic intoxication. Am Fam Physician 40(6):93–97

Chowdhury S, Yanful E (2011) Arsenic removal from aqueous solutions by adsorption on magnetite nanoparticles. Water Environ J 25:429–437

Cobo JM, Castineira M (1997) Oxidative stress, mitochondrial respiration, and. glycemic control: clues from chronic supplementation with Cr^{3+} or As^{3+} to male Wistar rats. Nutrition 13:965–970

Cui H, Li Q, Gao S, Shang JK (2012) Strong adsorption of arsenic species by amorphous zirconium oxide nanoparticles. J Ind Eng Chem 18:1418–1427

Deng S, Li Z, Huang J, Yu G (2010) Preparation, characterization and application of Ce-Ti oxide adsorbent for enhanced removal of arsenate from water. J Hazard Mater 179:1014–1021

Feng Q, Zhang Z, Ma Y, He X, Zhao Y, Chai Z (2012) Adsorption and desorption characteristics of arsenic onto ceria nanoparticles. Nanoscale Res Lett 7(1):84

Fu D, He Z, Su S, Xu B, Liu Y, Zhao Y (2017) Fabrication of α-FeOOH decorated graphene oxide-carbon nanotubes aerogel and its application in adsorption of arsenic species. J Colloid Interface Sci 505:105–114

Ghariani M, Adrien ML, Raucoules M, Bayle J, Jacomet Y, Grimaud D (1991) Subacute arsenic poisoning. Ann Fr Anesth Reanim 10:304–307

Gupta K, Bhattacharya S, Chattopadhyay D, Mukhopadhyay A, Biswas H, Dutta J, Ray NR, Ghosh UC (2011) Ceria associated manganese oxide nanoparticles: synthesis, characterization and arsenic(V) sorption behaviour. Chem Eng J 172:219–229

Haq I, Singh A, Kalamdhad AS (2020) Bioremediation for environmental sustainability toxicity, mechanisms of contaminants degradation, detoxification, and challenges. In: Arsenic: environmental contamination, health hazards, and bioremediation approaches for detoxification. Elsevier, pp 121–142

Karim M (2000) Arsenic in groundwater and health problems in Bangladesh. Water Resour 34:304–310

Katsoyiannis IA, Zouboulis AI (2004) Application of biological processes for the removal of arsenic from groundwaters. Water Res 38:17–26

Katz SA (2019) On the use of hair analysis for assessing As intoxication. Int J Environ Res Public Health 16:977

Lee SH, Kim KW, Lee BT, Bang S, Kim H, Kang H, Jang A (2015) Enhanced arsenate removal performance in aqueous solution by yttrium-based adsorbents. Int J Environ Res Public Health 12:13523–13541

Li R, Li Q, Gao S, Shang JK (2012b) Exceptional arsenic adsorption performance of hydrous cerium oxide nanoparticles: part A. Adsorption capacity and mechanism. Chem Eng J 185–186:127–135

Li Y, Jia SY, Guo JW, Zhuo J, Na P (2012a) TiO_2 pillared montmorillonite as a. photoactive adsorbent of arsenic under UV irradiation. Chem Eng J 191:66–74

Liu D, Deng S, Maimaiti A, Wang B, Huang J, Wang Y, Yu G (2018) As(III) and As(V) adsorption on nanocomposite of hydrated zirconium oxide coated carbon nanotubes. J Colloid Interface Sci 511:277–284

Liu H, Zuo K, Vecitis CD (2014) Titanium dioxide-coated carbon nanotube network. filter for rapid and effective arsenic sorption. Enviorn Sci Technol 48:13871–13879

Ma Y, Zheng YM, Chen JP (2011) A zirconium based nanoparticle for significantly enhanced adsorption of arsenate: synthesis, characterization and performance. J Colloid Interface Sci 354:785–792

Malana MA, Qureshi RB, Ashiq MN (2011) Adsorption studies of arsenic on nanoaluminium doped manganese copper ferrite polymer (MA, VA, AA) composite: kinetics and mechanism. Chem Eng J 172:721–727

Manna B, Dasgupta M, Gosh UC (2004) Crystalline hydrous titanium (IV) oxide (CHTO):an arsenic (III) scavenger from natural water. J Water Supply Res Technol 53(7):483–495

Mayo J, Yavuz C, Yean S, Cong L, Shipley H, Yu W, Colvin V (2007) The effect of nanocrystalline magnetite size on arsenic removal. Sci Technol Adv Mater 8:71–75

Miller SM, Zimmerman JB (2010) Novel, bio-based, photoactive arsenic sorbent: TiO2-impregnated chitosan bead. Water Res 44:5722–5729

Nicomel NR, Leus K, Folens K, Voort PVD, Laing GD (2016) Technologies for arsenic removal from water: current status and future perspectives. Int J Environ Res Public Health 13:62

Niu HY, Shi YL, Cai YQ, Wei FS (2009) Adsorption behavior of arsenic onto protonated titanate nanotubes prepared via hydrothermal method. Micropor Mesopor Mat 122:28–35

Ntim SA, Mitra S (2012) Adsorption of arsenic on multiwall carbon nanotube–zirconia nanohybrid for potential drinking water purification. J Colloid Interface Sci 375:154–159

Önnby L, Pakade V, Mattiasson B, Kirsebom H (2012) Polymer composite adsorbents using particles of molecularly imprinted polymers or aluminium oxide nanoparticles for treatment of arsenic contaminated waters. Water Res 46:4111–4120

Peng X, Luan Z, Ding J, Di Z, Li Y, Tian B (2005) Ceria nanoparticles supported on carbon nanotubes for the removal of arsenate from water. Mater Lett 59:399–403

Ratnaike RN (2003) Acute and chronic arsenic toxicity. Postgrad Med J 79(933):391–396

Saha S, Sarkar P (2012) Arsenic remediation from drinking water by synthesized nano-alumina dispersed in chitosan-grafted polyacrylamide. J Hazard Mater 227–228:68–78

Saleh TA, Agarwal S, Gupta VK (2011) Synthesis of MWCNT/MnO_2 and their application for simultaneous oxidation of arsenite and sorption of arsenate. Appl Catal Environ 106:46–53

Schoolmeester WL, White DR (1980) Arsenic poisoning. South Med J 73:198–208

Singh A, Haq I, Kumar A (2014) Isolation, screening and characterization of arsenic resistant bacteria from arsenic resistant contaminated soils of Hoogly, West Bengal. Ann Plant Soil Res 3:268–270

Tamaki S, Frankenberger WT Jr (1992) Environmental biochemistry of arsenic. Rev Environ Contam Toxicol 124:79–110

Tamilarasan P, Ramaprabhu S (2012) Iron-manganese binary oxide coated functionalized multiwalled carbon nanotubes for arsenic removal. AIP Conf Proc 1447:321–322

Tang W, Li Q, Gao S, Shang J (2011) Arsenic (III,V) removal from aqueous solution by ultrafine α-Fe_2O_3 nanoparticles synthesized from solvent thermal method. J Hazard Mater 192:131–138

Thanh DN, Singh M, Ulbrich P, Strnadova N, Šteˇpanek F (2011) Perlite incorporating γ-Fe_2O_3 and α-MnO_2 nanomaterials: preparation and evaluation of a new adsorbent for As(V) removal. Sep Purif Technol 82:93–101

USEPA (2001) Arsenic in drinking water. http://www.epa.gov/safewater/arsenic/history.html-expertpanel. Accessed May 2005

USEPA (2002) National primary drinking water regulations. Current drinking water regulations. U.S. Environmental Protection Agency. Office of Water. www.epa.gov/safewater/mcl.htmlmcls. Accessed May 2005

Vadahanambi S, Lee SH, Kim WJ, Oh IK (2013) Arsenic removal from contaminated water using three-dimensional graphene-carbon nanotube-iron oxide nanostructures. Environ Sci Technol 47:10510–10517

Velickovic Z, Vukovic GD, Marinkovic AD, Moldovan MS, Grujic AAP, Uskokovic PS, Ristic MD (2012) Adsorption of arsenate on iron(III) oxide coated ethylenediamine functionalized multiwall carbon nanotubes. Chem Eng J 181:174–181

Wang Y, Liu W, Wang T, Ni J (2015) Arsenate adsorption onto Fe-TNTs prepared by a novel water–ethanol hydrothermal method: mechanism and synergistic effect. J Colloid Interface Sci 440:253–262

Xu Z, Gao S, Shang JK (2010) As(III) removal by hydrous titanium dioxide prepared from one-step hydrolysis of aqueous $TiCl_4$ solution. Water Res 44:5713–5721

Zhang M, Gao B (2013) Removal of arsenic, methylene blue, and phosphate by biochar/AlOOH nanocomposite. Chem Eng J 226:286–292

Zhou W, Pan K, Tian C, Qu Y, Lu P, Sun CC (2008) Mesoporous $TiO_2/\alpha\text{-}Fe_2O_3$:.bifunctional composites for effective elimination of arsenite contamination through. simultaneous photocatalytic oxidation and adsorption. J Phys Chem 112(49):19584–19589

Emerging Contaminants in Wastewater: Sources of Contamination, Toxicity, and Removal Approaches

5

Mudita Chaturvedi, Ambuj Mishra, Krati Sharma, Geetansh Sharma, Gaurav Saxena, and Amit Kumar Singh

5.1 Introduction

We are moving very fast on the roads of development, innovating in the desire of becoming more and more efficient with resources. Day by day mankind is developing complex methods and chemicals for simplifying life. However, several incidents have been recorded in the history of, chemical compounds that were developed to improve the standard of living left a hazardous impact on the environment and human health. Contamination of air, water, and soil is not a problem that developed in recent years, contaminants like arsenic and lead are age-old contaminants, with the advancement in research we can identify new problems associated with their increasing concentrations. Like aggravation of arsenic pollution problem in recent years in South East Asian region (Fendorf et al. 2010). The insatiable quest for economic growth led to the quantitative expansion of manufacturing. Every industrial setup generates wastewater, composed of chemicals developed and used as part of technological advancement in manufacturing. Variety of organic hydrocarbons, heavy metals, nutrient enrichment by human waste, and microbes that are pathogenic for human health are primary constituents of wastewater. These compounds are difficult or impossible to be degraded by natural processes or their insertion is so huge that it persists in significant quantity for a long time and adversely affects human health and the environment. Pollutants that are present in trace amounts in the environment are micropollutants, these can be synthetic in origin such as strongly halogenated molecules (fluorinated surfactants, etc.) or can be natural compounds

M. Chaturvedi · A. Mishra (✉) · K. Sharma · A. K. Singh
School of Environmental Sciences, Jawaharlal Nehru University, New Delhi, India

G. Sharma · G. Saxena
EMBL - Environmental Microbiology and Biotechnology Laboratory, EERG - Ecotoxicology and Environmental Remediation Group, School of Biotechnology, Shoolini University of Biotechnology and Management Sciences, Solan, Himachal Pradesh, India

I. Haq, A. S. Kalamdhad (eds.), *Emerging Treatment Technologies for Waste Management*, https://doi.org/10.1007/978-981-16-2015-7_5

like antibiotics (penicillin) or estrogens. Their presence in trace amounts made it difficult to identify their effects; however, with advancements in research there possible negative impacts have been highlighted which raised the concern of stakeholders towards these trace compounds.

Emerging contaminant reaches the environment via many anthropogenic sources and spread in environmental matrices. Most of the emerging contaminants (ECs) present in the aquatic system originate from direct and diffused pollution sources, posing a challenge to the environment and society (Haq and Raj 2018). ECs are synthetic or natural chemical compounds, molecules of which have the potential to disturb the ecological cycle and cause a threat to the sanctity of the environment and human health. These contaminants are mobile but persistent in air, water, and soil sediments.

Table of DDT usage in agriculture is a horrendous example of how devastating impact a contaminant can pose on ecology and the environment. DDT was developed as an insecticide and was used for medicinal purposes during World War II. Its usage as a pesticide was started in 1945, as food demand was increased after the world war and the chemical proved itself effective during a crisis. It was in 1962 when negative impacts of DDT were highlighted (Carson 2002). This whole situation raised concern about bringing synthetic drugs for general public use without prior investigation of their effects on the environment, biodiversity, and human health. With growing innovations in manufacturing, humans are bringing out innumerable contaminants in nature. Most of these emerging contaminants can be categorized as flame retardants, surfactants, pharmaceuticals, personal care products, gasoline additives, and their degradation products, biocides, polar pesticides, and their degradation products, and various proven or suspected endocrine-disrupting chemicals (EDCs). In recent years problem of nanomaterials, plastics, and microplastics has become the center of attention for researchers. Along with these man-made complex compounds, well-known traditional trouble of heavy metal ions like mercury, cadmium, thallium, arsenic, lead, etc. has aggravated with time. Advances in epidemiological and toxicological studies have discovered emerging problems regarding these old pollutants.

Studies suggest that a significant amount of contaminants remains in treated water via sewage treatment plants, this treated water when discharged in natural resources of water like rivers and lakes (Stasinakis et al. 2013). These unfiltered contaminants cause severe damage to the natural aquatic ecosystem and biodiversity. When contaminated water or partially treated water is used for agricultural and for other secondary uses. Studies suggest their acute hazards effect on soil and agricultural produce (Boots et al. 2019).

5.2 Emerging Contaminants

Chemicals of synthetic or natural origin, which cause known or suspected adverse ecological or human health effects, are contaminants. Contaminants from a chemical class that so far has not been studied extensively, where there is either a concern from stakeholders (scientists, regulators, NGOs, etc.), that the contaminant class may be having an impact on the environment or human health are emerging contaminants; also, where there is a concern that existing environmental assessment paradigms are not appropriate for the contaminant class are contaminants of emerging concern (CEC). The ecotoxicological perspectives typically used to characterize environmental contaminants have fallen short when attempting to classify the complex network of emerging contaminants. The typical approach taken by toxicologists of characterizing a compound by its persistence, lipophilicity (preference for fat-tissue), and toxicity (both acute and chronic) has proven to be inadequate with regard to emerging contaminants. An emerging contaminant can exhibit what is known as "pseudo-persistence" whereby a contaminant is continually present in the environment because it is continually being released by sewage treatment plants. Lipophilicity, typically used to determine how easily a contaminant will cross cell membranes or enter tissues, is an incomplete characterization tool because many emerging contaminants are pharmaceutically engineered to be actively transported into cells and tissues. There are even more characteristics of emerging contaminants that limit our reliable characterization of their risk to humans and the environment. Our power to detect and measure these compounds currently surpasses our knowledge of their toxic effects on humans or wildlife. Besides, to establish a cause–effect relationship, toxicologists cannot simply cite the co-occurrence of contaminant and adverse effects. Rather, the uptake, mode of action, and biological endpoints of each emerging contaminant must be researched and documented to establish a correlation between contaminant and consequence.

These contaminants have been incorporated by anthropogenic activities in almost all spheres, i.e. biosphere, hydrosphere, lithosphere, atmosphere, and cryosphere, all spheres are affected and contaminated by these compounds. Emerging contaminants are not only ones whose existence is recently noticed by the scientific community, but it is also about new concerns raised about old contaminants due to advancement in research (Sauve and Desrosiers 2014). It is important to note that the majority of emerging contaminants are no pollutants that are new or have just gained entry into the environment. Rather, most emerging contaminants are well-established pollutants with a newly demonstrated toxic effect or mode of action. Thus, the word "emerging" refers not only to the contaminant itself but also to an emerging concern about the contaminant. As such, emerging contaminants are often referred to as "chemicals of emerging concern" or "contaminants of emerging concern." Since there is little toxicological information for the vast majority of the chemicals we use, particularly with regard to long-term, low-level exposure, it is possible that many chemicals with a long history of use (such as MTBE) could contribute to the future burden of emerging contaminants.

Bioaccumulation, toxicity, and resistance to degradation are key properties that can make any contaminant disastrous. With increasing knowledge of the negative impacts of naturally available chemical compounds, we are adding contaminants to the list of emerging contaminants. Environmental concerns can be sorted into three classes of contaminants. (1) a new compound, a molecule which is recently appeared to scientific research or not studied yet fully, (2) the contaminant is common or known to exist but the research of contaminant still going on, (3) old contaminant such as mercury, arsenic, DDT where new research is pushing our understanding of environmental and human health risk of such contaminant.

Proper knowledge of the type, source, and effects of any contaminant is a must to monitor and mitigate their harmful effects. In recent times, wastewater is loaded with emerging micro-contaminants, these contaminants are present in trace quantities in natural waters, it is because of advancement in research technology that quantitative analysis of these compounds is possible.

5.2.1 Pesticides

Any substance or mixture of substances that are intended for preventing, destroying, repelling, or mitigating any pest is pesticides. Pesticides are classified into different classes according to their applications such as herbicides, insecticides, fungicides, rodenticides, nematicides, and microbicides. Excessive use of pesticides in agricultural fields will result into leach out of chemicals in soil/sediments or through runoff it will infuse with streams can be a potential concern to aquatic life and ecosystem, as well as human health if they affect the drinking water source and recreational uses areas. There has been a steady growth in the production of technical-grade pesticides in India, from 5000 metric tons in 1958 to 102,240 metric tons in 1998. In 1996–1997 the demand for pesticides in terms of value was estimated to be around Rs. 22 billion (USD 0.5 billion), which is about 2% of the total world market (Aktar et al. 2009). Globally, around three million tons of pesticide is used equivalent to a market value of USD 40 million. Regardless of the benefits of pesticides in the crop field and their significance for the economy, overexploitation of pesticides is raising a serious threat to the environment and humans. The most widely used conventional pesticides are herbicide 2,4-dichloro phenoxy acid, herbicide glyphosate, the herbicide atrazine, fumigant metam sodium, herbicide acetochlor, inorganic chemical insecticides such as calcium arsenate, lead arsenate, and fluoro-organic compounds. Most conventional inorganic pesticides are permanently banned or for other pesticides countries have developed their standard concentration guidelines.

Around the world an enormous number of persistent organic contaminants are present in wastewater effluent, surface water, and drinking water. The pesticides alachlor, dieldrin, and DBCP are each classified under the USEPA designation "probable human carcinogen." Perhaps the greatest threat posed by pesticides, however, is their endocrine-disrupting potential. Pesticides such as atrazine, malathion, methoxychlor, and many others have been shown to disrupt endocrine systems, even at very low concentrations. Exposures can cause reproductive system damage including sterility, decreased fertility, and birth defects, as well as impaired development, immuno-suppression, and metabolic disorders. It represents the ineffectiveness of present and conventional technologies. In US EPA rainbow report >1000 pesticide is listed such as borax, nitrogen, warfarin, phorate, zinc phosphide. Pesticides are used in the excess amount in agricultural fields, golf courses, and residential areas to prevent insects, mosquitoes, and other pests (Ikehata et al. 2008). Surface water in the urban and agricultural areas is more prone to pesticide contamination which can pose threat to the aquatic ecosystem even present in low concentrations (ng/L) (Moschet et al. 2014). Pesticide can develop hemopoietic tissue of anterior kidney and immune-depressive effects in fishes, mammals (Dunier et al. 1993).

Pesticides can harm aquatic life, along with the probable harm to human encounter with them can have. Manufacturing to pesticide application in the field the people involved are high-risk groups for pesticide exposure. Pesticide can pose serious effects to human health when exposed in low doses like immune suppression, hormone disruption, diminished intelligence, reproductive abnormalities, and cancer. In India the first report of poisoning due to pesticides was from Kerala in 1958, where over 100 people died after consuming wheat flour contaminated with parathion (Karunakaran 1958).

5.2.2 Persistent Organic Pollutants (POPs)

POPs are chemical substance persist in the environment and cause the risk to human health and the environment. POPs have high lipid solubility, semi-volatile in nature so they tend to travel a long distance through the atmosphere and its travel through the food web not directly and it starts bio-accumulating. The 12 initially, targeted POPs include eight pesticides (aldrin, chlordane, DDT, dieldrin, endrin, heptachlor, mirex, and toxaphene), two types of industrial chemicals (polychlorinated biphenyls or PCBs and hexachlorobenzene), and two chemical families of unintended by-products of the manufacture, use, and/or combustion of chlorine and chlorine-containing materials (dioxins and furans). All 12 targeted POPs are also endocrine disruptors chemicals that can interfere with the body's hormones. Endocrine-disrupting chemicals can be hazardous at extremely low doses and pose a particular danger to those exposed in the womb. During prenatal life, endocrine disruptors can alter development and undermine the ability to learn, fight disease, and reproduce (Table 5.1).

In past, the widespread usage of DDT in the agriculture field to eliminate mosquitoes and other pests was a trend that led to the death and disappearance of many birds (bald eagles), later the government banned the usage of DDT. POPs are produced consciously by industries such as pesticides, PCBs, PCNs, etc. and some are produced as a by-product through industrial or combustion processes such as PCDDs, PCDFs, PAHs. After 1995, analyzing the impact of POPs on the environment many international societies are working legally to the binding instrument and eliminate POPs (Eljarrat et al. 2003). Present research studies suggested that emerging contaminates such as polybrominated diphenyl ether (PBDEs) are also tend to transport in remote areas of the aquatic system including deep-sea fish. PCB and PBDEs are structurally identical and widely used in flame retardants, plastic, textile, and electronics (Koenig et al. 2013).

United Nations Environment Program (UNEP) adopted the Stockholm convention in May 2001. Initially, 12 POPs were identified which were named as "dirty dozens." The treaty laid to eliminate or restrict the use of POPs enforced in May

Table 5.1 Persistent organic pollutants (POPs) and their sources of contamination, toxic effects with their current legal and environmental status

	POP	Source	Effects	Recent status of usage
1.	Aldrin and dieldrin	Pesticide (through dairy products, animal meat, insect repellents)	Major motor convulsions, malaise, incoordination, headache, dizziness, and gastrointestinal disturbances, and acute intoxication including fatalities	The use of both chemicals is severely banned in many parts of the world since the 1970s (since 2003 in India) but in use in many countries and the countries use is banned in still present in the environment due to excessive early use
2.	Chlordane	Pesticide (through the crop, dairy products, animal meat, and direct exposure through air pollution)	Affects the immune system and poses carcinogenic effects	Use is restricted in many countries including India
3.	DDT	Pesticide	Increased risk of cancer and diabetes, reduced reproductive success, and neurological disease	Agriculture use banned, still in use against malarial
4.	Endrin	Through food (insecticide and rodent repellent sprayed on the surface of leaves)	Highly toxic for aquatic animals and humans as a neurotoxin	Banned in many parts of the world including India
5.	Hexachlorobenzene	Food grains (food grains)	Skin disease, debilitation, causes porphyria turcica, a lethal metabolic disorder	Globally banned under Stockholm Convention on POP, including India
6.	Heptachlor	Food (used as soil pesticide)	Low doses result in adverse behavioral changes and declined reproductive success	Globally banned under Stockholm Convention on POP, including India, but still reported being present in rainwater
7.	Polychlorinated biphenyls	Electrical circuits, capacitors, as adhesive in paints, plastics, and	Immune dysfunction, reproductive failure; acute	Globally banned under Stockholm Convention on

(continued)

Table 5.1 (continued)

	POP	Source	Effects	Recent status of usage
		carbonless copy papers, etc.	effects include swelling of the eyelids, along with fatigue, nausea, and vomiting	POP, including India
8.	Toxaphene	Used as an insecticide on crops and livestock, this is present in the environment as residue and often as present in food	Exposure to humans is at a low level, reported as a possible carcinogen; highly toxic to aquatic animals	Globally banned under Stockholm Convention on POP, including India, but present in environment residue due to long half-life
9.	Mirex	As flame retardant in plastic, rubber, and electronic stuff. As insecticide for ants and termites. Environmental residues are a possible source	Toxic to many plant and aquatic species; possible carcinogen for humans	Globally banned under Stockholm Convention on POP, including India, but present in environment residue due to long half-life (one of the most persistent POP)
10.	Chlordecone	Agricultural crops (used as a pesticide)	Toxic to aquatic animals and carcinogenic to humans	Banned throughout the world in the Stockholm list later (extended)
11.	Lindane	Widely used as a pesticide in soil, crops, and as wood treatment	Highly toxic as immunotoxic, neurotoxic, carcinogenic effects to humans	Banned throughout the world in the Stockholm list later (extended) Still present in the environment as one of the most persistent POPs
12.	Endosulfans	Food crop is the most prominent source as used as a pesticide for the crop	It acts as ant antiandrogen, thus disrupts the endocrine system. Causes congenital physical disorders, mental retardation, and lethal in high dose	Banned worldwide under Stockholm Convention later in 2011

2004. There are additions to the list of POPs since then and today number has reached 26 persistent organic pollutants. A class of persistent organic compounds (POPs) are used in the textile industry for the dyeing and bleaching of cloths. Dyes get transported to domestic wastewater during the daily washing of clothes. Dyes are different from pigments as they are soluble in water or any organic solvent; however, pigments are insoluble in both liquid media. Nowadays synthetic dyes are used in large amounts in industries, the untreated effluent is directly transported to the aquatic system. Dyes and their by-products are toxic, carcinogenic, and mutagenic, and exposure to organic dyes can irritate the respiratory tract, skin, eyes, throat, asthma, and allergy to skin (Siyal et al. 2018). A fraction of new textiles releases a high concentration of polychlorinated dibenzo-p-dioxins (PCDDs) and polychlorinated dibenzofurans (PCDFs). These dioxins and furans transfer to distant matrices via distinct pathways from where it enters in the food chain (Križanec and Majcen Le Marechal 2006). Other persistent organic pollutant dyes are aromatic amines, oxazines, anthraquinone, pentachlorophenol, chloranil, phthalocyanine, and phenolic compounds all of these compounds are carcinogenic (Mustereţ et al. 2007).

Another chemical of emerging concern is from the class of perfluorooctanoic acid which is a listed persistent organic pollutant. Perfluorinated alkylated substance and poly-fluoroalkyl substances (PFAS) have been found ubiquitously around world-wide in water, air, sediments, organism, because of their use vastly in all household related processes (e.g. clothing, cooking, furniture, etc.) and its physicochemical properties PFAS residues are lipophobic and hydrophobic and this compound shows strong surface-active properties and usually attach to the natural surface, and due to C-F bond, most of these compounds are virtually indestructible and persist for longer duration in environment. Many studies identified the source of PFAS in the surface water as fluorochemical manufacturing industries, discharge of firefighting foam, landfill leachate, and degradation of precursor compound (Moller et al. 2010). PFAS and their by-products are considered as highly toxic, extraordinarily persistent chemicals and contaminate human blood and wildlife. And due to their unique physio-chemical property, they are still a considerable challenge to environmental scientists and also for regulatory authorities (Roland Kallenborn).

5.2.3 Endocrine-Disrupting Chemicals (EDCs)

EDCs represent a broad class of molecules such as organochlorinated pesticides and industrial chemicals, plastics and plasticizers, fuels, and many other chemicals that are present in the environment or are in widespread use (Haq and Raj 2018). We make several recommendations to increase understanding of the effects of EDCs, including enhancing increased basic and clinical research, invoking the precautionary principle, and advocating involvement of individual and scientific society stakeholders in communicating and implementing changes in public policy and awareness. EDC is a modified chemical or a mixture of chemicals that interferes with any aspects of hormonal activity. EDC disrupts the endocrine system either by forming an enantiomer of the natural hormone or by hindering its functional activity.

Environment Protection Agency (EPA) elaborates EDC as an external factor that might interfere with the development, release, transport, activity, or replacement of natural hormones which help in maintain bodily functions like homeostasis, reproduction, development, and behavior. EDCs include xenoestrogens, phthalates, PCBs, PFAS, etc. chemicals that can cause metabolic (obesity), biological, and reproductive disorders. The group of molecules identified as endocrine disruptors is highly heterogeneous and includes synthetic chemicals used as industrial solvents/lubricants and their by-products [polychlorinated biphenyls (PCBs), polybrominated biphenyls (PBBs), dioxins], plastics [bisphenol A (BPA)], plasticizers (phthalates), pesticides [methoxychlor, chlorpyrifos, dichlorodiphenyltrichloroethane (DDT)], fungicides (vinclozolin), and pharmaceutical agents [diethylstilbestrol (DES)]. Almost all POPs are endocrine disruptors. Nevertheless, in very broad terms, EDCs such as dioxins, PCBs, PBBs, and pesticides often contain halogen group substitutions by chlorine and bromine. They often have a phenolic moiety that is thought to mimic natural steroid hormones and enable EDCs to interact with steroid hormone receptors as analogs or antagonists. Even heavy metals and metalloids may have estrogenic activity, suggesting that these compounds are EDCs as well as more generalized toxicants. Several classes of EDCs act as antiandrogens and as thyroid hormone receptor agonists or antagonists, and more recently, androgenic EDCs have been identified. EDCs do not have any structural similarity; therefore, it is difficult to classify a set of chemicals as EDCs (Diamanti-Kandarakis et al. 2009). These compounds are present in very fewer amounts ranging from ng/L to μg/L in wastewater. However, long-term exposure to such low levels is hazardous for human health. The toxic effect of these molecules is not fully identified.

5.2.4 Pharmaceutical Personal Care Products (PPCPs)

Pharmaceutical personal care products are broadly divided into multiple categories based on their characteristics and usage. Pharmaceutical is divided into antibiotics, analgesics, steroid hormones, anti-inflammatory drugs, cytostatic drugs blood lipid, regulators, etc., However, personal care products (PCPs) are classified as fragrance, disinfectants, insect repellents, preservative, and sunscreen ultraviolet (UV) filters. At present more than 3000 PPCPs have been used in the medical field for humans and animals for the betterment of human standards (Muthanna et al. 2008). Humans are consuming PPCPs in large amount which releases in domestic wastewater as sewage, landfill leachate, biomedical waste, municipal waste, PPCPs may enter into the aquatic environment. Even an individual household is also contributing to PPCPs contamination of water through the addition of expired and unused medicine in sinks and drains as reported by Kummerer et al. (2009). Studies suggest the presence of low, subtoxic concentrations of PPCPs can pose threat to human beings and the environment. Many studies suggested that chemical of pharmaceutical origin is often not eradicated from the wastewater treatment plant and not biodegradable in the environment.

Biomedical waste often contains a high frequency and number of pharmaceutical effluents. Antibiotics are extensively used worldwide in the medical field. In surface and wastewater treatment plant both metabolized and unmetabolized antibiotics have been found, they are partially degraded in the environment and as a consequence, it starts accumulating in water bodies. The prolongation of antibiotics can stimulate the antibiotic-resistant bacteria in aquatic bodies and chances to develop drug-resistant microorganisms (Sui et al. 2015; Ahmed et al. 2015). PPCPs are present in groundwater, surface water as well in effluents coming from different sources, where the concentration of drugs varies from ppt to ppb. Research reveals that approximately 65% of pharmaceutical products are never ingested (Ruhoy et al. 2008). DEET has been used as an active constituent in insect repellents on a broader scale and it tends to persist in the aquatic system for a longer duration. Parabens are the ester of para-hydroxybenzoic acid-containing functional group either alkyl or benzyl group. Influents of sewage treatment plant (STP) paraben especially methylparaben (MEP) and propylparaben (PrP) are present with concentration up to 30–20μg/L, respectively (Yang et al. 2014). Triclosan (TCS) is mainly used in health care products such as cosmetics, deodorant, shampoo, mouth rinses, soap, and toothpaste at a concentration of 0.1–0.3% (w/w) (Thompson 2005). TCS is mainly of two types: dissociated and non-dissociated/methyl TCS, dissociated TCS degrades under sunlight which has less than one-hour lifespan, while non-dissociated/methyl TCS is comparatively more stable to photo-degradation (Subedi et al. 2015).

5.2.5 Naturally Occurring Emerging Contaminants

Plants, humans, and animals contribute toxins in wastewater systems, although domestic wastewater is not the major source of plant toxin pollution, analyzing agricultural runoff is momentous when we track down the presence of mycotoxin in surface waters (Gatidou et al. 2007). There are contradictory theories about the source of mycotoxin or other such naturally produced toxins. Some studies reported the occurrence of zearalenone in bottled water. The presence of these toxins in food items is of major concern to food regulating bodies, most common toxins are phytotoxin (plant origin), mycotoxin (fungal origin), phycotoxin (algal origin), zootoxins (animal origin), and bacterial toxin (bacterial origin). In recent times stakeholder has raised concern about presence and effects of mycotoxins on human health and livestock. Most commonly observed are aflatoxins, ochratoxin A, patulin, zearalenone (ZEA), fumonisins, and nivalenol/deoxynivalenol. Mycotoxins are produced by three infamous fungal genera: Aspergillus, Penicillium, and Fusarium. Mycotoxins can be produced before or after harvest on mold-infected crops, some other natural toxins are produced by organisms as the response to their defense mechanisms. Studies on the prevalence of mycotoxins in the aquatic environment are mainly focused on ZEA due to its strong oestrogenic activities. Phycotoxins impact on human exposure is sparsely explored in the scientific community. Several mycotoxins are neurotoxic, potentially lethal, and associated with chronic adverse health effects. Certain cyanobacteria and marine algal species are well known for causing discoloration of water red tides and green tides, this phenomenon is commonly called "harmful algal blooms." These toxic or sometimes lethal blooms effects native aquatic life badly. Both healthy and unhealthy humans have high viral loads in their feces and their urine, contributing to high viral loads in municipal sewage. These viruses are only partially eliminated by wastewater treatment systems and are now being considered an emerging contaminant. Some of the most prominent emergent viruses being found in water supplies are human polyomaviruses, hepatitis E virus, and human adenoviruses (Smital 2008).

5.2.6 Microplastic

Plastic pollution is pervasive, it affects the aquatic ecosystem, soil ecosystem, and widely affects natural flora and fauna. According to the recent estimation of plastic production done till 2015, the world has produced 6300 million metric tons of plastic waste; out of which 9% has been recycled, 12% was incinerated and 79% was accumulated in landfills or the natural environments (Geyer et al. 2017) Although we are aggravating the rate of production daily, if we continue with the same rate of production and waste management trends by 2050, we will have 12,000 million tons of plastic waste accumulated in our natural environment. The size continuum of plastic pollutants has been defined from macroplastic (>5 mm), through microplastic (5 mm to 1μm) to neoplastic (<1μm). (Gigault et al. 2018). Plastics are categorized into two main polymer families, i.e. thermoplastics and thermosets. Plastics are synthetic organic polymers derived from natural organic materials such as cellulose, coal, natural gas, salt, and crude oil which is the most common source of naphtha, feedstock for plastic manufacturing. Plastic can be of distinct origin, composition, and properties such as polyethylene (PE), polypropylene (PP), polystyrene (PS), nylon (PA), polytetrafluoroethylene (PTFE), thermoplastic polyester (PET), poly(vinyl chloride) (PVC), cellulose acetate (CA), polyester polystyrene, polycarbonate, polyethylene terephthalate, foamed polystyrene and biodegradable polylactic acid (PLA). Polystyrene is most commonly used in packaging and industrial insulation, polyethylene is used in personal care products (PPCPs) as in facials and scrubs, polypropylene is used in fishing gears, these are some of the most commonly used plastics. The most prominent type of microplastic present or identified in the aquatic environment are spheres, pellets, irregular fragments, and fibers. In residential areas, machine washing can produce more than 1900 fibers (100 particles L^{-1}) in one wash cycle. Microplastic is categorized into primary and secondary based on their process of formation: primary microplastic are generated intentionally in microscopic scale through the process of grinding or extrusion, either precursor to other products and integrated into a variety of activities such as facial exfoliating cleanser, air blasting boat cleaning media and used in medicines as a vector for drugs. Secondary microplastics are the by-products of degraded and fragmented macroplastic due to mechanical, photolytic, or chemical processes. Plastic having a density greater than 1.02 g/cm^3 starts sinking and accumulates over sea beds, while low-density plastic floats on the sea-surface or suspension in the water column, however, low-density plastic tends to sink through biofouling (Alomar et al. 2016). Therefore, via multiple passage microplastic enters into the aquatic system such as water, wastewater (industrial, landfill, municipal, biomedical waste) being discharged into the sea, rivers, and water bodies. Thomson et al. (2004) identify microplastic fragments around UK beaches materials used for clothing (polyester, acrylic), packaging (polyethylene, polypropylene), rope (polyamide). RASFF and ESFA reported the presence of microplastic in food products microplastic or synthetic microplastic have found in canned sardines, sprats, salt, beer, honey, sugar, and the sealed plastic drinking bottle also contain plastic (Rist et al. 2018). Microplastic is especially concerning the ecosystem because of its small size and lack of technology available to quantify the presence of microplastic in the environment and its potential to hurt marine biota and humans.

5.3 Source of Contamination of Emerging Contaminants

The water which becomes unfit for human use after being used for industrial processes and household activities is wastewater. These anthropogenic activities are responsible for the amalgamation of toxic contaminants in natural water bodies. Aquatic systems suffer pollution from many sources including water leakage from septic and sewage tanks, fertilizers and agrochemicals runoff, seepages from landfill sites. Studies suggest that even after being treated in conventional wastewater treatment plants (WWTPs), a significant quantity of emerging contaminants remains untreated in wastewaters (Gogoi et al. 2018). Trace amounts of pharmaceuticals, endocrine-disrupting chemicals (EDCs), personal care products (PCPs), and their by-products have been identified in effluents from WWTPs, these form a group of micro-contaminants that bioaccumulates in the food chain. This contaminant reaches groundwater via surface leakages and groundwater recharging points. Groundwater is highly contaminated due to the continuous mixing of waste coming from wastewater treatment plants (WWTP) in the aquatic system, and it contains multifarious emerging contaminants.

Domestic and industrial wastewater are distinctive in their composition and properties. Due to their distinct properties and composition domestic and industrial waste are generally collected and treated separately, which allows specific pollution mitigation strategies. ECs generated from industries and urban wastewater treatment plants are directly discharged into rivers, lakes, ponds and where their environmental fate is of concern such as leach out in groundwater, sorption in sediments, decreased fertility of the soil. The life and passage of emerging contaminant from the origin point to sinking point (aquatic system) are highly dependent on the ECs properties such as volatility, polarity, adsorption properties, persistence (Farré et al. 2011).

5.3.1 Domestic and Hospital Effluents

Domestic wastewater is rich in a pathogenic microorganism, biodegradable organic matter, and nutrients which are responsible for the aggravation of eutrophication processes in natural water bodies. It poses long-term damage to biodiversity by incorporating nondegradable organic compounds like oil, grease, phenols, cyanide, pesticides, and detergents along with metals that bioaccumulate in the food chain. Domestic wastewater treatment was considered to be more efficient compared to industrial wastewater. There is a significant change in domestic wastewater composition for the past 10–15 years. Increasing the use of personal care products and other strong chemicals in daily household needs is becoming a tough chore to handle (Tjandraatmadja et al. 2010). Domestic wastewaters are a major contributor of microbial contaminants, phosphorus, and nitrogen in natural water bodies. Excess amounts of phosphorus and nitrogen induce eutrophication in water resources, such conditions in the environment may also favor the growth of toxin-producing cyanobacteria. This makes untreated domestic wastewater hazardous for the health of human and aquatic life. Constituent contaminants in wastewater vary geographically based on distinct domestic usage patterns (Tran et al. 2018). Major contaminants from domestic wastewater and their effects are:

- Antibiotics: the concentration level of antibiotics in influent and effluent of WWTPs has been recorded in the range of few tens of microgram per liter. Sulfonamide, macrolides, fluoroquinolones, and trimethoprim are the most common antibiotics found worldwide in wastewater. One of the highly prescribed drugs is β-lactam. However, its insignificant presence in wastewater is due to chemical hydrolysis or chemical transformation in alkaline/acidic conditions. Antifungal and antimicrobial agents like miconazole, thiabendazole, triclocarban, and triclosan are extensively used in household products (shampoos, soaps, shower gels, toothpaste, and dermal creams). Generally, concentration levels of triclocarban and triclosan are higher in wastewater, whereas miconazole and thiabendazole which are used as therapeutics reach the aquatic system in comparatively lower concentrations.
- Therapeutic drugs: beta-adrenoreceptor blocking agents, antiepileptic and antipsychotic drugs, lipid regulators, anti-inflammatory drugs, lipid regulators, steroids, and synthetic hormones like estrogen are all used as medication, some of these are excreted from the human system in non-metabolized form and reach aquatic systems. in long term, these drugs cause a serious threat to aquatic life.
- UV filters are part of personal care products that are used to block UV rays, these are used in sunscreens, beauty creams, dyes, and other cosmetics. These can be introduced in aquatic systems in two ways. Firstly, by water recreational activities like swimming in lakes and rivers. Another is from used domestic water via WWTPs.
- Artificial sweeteners: these are mainly used as food additives (low-calorie dietary products), like sugar substitutes in substantial concentrations. These come out of the human system in unchanged form. Few of these like cyclamate and saccharin

can be biodegraded, whereas, sucralose, which is a chlorinated compound, and acesulfame are persistent and may disrupt the aquatic environment.

5.3.2 Industrial Wastewater

There are several industrial compounds which can be released into the environment, causing adverse effect to the environment. Effluent released from industries including pharmaceutical industries, tannery, iron smelters, textile industry, cellulose production, dishes factory, paper, and pulp factory contains several emerging contaminants. Along with primary compounds that are released in water bodies from industries, there are many degraded by-products from primarily infused contaminants that contribute as emerging contaminate. Industrial emerging contaminants are as follows:

- 1,4-Dioxane: it is used as a solvent in the manufacturing of chemicals and laboratory reagents. When combining with 1,1,1-trichloroethane it can resist natural biodegradable processes and have the ability to persist in groundwater and directly leach out from the soil. 1,4-Dioxane is also present in PPCPs (Abe 1999).
- Benzotriazole derivatives: it is present in many pharmaceutical antibiotics as antifungal, antibacterial, and antihelmintic drugs and tends to persist in an aqueous environment.
- Polybrominated biphenyl ether (PBDEs): PBDEs are the synthetic chemicals used as an additive retard fire and flame for multiple household and commercial purposes. In a coastal and estuarine ecosystem, the presence of PBDEs has been identified. From industrial manufacturing and household products (furniture, electronics, bedding) sites it is reaching the environment. Exposure to PBDEs can potentially cause neurodevelopmental and thyroid dysfunctions (Brown and Nixon 1979) and have potential endocrine-disrupting properties (Rahman et al. 2001).

5.3.3 Agriculture Runoff

The ECs where agriculture is currently the predominant source of surface water contamination, compared to non-agricultural sources, are the natural toxins, veterinary medicines, hormones, and transformation products of man-made chemicals used in agriculture. Emerging contaminant loads of agricultural land soil get transported to aquatic systems by leaching, runoff, and drainage processes. ECs are released in the agricultural environment directly via the application of biosolids, pesticides, or other solid waste materials, and indirectly from veterinary medicines through animal bio-waste. Multifarious ECs have been detected in agricultural systems such as human pharmaceuticals, personal care products, veterinary drugs, hormones, and transformation products of man-made chemicals in a very low ng/l range. Prospective of developing new technology for effective and efficient use of pesticides is high, the existing pesticide may be made into nanoparticulate (1–100 nm particles) form to give the active ingredient beneficial properties for pest control, such as increased solubility, increased stability, the capacity for absorption into plants or increased toxicity to pests (Lyons et al. 2009). Although there are speculations that these nanoparticles will be easily transferred to surface water and contaminate aquatic systems, many man-made substances that enter agricultural systems will be degraded by chemical, physical, and biological processes. While these reactions may result in the complete breakdown of a chemical (i.e. they convert the chemical to carbon dioxide, water, and methane), in many instances the processes may result in the formation of stable intermediate chemicals which we call degradation or transformation products (e.g. Roberts 1998; Roberts and Hutson 1999).

Typical wastewater systems involve a primary treatment phase, where solid material is settled out and removed, followed by a secondary treatment where microbes are used to break down organic matter (including chemicals). One common secondary treatment process is activated sludge treatment (AS). In AS, chemical contaminants may be degraded but can also stick to the sludge particles in the system. At the end of the process, the sludge is removed and, as it is high in nutrients, in many regions, it is then applied to agricultural land as a fertilizer. ECs that stick to sludge can therefore enter the environment when sludge (biosolids) from wastewater treatment works is applied to land (Kinney et al. 2006; Topp et al. 2008).

5.4 Pollution and Toxicity of Emerging Contaminants

Emerging contaminants, such as persistent organic pollutants (POPs), endocrine-disrupting chemicals (EDCs), pharmaceuticals and personal care products (PCPs), etc., and trace levels of their transformation products are major disruptors for the aquatic ecosystem, human health, and biodiversity (Haq and Raj 2019). Lack of advanced technology and health standards regarding newly detected contaminants is another complication in wastewater treatment. The concerning levels of contaminants are released in soil and water via treated water from wastewater treatment plants. The efficiency and mechanism of functioning of current WWTPs are not sufficient to remove all contaminants from wastewaters.

5.4.1 Adverse Impacts on Human Health and Biodiversity

The presence of contaminants implicates acute harmful effects on natural flora and fauna in surface waters, seawater, and soils. In the recent decade, plastic pollution due to microplastic (1–5 mm) and nanoplastic (<1 mm) size particles is of emerging concern for aquatic fauna. Microplastic ingestion has been widely observed in the range of marine animals such as fish, bivalves, and crustaceans which are consumed by humans, along with concern to aquatic life, it raises the concern of microplastic ingestion in humans also. These microplastic get chemicals to adhere to their surfaces and they significantly cause chemical damage to ambient life. These ingested contaminants block the digestive system, alter feeding behavior, disrupt the endocrine system, retard growth and reproductive capabilities. Even artificial aquaculture systems are also affected by these contaminants as a fish meal which is given to cultured fishes, shrimps, or other seafood animals has been taken from natural contaminated sources. Experiments performed on Fulmarus glacialis (Northern fulmar) showed the persistence of ingested plastics in the gizzard, mixed with stomach oil which is used by the bird as an energy source during long flights (Kühn et al. 2020). Due to their small size microplastic is ingested by a benthic organism and has ecological impacts. Rochman suggested that microplastic of range 2–5 mm could take a longer time to digest because of durability and retain for longer duration in organism body as a consequence increase the exposure time to toxins (Rochman et al. 2013). Microplastic abundantly emerging in sea beds and cause a major threat to the marine food web. Some organic contaminants associated with plastic after inhalation it starts interfering with hormone regulation in animal (Teuten et al. 2009). Ingestion of microplastic especially in fiber form poses threat to the organism as it can cause blockage in the digestive tract, be translocated in different tissues within the organism, and undergo accumulation (Wright et al. 2013). Knowledge about the effects of ingestion of microplastic by a human being requires more research and investigation.

Bioaccumulation aggravates the tendency of abnormal hormonal regulators causing reproductive impairments, decreased fecundity, an increase of breast and testosterone cancers, and persistent antibiotic resistance (Tijani et al. 2013). Another similar example is of alternation in the endocrine and reproductive system of Japanese medaka fish because of harmful exposure to the UV filters and oxybenzone (Coronado et al. 2008).

Much research has been performed on the toxicity of pesticides. Many pesticides are carcinogenic to animals and, for this reason, are likely human carcinogens. For example, the pesticides alachlor, dieldrin, and DBCP are each classified under the USEPA designation "probable human carcinogen." Perhaps the greatest threat posed by pesticides, however, is their endocrine-disrupting potential. Exposures can cause reproductive system damage including sterility, decreased fertility, and birth defects, as wells as impaired development, immuno-suppression, and metabolic disorders. Studies have linked low concentrations (0.1 ppb) of the pesticide atrazine to developmental deformities in frogs including the development of multiple sex organs and small feminized larynxes. In another study, frogs exposed to even trace amounts of malathion experienced a near-total collapse of their immune systems, with antibody production limited to as little as 1–2% of normal. Many studies and research have suggested that many emerging organic contaminants (EOC) are endocrine disruptor chemicals (EDCs) (Liu et al. 2009). EDCs are present in a wide range of compounds, include plastic bottles, detergents, flame retardants, toys, cosmetics, pesticides, etc. and they might have adverse developmental and reproductive effects on humans and wildlife. EOCs are toxic and persistent and even it presents in low concentration; it might produce potentially harmful effects on the ecosystem and human health (Jurdo et al. 2012).

Some case studies associated with the chronic effect of synthetic chemicals usage on fauna are, the veterinary use of diclofenac, which is a human pharmaceutical used as an anti-inflammatory treatment, was found to be responsible for the massive decline in populations of vulture species in certain areas of Asia (Oaks et al. 2004). The veterinary drug ivermectin, which is used to treat parasitic infections in livestock, has been shown to affect the growth of aquatic invertebrates at a concentration lower than those that are expected to occur in the aquatic environment (Garric et al. 2007). Ethinylestradiol, one of the active ingredients in the contraceptive pill, has been associated with endocrine disruption in fish (Lange et al. 2001); and there is concern that long-term exposure to antibiotic pharmaceuticals, used in human and veterinary medicine, may be contributing to the selection of resistant bacteria in the environment which may have significant implications for human health (Boxall et al. 2003).

5.4.2 Water Pollution

In recent year the jeopardy of groundwater contamination from emerging contaminants has intensified mainly due to multifarious anthropogenic activities which significantly affect the natural water cycle as well as the passage of contaminants in the aquatic system (Díaz-Cruz and Barceló 2008). Groundwater contamination can be resultant from numerous potential sources, including leakage from sewers, septic tanks, landfills, urban runoff, municipal waste, overuse of fertilizers, gasoline spills, aquaculture, and many more. Furthermore, it is difficult to eliminate contaminants from groundwater because of the slow flow rate movement, long residence times, low microbial populations, and redox controls in the saturated zone. Thus, the emerging contaminates remain stranded in groundwater for a longer period. Compared to other aquatic resource the existence of emerging contaminate is poorly categorized in groundwater. The source of emerging contaminate in groundwater is divided into a point and diffuse source—point source originated from particular locations and the amount of pollution is generally more constrained such as from industrial effluents, municipal solid waste treatment plant, hazardous waste effluent, waste disposal site, biomedical waste site while diffuse pollution, in contrast, originated from scattered sources and occurs over broad geographical scales such as from agricultural runoff, urban-storm runoff leakage from urban sewerage system (Buerge et al. 2011).

One of the major sources of groundwater pollution is pesticides and their metabolites. Present studies have been shifted towards pesticide metabolites also on degraded and reaction products. By their nature, this product is biologically active many are toxic. Research has shown that pesticide metabolites are frequently perceived in groundwater in higher concentrations compared to the parent compound. An extensive series of pharmaceutical products have been spotted in surface and groundwater including wastewater disposal. These have included:

- Veterinary and human antibiotics—e.g. ciprofloxacin, erythromycin, lincomycin, sulfamethoxazole, tetracycline.
- Personal care products—e.g. DEET, parabens, antifungal and bactericide agents (triclosan), polycyclic musk (tonalite, galactoside).
- UV filters/sunscreen—organic filters include benzophenones and methoxycinnamates.

Worldwide freshwater resources are facing challenges of thousands of chemical compounds due to anthropogenic activities. Yearly around 300 million tons of synthetic compounds used in industrial and consumer products generally find their way to natural waters (Schwarzenbach et al. 2006). Due to chemical compounds released from wastewater discharge, incorporated with the agricultural runoff of fertilizers/pesticide, leach out from landfill sites (biomedical waste, hazardous waste), and atmospheric deposition, that majorly affecting the surface water quality and such contamination can increase the problem for drinking supplies also. Streams disperse the emerging contaminates into different water bodies including aquifers, estuaries, and marine systems. ECs are unable to remove from conventional wastewater treatment plants (WWTPs); therefore, many of these chemical or synthetic compounds are directly released into surface waters (Pal et al. 2010). The overapplication of fertilizers and pesticides has increased the discharge rate of nutrients and pollutants into the aquatic system. However, the discharge of nutrients from a point source (WWTP and industry) still contributes to the significant pollution load to the aquatic system (Kronvang et al. 2008). Increased levels of the nutrient in the aquatic system may result in eutrophication, increased water turbidity, oxygen depletion, coral reef destruction, vulnerability increase to marine life (Kalff 2002). Antibiotics can result in the development of resistant bacterial strains and are associated with a public health concern. In China antibiotic frequently detected in surface water has been observer such as macrolides, sulfonamides, and fluoroquinolones (Yang et al. 2014).

5.4.3 Soil Pollution

It is not quite astonishing that contaminants that had spread across the aquatic system have made to retain in soil and sediments. Contaminates of soil reach human, animal, and plant systems via distinct routes of exposure like ingestion of soil and dust particles, uptake in food crops which has been absorbed by plants through roots, inhalation, leaching to groundwater which is commonly used for drinking. Most discussed soil contaminants in the scientific community are lead, petroleum, radon, chromate copper arsenate, and creosote. Anthropogenic activities such as industrial dumping, mining, manufacturing, local waste disposal, pesticide usage in agriculture are major contributors to soil pollution. The type of soil determines the probability of groundwater contamination and can allow more contaminants to pass in groundwater than clay. The presence of organic matter holds contaminants tightly which leads to the accumulation of contaminants in soil. Studies suggest that microplastic pollution can alter the coupling between carbon and nutrient cycling through a significant increase in nutrients in dissolved organic matter and CO_2 fluxes (Geyer et al. 2017). Some emerging contaminants like petroleum are harmful to soil microorganisms and their activity. These microbes play important role in providing nutrients to plant, so plant nutrient availability also gets affected by these contaminants. Petroleum hydrocarbons like benzene, toluene, ethylbenzene, and xylene can cause cancer. There are regulatory standards for safe levels in drinking water.

The challenges posed by emerging contaminants in soils are crucial and require rigorous actions and collaboration. There is a need for monitoring data and risk assessment models, but also for awareness-raising and new guidelines and authority models to deal with emerging contaminants in soils. Detection of these contaminants in soil and sediment is particularly challenging due to the low detection limits required, their intricate nature, and the difficulty in separating these compounds from interfering.

5.5 Removal Approaches

Since the last decades several physical, chemical, and biological technologies have been developed and studied to eliminate or degrade the residues of ECs (Zhang et al. 2008). Biological treatment has been the preferred technology for the removal, including activated sludge, constructed wetlands, membrane bioreactor (MBR), aerobic bioreactor, trickling filter, enzyme treatment, biosorption, and plant-based (Phyto) technologies (Haq et al. 2016a, b; Haq et al. 2017, 2018; Kishor et al. 2018; Mulla et al. 2019; Bharagava et al. 2017, 2018, 2020; Mulla et al. 2020; Kumar et al. 2020; Bharagava and Saxena 2020; Deb et al. 2020; Haq et al. 2020; Saxena et al. 2020a, b, c, d, e, f). Chemical technologies also have been broadly used for the degradation of these micropollutants because some non-biodegradable organic compounds cannot be removed appropriately from biological treatment, it includes conventional oxidation methods such as photocatalysis, photolysis, ozonation, advanced oxidation processes, etc. moreover each method has advantages and challenges (Goutam et al. 2020). Several pesticides and pharmaceutical are highly water-soluble will remain in aqueous system so, physical treatment processes as coagulation and flocculation followed by sedimentation, sand filtration, and disinfection. These processes are not that effective as a comparison to chemical and biological processes it removes around 20–30% of contaminant (Kastl et al. 2004; Stackelberg et al. 2007).

5.5.1 Physical Treatments

Physical treatment may involve the technologies to remove the emerging contaminants without changing their chemical state and may be classified as phase partitioning processes (adsorption processes) or membrane processes. In the adsorption process several adsorbing agents, which may be natural or synthetic or waste materials, are used to adsorb the target contaminant to adsorb on the surface, converting one face transfer to another face transfer is involved (Dhangar and Kumar 2020). In this technology, sore surface area and more porosity may be the good indicator of the optimum success of the technology (Rivera-Utrilla et al. 2013). The removal efficiency may be depending on the selected material as an adsorbent (Rodriguez-Narvaez et al., 2017). These may include activating charcoal from wood or herbaceous plant, biochar, etc.

In the membrane technologies, solutions are filtered by retaining contaminants on the membrane, with variable filtration features (pore size, hydrophobicity, and surface charge as per the material from it has been obtained, determines the efficiency of the removal of the contaminants (Dhangar and Kumar 2020). Hydrostatic pressure performs a crucial part in the membrane processes, allows water and low molecular weights to pass through the membrane by retaining the suspended particles and high molecular weight solutes. Based on the size of the pores and the types of membrane, the processes may be classified into microfiltration (MF), nanofiltration (NF), ultrafiltration (UF), forward osmosis (FO), and reverse osmosis (RO).

5.5.2 Chemical Treatment

Some of the ECs that may not be appropriately removed by biological methods may be remediated by chemical methods which include several oxidation methods like photocatalysis, photolysis, ozonation, advanced oxidation processes. In this method, the ECs may be remediated either by complete mineralization or by their alteration to inorganic molecules such as water, carbon dioxide, and nitrogen. To achieve this goal, these techniques use various chemical agents like ozone, hydrogen peroxide, and chlorine; metal oxides and transition metal-based catalysts; and some energy or radiation sources such as ultrasound, electric current, gamma, solar, and UV-radiation (Dhangar and Kumar 2020).

For example, removal of contaminants such as sulfamethoxazole (SMX), diclofenac (DCF), and carbamazepine (CBZ) may be possible by such above cited (ozonation, sonolysis, and photocatalytic oxidation) chemical methods in synthetically prepared aqueous solutions. To achieve the optimum efficiency for remediation of such ECs may be preferred one of the methods or a different combination of the same, which may depend on the characteristics of the chemical (Fraiese et al. 2019).

The chemical treatments are further classified as conventional oxidation processes and advanced oxidation processes (AOPs). In conventional oxidation processes chlorination, ozonation, Fenton process, and photolysis, etc. may be counted (Dhangar and Kumar 2020). But these conventional methods have some disadvantages, which could be fulfilled by advanced oxidation methods. In these methods high oxidizing state of a chemical or radiation is used to remove the target ECs. This process may include photocatalysis, electro-Fenton process, solar-Fenton process, and ultrasound irradiation (Salami et al. 2017). These chemical processes may be improvised or combined as per the required efficiencies but may also provide few disadvantages as well.

5.5.3 Biological Treatments

Tertiary treatment is widely used for the removal of contaminants from an aqueous system it works on biodegradation and is also known as effluent polishing treatment. Biodegradation is the process of conversion of a substance into new compounds through biochemical actions of microorganisms including bacteria, fungus (Rodriguez-Narvaez et al. 2017). Tertiary treatment mainly divided into aerobic and anaerobic processes-activated sludge, membrane bioreactor, and anaerobic sludge reactor, anaerobic film reactor, respectively. Conventionally biodegradation method has been used for the treatment of wastewater for the removal of EC. The rate of removal of the contaminant is strongly dependent on treatment technology, target contaminant, and treatment conditions. Recognition of degradation products in environmental samples is challenging because it is not present in very low concentration but also present in complex matrixes that may interfere in the detection process (Zhou et al. 2009).

5.5.3.1 Constructed Wetlands

Wetlands have been receiving much attention worldwide due to their intrinsic tendency to remove pesticides, surfactant, PPCPs, and other micro contaminate, hence wetlands offer important ecological benefits (Gregoire et al. 2009). To reduce sediments, nutrients load, and pollution due to anthropogenic activities entering into the surface water and groundwater, some countries as Sweden, Denmark, and the USA has adopted a technique to develop or restored wetlands to revive the ecosystem services that were vanished due to anthropogenic activities such as the conversion of wetlands into the agricultural field, aquaculture (Hoffmann and Baattruo-Pedersen 2007; Thiere et al. 2009; Hoffmann et al. 2011). European Union (EU) water framework directive which states that a good ecological and chemical quality should be reached in water bodies has been adopted by EU members by the end of 2015 (European Commission 2000). Denmark government through Danish Action Plans on a national scale has set the target of increasing the wetlands area and reduce the nutrient load in the sea.

5.5.3.2 Biological Trickling Filter

Bio-trickling filter is used as a conventional treatment method in WWTPs for the removal of biochemical oxygen demand (BOD), chemical oxygen demand (COD), pathogen decontamination, odor, and air pollution control. The efficiencies of treatment methods such as activated sludge, aerated lagoon, trickling filter vary according to the presence of a contaminant in wastewater. Trickling filters or biobeds were used either alone or with association with other treatment methods. Kasprzyk-Horden et al. (2009) run a 5-month monitoring program in South Wales in the UK to analyze the efficiency of two different wastewater treatment technologies which is activated sludge and trickling filter beds, with two distinct wastewater treatment plant (WWTP Cilfynydd and WWTP Coslech) discharging treated water in two different rivers (Taff and Ely). Their outcome was activated sludge treatment which was much more efficient than trickling filter beds. For WWTP using trickling filter, the average came less than 70% removal for 55 PPCPs studied. However WWTP using activated sludge gave efficiency above 85%.

5.5.3.3 Biologically Activated Carbon

Several studies indicated that biologically activated carbon adsorption tends to remove contaminant in a broad spectrum of EC. Eliminate the effluent organic matter from treated water so it can be reused in different activities. To eliminate emerging organic contaminant (EOC) from wastewater the use of adsorbents is a very common method. Activated carbon is highly porous and effective absorbent and used to remove repugnant taste and odor and EOC worldwide (Moreno-Castilla 2004; Quinlivan et al. 2005). Adsorption depends on surface area, pore structure, and surface structure. On the surface of activated carbon, the immobilization of microorganism (biofilm) under appropriate temperature and nutrient condition produces the biological activated carbon and plays a significant role in adsorption and biodegradation simultaneously. This active biofilm effectively removes biodegradable pollutant such as EOC and organic pollutants from the water and increases the lifespan of carbon bed (Hijnen et al. 2014; Korotta-Gamage et al. 2017). Nowadays biologically activated carbon treatment is used after the ozonation process to enhance the process efficiency of treatment and quality of water. When ozonation has not used before biological activated carbon treatment showed lower efficiency in the removal of EDC (e.g. BPA, octylphenol), However, biological activated carbon treatment in combination with ozonation used in the removal of emerging contaminant mainly pesticide (e.g. atrazine and triclosan) and pharmaceuticals (analgesics, antibiotics, lipid regulator and anti-depressant) shows more efficiency (Korotta-Gamage and Sathasivan 2017).

5.5.3.4 Biosorption

Biosorption is a mixture of physicochemical and biological techniques, used to remove pollutants from industrial wastewaters mainly which are not easily biodegradable such as dyes, metal, etc. before discharge into an aqueous system (Gadd 2009). In biosorption-sorption and bio-oxidation occur when microorganism traps/immobilizes into adsorbents. Both living and non-living microorganism and their components have the tendency and capabilities to detoxify the organic and inorganic pollutants (Gadd and White 1993; Gadd 2007). However, a product such as all organic materials including macroalgae (seaweeds), plants and animal biomass, and derived products (chitosan) are capable of biosorption. Due to their conventional ion-exchange property and deceptive efficiency and availability of biomass and waste bio-products, biosorption has been established as a capable technique for the removal of a pollutant from the environment. Nguyen et al. (2014) by using live cultured and harvested white-rot fungus (Trametes Versicolor) compare the biosorption and biodegradation technique for the removal of trace organic carbon

(TOC) from wastewater. Studies conclude that live white-rot fungi treatment has more efficiency than inactivated white-rot fungi (Biosorption). The removal of ECs such as 4-start-octylphenol, triclosan, pentachlorophenol, etc. had been achieved by more than 80%. And some pharmaceuticals removed completely such as ibuprofen, naproxen, and gemfibrozil.

5.5.3.5 Membrane Bioreactor (MBR)

MBR is the combination of a membrane process (microfiltration and ultrafiltration) and suspended growth bioreactor. Now worldwide MBR is used for industrial and municipal waste treatment and reclamation. Studies have found that MBR is a more effective remover of a wide range of ECs including that compounds are resistant to activated sludge process and constructed wetlands (Radjenovic et al. 2009; Luo et al. 2014) because sludge and physical retention time on the membrane which enhances the microbial degradation of all molecules according to carrying capacity of membranes. The efficiency of MBR treatment depended on sludge age, concentration, and presence of anoxic and anaerobic compartments, the composition of wastewater, operating temperature, pH, and conductivity (Kovalova et al. 2012). Trinh et al. (2012) examined the 48 EC removal efficiency through full-scale MBR. The result showed above 90% of elimination of ECs through the MBR. However, some compounds were partially removed (24–68%) such as amitriptyline, carbamazepine, diazepam, diclofenac, fluoxetine, gemfibrozil, omeprazole, sulfamethoxazole, and trimethoprim.

5.5.3.6 Phytoremediation

It has been observed in the last two decades, varied chemicals and emerging contaminants, such as heavy metals, volatile organic compounds, pesticides, personal care products (PCPs), pharmaceuticals, organics, and pathogenic microbes trouble the water and environment (Rai 2018). Plants are the interface for soil air and water, thus can emerge as eco-friendly, cost-effective, and better tool for removal of emerging contaminants with the technology named phytoremediation (Mishra 2021; Rai 2018). Some research has been done on the elimination of emerging environmental contaminants (volatile organic carbons, pesticides, PPCPs, organics, and pathogenic microbes) and heavy metals with help of worldwide spread plants species particularly wetland plants (Pandey et al. 2020; Rai 2018; Kamusoko and Jingura 2017), but still lacks plenty of work to establish plant-based technology as game-changer. In this technology, pollutants may be absorbed by roots from the medium (soil or water) into the plants and either accumulated or degraded by plant tissues (Mishra 2021). Several plant species have been identified with years of studies, which remediate emerging contaminants of concern. As an example, Typha latifolia, Populus deltoids, Populus trichocarpa, etc. have been reported to remediate trichloroethylene (Amon et al. 2007; Rai 2018). ECs, like pesticides, plants, such as Ceratophyllum demersum, Elodea canadensis, and Lemna minor, have been evidenced to perform rapid biotransformation of metolachlor and atrazine from herbicide-contaminated aquatic bodies (Williams 2002). Scirpus Validus remediate simazine and metolachlor from wastewaters (Stearman 2003). For petroleum and other hydrocarbon contaminated land could be efficiently phytoremediation by Juncus roemerianus (Syranidou et al. 2017).

5.6 Conclusion

In the changing and developing world emerging contaminants unknowingly became an inextricable part of our life, which we could identify with time. Saying this may not be enough as regularly new emerging pollutant comes inexistent as the result of evolution and emergence of new technology developed in the favor and welfare of humans. Some of them are being identified earlier and some are later with their acerbating consequences. A crucial point is that we are not ready yet, for the expected ECs in the coming future; to overcome the situation many vicissitudes can be expected regarding future emerging contaminants. The emergence of ECs would lead to more changing and troublesome political, economic, ecological, and health consequences. Only working on the extenuating counters may not be enough to fight the troublesome consequences of emerging contaminants. It needs careful observation over the developing world to develop more sustainable technologies.

Acknowledgments The authors are thankful to the University Grants Commission for providing financial assistance for our research works.

References

Abe A (1999) Distribution of 1, 4-dioxane in relation to possible sources in the water environment. Sci Total Environ 227(1):41–47

Ahmed MB, Zhou JL, Ngo HH, Guo W (2015) Adsorptive removal of antibiotics from water and wastewater: progress and challenges. Sci Total Environ 532:112–126

Aktar W, Sengupta D, Chowdhury A (2009) Impact of pesticides use in agriculture: their benefits and hazards. Interdiscip Toxicol 2(1):1–12

Alomar C, Estarellas F, Deudero S (2016) Microplastics in the Mediterranean Sea: deposition in coastal shallow sediments, spatial variation and preferential grain size. Mar Environ Res 115:1–10

Amon JP et al (2007) Development of wetland constructed for the treatment of groundwater contaminated by chlorinated ethenes. Ecol Eng 30:51–66

Bharagava RN, Saxena G (2020) Progresses in bioremediation technologies for industrial waste treatment and management: challenges and future prospects. In: Bioremediation of industrial waste for environmental safety. Springer, Singapore, pp 531–538

Bharagava RN, Saxena G, Mulla SI, Patel DK (2017) Characterization and identification of recalcitrant organic pollutants (ROPs) in tannery wastewater and its phytotoxicity evaluation for environmental safety. Arch Environ Contam Toxicol 75(2):259–272

Bharagava RN, Purchase D, Saxena G, Mulla SI (2018) Applications of metagenomics in microbial bioremediation of pollutants: from genomics to environmental cleanup. In: Das S, Dash H (eds) Microbial diversity in the genomic era, 1st edn. Academic Press, Elsevier, Cambridge, MA

Bharagava RN, Saxena G, Mulla SI (2020) Introduction to industrial wastes containing organic and inorganic pollutants and bioremediation approaches for environmental management. In: Bioremediation of industrial waste for environmental safety. Springer, Singapore, pp 1–18

Boots B, Russell CW, Green DS (2019) Effects of microplastics in soil ecosystems: above and below ground. Environ Sci Technol 53(19):11496–11506

Boxall ABA, Kolpin DW, Halling-Sorensen B, Tolls J (2003) Are veterinary medicines causing environmental risks? Environ Sci Technol 37:286A–294A

Brown GG, Nixon R (1979) Exposure to polybrominated biphenyls: some effects on personality and cognitive functioning. JAMA 242(6):523–527

Buerge IJ, Keller M, Buser HR, Müller MD, Poiger T (2011) Saccharin and other artificial sweeteners in soils: estimated inputs from agriculture and households, degradation, and leaching to groundwater. Environ Sci Technol 45(2):615–621

Carson R (2002) Silent spring. Houghton, Mifflin Harcourt

Coronado M, De Haro H, Deng X, Rempel MA, Lavado R, Schlenk D (2008) Estrogenic activity and reproductive effects of the UV-filter oxybenzone (2-hydroxy-4-methoxyphenyl-methanone) in fish. Aquat Toxicol 90(3):182–187

Deb VK, Rabbani A, Upadhyay S, Bharti P, Sharma H, Rawat DS, Saxena G (2020) Microbe-assisted phytoremediation in reinstating heavy metal-contaminated sites: concepts, mechanisms, challenges, and future perspectives. In: Microbial technology for health and environment. Springer, Singapore, pp 161–189

Dhangar K, Kumar M (2020) Tricks and tracks in removal of emerging contaminants from the wastewater through hybrid treatment systems: a review. Sci Total Environ 738:140320

Diamanti-Kandarakis E, Bourguignon JP, Giudice LC, Hauser R, Prins GS, Soto AM, Zoeller RT, Gore AC (2009) Endocrine-disrupting chemicals: an Endocrine Society scientific statement. Endocr Rev 30(4):293–342

Díaz-Cruz MS, Barceló D (2008) Trace organic chemicals contamination in ground water recharge. Chemosphere 72(3):333–342

Dunier M, Siwicki AK (1993) Effects of pesticides and other organic pollutants in the aquatic environment on immunity of fish: a review. Fish Shellfish Immunol 3(6):423–438

Eljarrat E, Barcelo D (2003) Priority lists for persistent organic pollutants and emerging contaminants based on their relative toxic potency in environmental samples. TrAC Trends Anal Chem 22(10):655–665

European Commission (2000) Directive 2000/60/EC of the European Parliament and of the Council of 23 October 2000 establishing a framework for Community action in the field of water policy (OJ L 327, 22.12.2000, p.1)

Farré M, Sanchís J, Barceló D (2011) Analysis and assessment of the occurrence, the fate and the behavior of nanomaterials in the environment. TrAC Trends Anal Chem 30(3):517–527

Fendorf S, Michael HA, van Geen A (2010) Spatial and temporal variations of groundwater arsenic in south and Southeast Asia. Science 328(5982):1123–1127

Fraiese A, Naddeo V, Uyguner-Demirel CS, Prado M, Cesaro A, Zarra T, Liu H, Belgiorno V, Ballesteros F Jr (2019) Removal of emerging contaminants in wastewater by sonolysis, photocatalysis and ozonation. Glob NEST J 21:98–105

Gadd GM (2007) Geomycology: biogeochemical transformations of rocks, minerals, metals and radionuclides by fungi, bioweathering and bioremediation. Mycol Res 111(1):3–49

Gadd GM (2009) Biosorption: critical review of scientific rationale, environmental importance and significance for pollution treatment. J Chem Technol Biotechnol 84(1):13–28

Gadd GM, White C (1993) Microbial treatment of metal pollution—a working biotechnology? Trends Biotechnol 11(8):353–359

Garric J, Vollat B, Duis K, Pery A, Junker T, Ramil M, Fink G, Ternes TA (2007) Effects of the parasiticide ivermectin on the cladoceran Daphnia magna and the green alga Pseudokirchneriella subcapitata. Chemosphere 69(6):903–910

Gatidou G, Thomaidis NS, Stasinakis AS, Lekkas TD (2007) Simultaneous determination of the endocrine disrupting compounds nonylphenol, nonylphenol ethoxylates, triclosan and bisphenol a in wastewater and sewage sludge by gas chromatography–mass spectrometry. J Chromatogr A 1138(1–2):32–41

Geyer R, Jambeck JR, Law KL (2017) Production, use, and fate of all plastics ever made. Sci Adv 3 (7):e1700782
Gigault J, Ter Halle A, Baudrimont M, Pascal PY, Gauffre F, Phi TL, El Hadri H, Grassl B, Reynaud S (2018) Current opinion: what is a nanoplastic? Environ Pollut 235:1030–1034
Gogoi A, Mazumder P, Tyagi VK, Chaminda GT, An AK, Kumar M (2018) Occurrence and fate of emerging contaminants in water environment: a review. Groundw Sustain Dev 6:169–180
Goutam SP, Saxena G, Roy D, Yadav AK, Bharagava RN (2020) Green synthesis of nanoparticles and their applications in water and wastewater treatment. In: Bioremediation of industrial waste for environmental safety. Springer, Singapore, pp 349–379
Gregoire C, Elsaesser D, Huguenot D, Lange J, Lebeau T, Merli A et al (2009) Mitigation of agricultural nonpoint-source pesticide pollution in artificial wetland ecosystems–a review. In: Climate change, intercropping, pest control and beneficial microorganisms. Springer, Dordrecht, pp 293–338
Haq I, Kumar S, Kumari V, Singh SK, Raj A (2016a) Evaluation of bioremediation potentiality of ligninolytic Serratia liquefaciens for detoxification of pulp and paper mill effluent. J Hazard Mater 305:190–199
Haq I, Kumari V, Kumar S, Raj A, Lohani M, Bhargava RN (2016b) Evaluation of the phytotoxic and genotoxic potential of pulp and paper mill effluent using Vigna radiata and Allium cepa. Adv Biol. 8065736
Haq I, Kumar S, Raj A, Lohani M, Satyanarayana GNV (2017) Genotoxicity assessment of pulp and paper mill effluent before and after bacterial degradation using Allium cepa test. Chemosphere 169:642–650
Haq I, Mazumder P, Kalamdhad AS (2020) Recent advances in removal of lignin from paper industry wastewater and its industrial applications–A review. Bioresour Technol 312:123636
Haq I, Raj A (2018) Endocrine-disrupting pollutants in industrial wastewater and their degradation and detoxification approaches. In: Bharagava RN, Chowdhary P (eds) Emerging and eco-friendly approaches for waste management. Springer Nature Singapore Pte Ltd, pp 121–142
Haq I, Raj A (2019) Pulp and paper mill wastewater: ecotoxicological effect and bioremediation approaches for environmental safety. In: Bharagava RN, Saxena G (eds) Biological agents and methods for industrial waste management. Springer Nature Singapore Pte Ltd, pp 333–356
Haq I, Raj A, Markandeya (2018) Biodegradation of Azure-B dye by Serratia liquefaciens and its validation by phytotoxicity, genotoxicity and cytotoxicity studies. Chemosphere 196:58–68
Hijnen WAM, Schurer R, Martijn B, Bahlman JA, Hoogenboezem W, van der Wielen PWJJ (2014) Removal of easily and more complex biodegradable NOM by full-scale BAC filters to produce biological stable drinking water. In: Progress in slow sand and alternative biofiltration processes. IWA, London, pp 413–421
Hoffmann CC, Baattrup-Pedersen A (2007) Re-establishing freshwater wetlands in Denmark. Ecol Eng 30(2):157–166
Hoffmann CC, Kronvang B, Audet J (2011) Evaluation of nutrient retention in four restored Danish riparian wetlands. Hydrobiologia 674(1):5–24
Ikehata K, Gamal El-Din M, Snyder SA (2008) Ozonation and advanced oxidation treatment of emerging organic pollutants in water and wastewater. Ozone Sci Eng 30(1):21–26
Jurado A, Vàzquez-Suñé E, Carrera J, de Alda ML, Pujades E, Barceló D (2012) Emerging organic contaminants in groundwater in Spain: a review of sources, recent occurrence and fate in a European context. Sci Total Environ 440:82–94
Kalff J (2002) Limnology: inland water ecosystems. Prentice Hall, Upper Saddle River
Kamusoko R, Jingura RM (2017) Utility of Jatropha for phytoremediation of heavy metals and emerging contaminants of water resources: a review. CLEAN–Soil, Air, Water 45(11):1700444
Karunakaran CO (1958) The Kerala food poisoning. J Indian Med Assoc 31:20
Kasprzyk-Hordern B, Dinsdale RM, Guwy AJ (2009) The removal of pharmaceuticals, personal care products, endocrine disruptors and illicit drugs during wastewater treatment and its impact on the quality of receiving waters. Water Res 43(2):363–380

Kastl G, Sathasivan A, Fisher IAN, Van Leeuwen J (2004) Modeling DOC removal enhanced coagulation. J Am Water Works Ass 96(2):79–89

Kinney CA, Furlong ET, Zaugg SD, Burkhardt MR, Werner SL, Cahill JD, Jorgensen GR (2006) Survey of organic wastewater contaminants in biosolids destined for land application. Environ Sci Technol 40(23):7207–7215

Kishor R, Bharagava RN, Saxena G (2018) Industrial wastewaters: the major sources of dye contamination in the environment, ecotoxicological effects, and bioremediation approaches. In: Recent advances in environmental management. CRC Press, Boca Raton, FL, pp 1–25

Koenig S, Huertas D, Fernández P (2013) Legacy and emergent persistent organic pollutants (POPs) in NW Mediterranean deep-sea organisms. Sci Total Environ 443:358–366

Korotta-Gamage SM, Sathasivan A (2017) A review: potential and challenges of biologically activated carbon to remove natural organic matter in drinking water purification process. Chemosphere 167:120–138

Kovalova L, Siegrist H, Singer H, Wittmer A, McArdell CS (2012) Hospital wastewater treatment by membrane bioreactor: performance and efficiency for organic micropollutant elimination. Environ Sci Technol 46(3):1536–1545

Križanec B, Majcen Le Marechal A (2006) Dioxins and dioxin-like persistent organic pollutants in textiles and chemicals in the textile sector. Croat Chem Acta 79:177–186

Kronvang B, Jensen JP, Hoffmann CC, Boers P (2008) Nitrogen transport and fate in European streams, rivers, lakes, and wetlands. In: Nitrogen in the environment. Academic press, New York, pp 241–270

Kühn S, Booth AM, Sørensen L, Van Oyen A, Van Franeker JA (2020) Transfer of additive chemicals from marine plastic debris to the stomach oil of northern fulmars. Front Environ Sci 8:138

Kumar V, Chandra R, Thakur IS, Saxena G, Shah MP (2020) Recent advances in physicochemical and biological treatment approaches for distillery wastewater. In: Combined application of physico-chemical & microbiological processes for industrial effluent treatment plant. Springer, Singapore, pp 79–118

Kümmerer K (2009) The presence of pharmaceuticals in the environment due to human use–present knowledge and future challenges. J Environ Manage 90(8):2354–2366

Lange R, Hutchinson TH, Croudace CP, Siegmund F, Schweinfurth H, Hampe P, Panter GH, Sumpter JP (2001) Effects of the synthetic estrogen 17 alpha-ethinylestradiol on the life cycle of the fathead minnow. Environ Toxicol Chem 20(6):1216–1227

Liu ZH, Kanjo Y, Mizutani S (2009) Removal mechanisms for endocrine disrupting compounds (EDCs) in wastewater treatment—physical means, biodegradation, and chemical advanced oxidation: a review. Sci Total Environ 407(2):731–748

Luo Y, Guo W, Ngo HH, Nghiem LD, Hai FI, Zhang J et al (2014) A review on the occurrence of micropollutants in the aquatic environment and their fate and removal during wastewater treatment. Sci Total Environ 473:619–641

Lyons K, Scrinis G (2009) Under the regulatory radar? Nanotechnologies and their impacts for rural Australia. In: Tracking rural change. ACT ANU E Press, Canberra, p 151

Mishra A (2021) Phytoremediation of heavy metal-contaminated soils: recent advances, challenges, and future prospects. In: Bioremediation for environmental sustainability. Elsevier, Amsterdam, pp 29–51. https://doi.org/10.1016/B978-0-12-820524-2.00002-X

Möller A, Ahrens L, Surm R, Westerveld J, van der Wielen F, Ebinghaus R, de Voogt P (2010) Distribution and sources of polyfluoroalkyl substances (PFAS) in the river Rhine watershed. Environ Pollut 158(10):3243–3250

Moreno-Castilla C (2004) Adsorption of organic molecules from aqueous solutions on carbon materials. Carbon 42(1):83–94

Moschet C, Wittmer I, Simovic J, Junghans M, Piazzoli A, Singer H et al (2014) How a complete pesticide screening changes the assessment of surface water quality. Environ Sci Technol 48 (10):5423–5432

Mulla SI, Bharagava RN, Belhaj D, Saratale GD, Bagewadi ZK, Saxena G, Kumar A, Mohan H, Yu C-P, Ninnekar HZ (2019) An overview of nitro group-containing compounds and herbicides degradation in microorganisms. Microb Metab Xenobiotic Compounds:319–335

Mulla SI, Ameen F, Talwar MP, Eqani SAMAS, Bharagava RN, Saxena G, Tallur PN, Ninnekar HZ (2020) Organophosphate pesticides: impact on environment, toxicity, and their degradation. In: Bioremediation of industrial waste for environmental safety. Springer, Singapore, pp 265–290

Mustereţ CP, Teodosiu C (2007) Removal of persistent organic pollutants from textile wastewater by membrane processes. Environ Eng Manage J 6:3

Muthanna TM, Plósz BG (2008) The impact of hospital sewage discharge on the assessment of environmental risk posed by priority pharmaceuticals: hydrodynamic modelling and measurements. In: International conference on urban drainage, Edinburgh, Scotland, pp 1–10

Nguyen LN, Hai FI, Yang S, Kang J, Leusch FD, Roddick F et al (2014) Removal of pharmaceuticals, steroid hormones, phytoestrogens, UV-filters, industrial chemicals and pesticides by Trametes versicolor: role of biosorption and biodegradation. Int Biodeter Biodegr 88:169–175

Oaks JL, Gilbert M, Virani MZ, Watson RT, Meteyer CU, Rideout BA, Shivaprasad HL, Ahmed S, Chaudhry MJI, Arshad M, Mahmood S, Ali A, Khan AA (2004) Diclofenac residues as the cause of vulture population decline in Pakistan. Nature 427:630–633

Pal A, Gin KYH, Lin AYC, Reinhard M (2010) Impacts of emerging organic contaminants on freshwater resources: review of recent occurrences, sources, fate and effects. Sci Total Environ 408(24):6062–6069

Pandey VC, Mishra A, Shukla SK, Singh DP (2020) Reed canary grass (Phalaris arundinacea L.): coupling phytoremediation with biofuel production. In: Phytoremediation potential of perennial grasses. Elsevier, San Diego, p 165. https://doi.org/10.1016/B978-0-12-817732-7.00007-9

Quinlivan PA, Li L, Knappe DR (2005) Effects of activated carbon characteristics on the simultaneous adsorption of aqueous organic micropollutants and natural organic matter. Water Res 39 (8):1663–1673

Radjenović J, Petrović M, Barceló D (2009) Fate and distribution of pharmaceuticals in wastewater and sewage sludge of the conventional activated sludge (CAS) and advanced membrane bioreactor (MBR) treatment. Water Res 43(3):831–841

Rahman F, Langford KH, Scrimshaw MD, Lester JN (2001) Polybrominated diphenyl ether (PBDE) flame retardants. Sci Total Environ 275(1–3):1–17

Rai PK (2018) Phytoremediation of emerging contaminants in wetlands. CRC Press, London

Rist S, Almroth BC, Hartmann NB, Karlsson TM (2018) A critical perspective on early communications concerning human health aspects of microplastics. Sci Total Environ 626:720–726

Rivera-Utrilla J, Sánchez-Polo M, Ferro-García MÁ, Prados-Joya G, Ocampo-Pérez R (2013) Pharmaceuticals as emerging contaminants and their removal from water. A review. Chemosphere 93(7):1268–1287

Roberts TR (1998) Metabolic pathway of agrochemicals. Part I. In: Herbicides and plant growth regulators. The Royal Society of Chemistry, Cambridge

Roberts TR, Hutson DH (1999) Metabolic pathway of agrochemicals. Part II. In: Insecticides and fungicides. The Royal Society of Chemistry, Cambridge

Rochman CM, Hoh E, Kurobe T, Teh SJ (2013) Ingested plastic transfers hazardous chemicals to fish and induces hepatic stress. Sci Rep 3:3263

Rodriguez-Narvaez OM, Peralta-Hernandez JM, Goonetilleke A, Bandala ER (2017) Treatment technologies for emerging contaminants in water: a review. Chem Eng J 323:361–380

Ruhoy IS, Daughton CG (2008) Beyond the medicine cabinet: an analysis of where and why medications accumulate. Environ Int 34(8):1157–1169

Salimi M, Esrafili A, Gholami M, Jafari AJ, Kalantary RR, Farzadkia M, Kermani M, Sobhi HR (2017) Contaminants of emerging concern: a review of new approach in AOP technologies. Environ Monit Assess 189(8):414

Sauvé S, Desrosiers M (2014) A review of what is an emerging contaminant. Chem Cent J 8(1):1–7
Saxena G, Purchase D, Bharagava RN (2020a) Environmental hazards and toxicity profile of organic and inorganic pollutants of tannery wastewater and bioremediation approaches. In: Bioremediation of industrial waste for environmental safety. Springer, Singapore, pp 381–398
Saxena G, Kishor R, Bharagava RN (2020b) Application of microbial enzymes in degradation and detoxification of organic and inorganic pollutants. In: Bioremediation of industrial waste for environmental safety. Springer, Singapore, pp 41–51
Saxena G, Thakur IS, Kumar V, Shah MP (2020c) Electrobioremediation of contaminants: concepts, mechanisms, applications and challenges. In: Combined application of physico-chemical & microbiological processes for industrial effluent treatment plant. Springer, Singapore, pp 291–313
Saxena G, Goutam SP, Mishra A, Mulla SI, Bharagava RN (2020d) Emerging and ecofriendly technologies for the removal of organic and inorganic pollutants from industrial wastewaters. In: Bioremediation of industrial waste for environmental safety. Springer, Singapore, pp 113–126
Saxena G, Kishor R, Saratale GD, Bharagava RN (2020e) Genetically modified organisms (GMOs) and their potential in environmental management: constraints, prospects, and challenges. In: Bioremediation of industrial waste for environmental safety. Springer, Singapore, pp 1–19
Saxena G, Purchase D, Mulla SI, Bharagava RN (2020f) Degradation and detoxification of leather tannery effluent by a newly developed bacterial consortium GS-TE1310 for environmental safety. J Water Process Eng 38:101592
Schwarzenbach RP, Escher BI, Fenner K, Hofstetter TB, Johnson CA, Von Gunten U, Wehrli B (2006) The challenge of micropollutants in aquatic systems. Science 313(5790):1072–1077
Siyal AA, Shamsuddin MR, Khan MI, Rabat NE, Zulfiqar M, Man Z et al (2018) A review on geopolymers as emerging materials for the adsorption of heavy metals and dyes. J Environ Manage 224:327
Smital T (2008) Acute and chronic effects of emerging contaminants. In: Emerging contaminants from industrial and municipal waste. Springer, Berlin, Heidelberg, pp 105–142
Stackelberg PE, Gibs J, Furlong ET, Meyer MT, Zaugg SD, Lippincott RL (2007) Efficiency of conventional drinking-water-treatment processes in removal of pharmaceuticals and other organic compounds. Sci Total Environ 377(2–3):255–272
Stasinakis AS, Thomaidis NS, Arvaniti OS, Asimakopoulos AG, Samaras VG, Ajibola A et al (2013) Contribution of primary and secondary treatment on the removal of benzothiazoles, benzotriazoles, endocrine disruptors, pharmaceuticals and perfluorinated compounds in a sewage treatment plant. Sci Total Environ 463:1067–1075
Stearman GK (2003) Pesticide removal from container nursery runoff in constructed wetland cells. J Environ Qual 32:1548–1556
Subedi B, Balakrishna K, Sinha RK, Yamashita N, Balasubramanian VG, Kannan K (2015) Mass loading and removal of pharmaceuticals and personal care products, including psychoactive and illicit drugs and artificial sweeteners, in five sewage treatment plants in India. J Environ Chem Eng 3(4):2882–2891
Sui Q, Cao X, Lu S, Zhao W, Qiu Z, Yu G (2015) Occurrence, sources and fate of pharmaceuticals and personal care products in the groundwater: a review. Emer Contam 1(1):14–24
Syranidou E, Christofilopoulos S, Kalogerakis N (2017) Juncus spp.—the helophyte for all (phyto)-remediation purposes? N Biotechnol 38:43. https://doi.org/10.1016/j.nbt.2016.12.005
Teuten EL, Saquing JM, Knappe DR, Barlaz MA, Jonsson S, Björn A et al (2009) Transport and release of chemicals from plastics to the environment and to wildlife. Philos Trans Roy Soc 364 (1526):2027–2045
Thiere G, Milenkovski S, Lindgren PE, Sahlén G, Berglund O, Weisner SE (2009) Wetland creation in agricultural landscapes: biodiversity benefits on local and regional scales. Biol Conserv 142(5):964–973
Thompson JN (2005) The geographic mosaic of coevolution. University of Chicago Press, Chicago
Thompson RC, Olsen Y, Mitchell RP, Davis A, Rowland SJ, John AW et al (2004) Lost at sea: where is all the plastic? Science 304(5672):838–838

Tijani JO, Fatoba OO, Petrik LF (2013) A review of pharmaceuticals and endocrine-disrupting compounds: sources, effects, removal, and detections. Water Air Soil Pollut 224(11):1770

Tjandraatmadja G, Cook S, Pollard C, Sheedy C, Gozukara Y, Marleni N, Diaper C (2010) Sources of priority contaminants in domestic wastewater. In: Ozwater 10 Conference. Australian Water Association, Brisbane. Electronic version https://publications.csiro.au/rpr/pub?list=BRO&pid=csiro:EP092360&sb=RECENT&n=20&rpp=25&page=119&tr=5467&dr=all&dc4.browseYear=2010

Topp E, Monteiro SC, Beck A, Ball Coehlo B, Boxall AB, Duenk PW, Kleywegt S, Lapen DR, Payne M, Sabourin L, Li H, Metcalfe CD (2008) Runoff of pharmaceuticals and personal care products following application of biosolids to an agricultural field. Sci Total Environ 396:52–59

Tran NH, Reinhard M, Gin KYH (2018) Occurrence and fate of emerging contaminants in municipal wastewater treatment plants from different geographical regions-a review. Water Res 133:182–207

Trinh T, Van Den Akker B, Stuetz RM, Coleman HM, Le-Clech P, Khan SJ (2012) Removal of trace organic chemical contaminants by a membrane bioreactor. Water Sci Technol 66 (9):1856–1863

Williams JB (2002) Phytoremediation in wetland ecosystems: Progress, problems, and potential. Crit Rev Plant Sci 21:607–635

Wright SL, Thompson RC, Galloway TS (2013) The physical impacts ofm microplastics on marine organisms: a review. Environ Pollut 178:483–492

Yang G, Fan M, Zhang G (2014) Emerging contaminants in surface waters in China—a short review. Environ Res Lett 9(7):074018

Zhang Y, Zhou JL (2008) Occurrence and removal of endocrine disrupting chemicals in wastewater. Chemosphere 73(5):848–853

Zhou JL, Zhang ZL, Banks E, Grover D, Jiang J (2009) Pharmaceutical residues in wastewater treatment works effluents and their impact on receiving river water. J Hazard Mater 166 (2–3):655–661

Application of Biochar for Sustainable Development in Agriculture and Environmental Remediation

6

Izharul Haq, Anshu Singh, Devki, and Ajay S. Kalamdhad

6.1 Introduction

The early developments have largely focused on the use of biochar as a soil amendment in agriculture, but other applications in environmental remediation. Thus, this chapter offers comprehensive and updated information related to production of biochar, its use for agriculture sustainability as well as environmental remediation. The special structure and greater surface area, the high load density of the biochar help to absorb various soil contaminants. It also stabilises biomass and native soil organic matter (SOM) which enhances soil aeration, improves microbial activity and immobilises nitrogen which together reduces the emission of major greenhouse gases, i.e. CH4, CO2 and N2O. The role of biochar in developing sustainable development in the agriculture system is immense, and so is its potential to moderate climate change, which is far beyond its use in agriculture.

Biochar, a by-product of the pyrolysis process, is a biomass-derived black carbon intended for use as a soil modification. As a soil change, it is mainly used to improve soil nutrient status, C storage and/or filtration of percolating soil water. Biochar has an intrinsic energy value that can be used to increase the energy output of pyrolysis. Research has shown, however, that the use of biochar in soil can be more beneficial as it can increase soil organic carbon (SOC), boost the supply of nutrients to plants and thus enhance plant growth and soil physical, chemical and biological properties.

I. Haq (✉) · A. S. Kalamdhad
Department of Civil Engineering, Indian Institute of Technology Guwahati, Guwahati, Assam, India
e-mail: izhar@iitg.ac.in

A. Singh
Defence Institute of Bio-energy Research-DRDO, Haldwani, Uttarakhand, India

Devki
Department of Microbiology, JECRC University, Jaipur, Rajasthan, India

I. Haq, A. S. Kalamdhad (eds.), *Emerging Treatment Technologies for Waste Management*, https://doi.org/10.1007/978-981-16-2015-7_6

Biochar is believed to support crop production through three primary mechanisms: direct alteration of soil chemistry through its intrinsic elemental and compositional structure (e.g. availability of nutrients and light organic molecules and decrease in soil acidity); provision of chemically active surfaces that modify soil nutrient dynamics or otherwise catalyse useful soil reactions (e.g. increasing the cation exchange capacity of the soil) and modifying physical character of the soil in a way that benefits root growth and/or nutrient and water retention and acquisition (e.g. reduction of soil bulk density, creation of stable macro-aggregates, improved tilth, provision of shelter for microorganisms).

Biochar has also been studied as a means of more sustainable catalytic converting (Kastner et al. 2012) activated carbon (Park et al. 2013), novel magnetic adsorbents (Chen et al. 2011), phosphate traps for soil remediation (Cao et al. 2009) and hydrogen storage (Sevilla and Mokaya 2014) using biorenewable materials. These green technological advances will increase the viability of pyrolysis-based bioenergy and foster economic growth in rural communities, while offering sustainable alternatives to meet consumer needs.

While increasingly recorded research has shown the beneficial effects of biochar on agricultural growth in recent years, the remedial aspect of biochar is missing. We, therefore, mentioned in this chapter the production of biochar and the use of biochar for sustainable development in agriculture and environmental rehabilitation.

6.2 Production of Biochar

Biochar processing from waste biomass is inexpensive and advantageous because this method offers renewable alternatives to fossil fuels, and the biochar commodity can also be used to combat climate change (Barrow 2012). Waste biomass has been commonly used to manufacture biochar from a number of sources (Cantrell et al. 2012), such as crop residues, forest waste, animal manure, food process (Li et al. 2013; Ahmad et al. 2014). Biochar processing can also produce oil and gas products that could be used as a source of renewable energy (Windeatt et al. 2014). Biomass thermal decomposition into biochar, oil and gaseous materials can be accomplished using a range of approaches including pyrolysis, gasification, hydrothermal carbonisation (HTC), torrefaction and conventional carbonisation methods (Stelt et al. 2011). Gasification is distinct from the general pyrolysis process (Ahmad et al. 2014). For gasification, biomass transforms into CO and H2 rich gases by reacting with biomass at high temperatures (N700 °C) in a regulated oxygen and/or steam system (Mohan et al. 2006). However, the yield of char from biomass gasification is very low, which is not suggested for the development of biochar. Hydrothermal carbonisation provided the final carbonaceous material (hydrochar) from a wet feedstock without an energy intensive pre-drying phase (Kumar et al. 2016), whereas extra thermal energy is usually needed for post-treatment of HTC streams, such as the separation of solid and liquid products (Kambo and Dutta 2015). Torrefaction is a pre-treatment process that has been extensively studied to transform biomass into carbon-rich solid fuel, such as biochar (Chen et al. 2015). It could also extract volatiles through different decomposition reactions in order to reduce the main limitations of biomass, improve the quality of biomass and adjust the combustion behaviour (Yu et al. 2017). However, the high inorganic metallic content of ash remains a significant challenge for the comprehensive use of biochar during the torrefaction process (Kambo and Dutta 2015). Biochar can be produced by thermochemical decomposition of biomass at 200–900 °C in the absence of oxygen, commonly known as pyrolysis (Ahmad et al. 2014; Demirbas and Arin 2002). Pyrolysis is generally divided into fast, intermediate and slow depending on the time of residence and temperature (Mohan et al. 2006). Fast pyrolysis with a very short retention time (<2 s) generates more liquid fuel, which is often used to produce bio-oil from biomass yielding around 75% (Mohan et al. 2006). Slow and intermediate pyrolysis processes with a residence time of a few minutes to several hours or even days are generally preferred for biochar production (25–35%) (Brown et al. 2009). The conventional way to synthesise biochar is through slow pyrolysis. Compared to HTC and torrefaction, it has higher carbon content (Wilk and Magdziarz 2017). In addition, slow pyrolysis results in lower liquid fuel yields and higher biochar yields compared to other thermal chemical processes (Windeatt et al. 2014; Yu et al. 2017). Biochar production is stated to be a complicated physicochemical process that affects the inherent inorganic substances and the pyrolysis mechanisms and interactions of major components such as cellulose, hemicellulose and lignin in biomass (Lian and Xing 2017). In general, the characteristics of biochar are determined by many variables, such as biomass feedstock, pyrolysis temperature, pyrolysis temperature residence and pyrolysis atmosphere (Windeatt et al. 2014). Preparation temperature and parent material (raw material, precursor, feedstock) are two major factors regulating the properties of biochars (Lian and Xing 2017; Tang et al. 2013). Lee et al. (2017) examined the ability of agricultural waste as a feedstock for slow pyrolysis biocharging (50 mL min − 1, N2 at 500 °C). The relationship between the thermochemical properties of the feedstocks and the biochar developed was calculated using various characterisation methods. In particular, several variables in the production of biochar, including pyrolysis conditions and feedstock types, can affect its exact function in environmental management (Ahmad et al. 2014; Safaei Khorram et al. 2016). The type of feedstock had a greater impact on the efficiency of biochars for metal removal than the pyrolysis temperature (Higashikawa et al. 2016). It has been presumed that the porous structure and sorption characteristics of the activated biochar are based on the physicochemical properties of the precursor biochar as well as the activation methods (Park et al. 2013). Actually, activated carbon, which is carbon dioxide treated with oxygen to increase microporosity and surface area, is the most widely used carbonaceous sorbent (Ahmad et al. 2014). In particular, the aromatic structures of the biochar can play an important role in the formation of porous networks during the activation process. In short, feedstock types and production processes may lead to biochars with different properties and therefore different effects, further studies on the effects of different types of biochars are required to systematically evaluate their effects on agriculture and environmental remediation (Saifullah et al. 2018).

6.3 Biochar and Microorganism

Bonanomi et al. (2015) reported that biochar is effective against both air-borne (e.g. *Botrytis cinerea* and different types of powdery mildew) and soil-borne pathogens (e.g. *Rhizoctonia solani* and species of *Fusarium* and *Phytophthora*). The use of biochar derived from citrus wood was capable of regulating air-borne grey mould, *Botrytis cinerea* on *Lycopersicon esculentum*, *Capsicum annuum* and *Fragaria ananassa*. While there is a lack of published data on the effects of biochar on soil-borne pathogens, evidence from Elmer et al. (2010) has shown that regulation of such pathogens may be possible. The addition of biochar in 0.32, 1.60 and 3.20% (w/w) of *Asparagus* soils infested with *Fusarium* increased the biomass of *Asparagus* plants and reduced *Fusarium* root rot disease (Elmer et al. 2010). In the same way, *Fusarium* root rot disease in *Asparagus* was also reduced by biochar inoculated with mycorrhizal fungi (Thies and Rillig 2009). A study of the suppression of bacterial wilt in tomatoes showed that biochar produced from municipal organic waste reduced the incidence of disease in *Ralstonia solanacearum* infested soil (Nerome et al. 2005). Ogawa (2006) promoted the use of modified biochars and biochars to manage soil diseases caused by bacteria and fungi. The mechanism of disease suppression has been due to the presence of calcium compounds, as well as changes in the physical, chemical and biological characteristics of the soil.

According to Mackie et al. (2015), a mixture of compost and biochar-compost has increased microbial biomass, phospholipid fatty acids (PLFAs) and various enzyme activities, i.e. phosphatase, arylsulphatase, as well as an increase in bacterial taxon, i.e. *Actinobacteria,* 5–007-*Proteobacteria,* β-*Proteobacteria, Firmicutes* and *Gemmatimonadetes.*

Diazotrophs, a specialised group of bacteria (and Archaea), have a common function: they possess the enzyme nitrogenase and the ability to reduce atmospheric N2 to NH3, which can be nitrified ($NO3^-$) prior to plant uptake. These diazotrophs act as either free-living N-fixing soil bacteria (e.g. *Azospirillum* sp.; *Azotobacter* sp.) or mutualists inside plants (e.g. *Rhizobia* forming legume nodules or *Actinorhizal* associations of *Frankia* sp.) (DeLuca et al. 2006). Free-living bacteria are less successful at N2 fixation than symbiont rhizobia, i.e. 5 compared to 3–206 kg N2 ha − 1 y − 1, respectively. Despite the ubiquitous presence of these free-living N2 soil bacteria, few studies have shown that the use of activated carbon (Berglund et al. 2004) and biochar (Gundale and DeLuca 2006) will improve nitrification. Biochar micro-environment can also provide a favourable niche (fine structural pores) in which the concentration of oxygen decreases; low oxygen voltages with Fe and Mo ions are needed in order for nitrogenase to work effectively (Thies and Rillig 2009). Biochars are typically low in inorganic-N2 and this can give diazotrophs a competitive advantage for the colonisation of large surface biochars. This element, combined with the capacity of biochar for NH4+ exchange with soil solution, could alter the availability of soil-N2 to the plant and stimulate nodulation and fixation.

The role of biochar in the adsorption and protection of chemical signalling molecules derived from plants, such as node factors that enhance root nodulation through *Rhizobia*, has been suggested (Thies and Rillig 2009). Evidence exists to show that increasing rates of biochar application to soil can increase the proportion of N2 derived from *Phaseolus vulgaris* fixation and this increases in yield (Rondon et al. 2007). These beneficial effects were related to increased availability of Mo and B (source not determined) with an increase in soil pH. *Rhizobia* has increased role in neutral pH soils, whereas increasing alkalinity in acid soil improves nodulation and fixation.

The different functional groups within 'soil fungi,' i.e. saprophytes, pathogens and mycorrhizae, react differently to biochar applications (Thies and Rillig 2009). Saprophytes theoretically alter the persistence of biochar soil through decomposition. Their invasive hyphal growth and extracellular enzymatic ability enable them to colonise biochar pores. Soil pathogenic fungi are extensive and especially important in the management of plant diseases. However, the impact of biochar on soil pathogens (population structure and function) appears to be minimal. Matsubara et al. (2002) have shown that the resistance of asparagus seedlings to *Fusarium oxysporum* has been improved by the use of biochar.

While biochars themselves can contain only small amounts of plant nutrients to support mycorrhizal fungi (Lehmann et al. 2003a, 2003b; Gundale and DeLuca 2006), it is proposed that biochars increase the availability of soil nutrients by altering the physical-chemical properties of the soil. These changes in themselves alter the supply of nutrients and possibly mycorrhizal abundance, and modify the local nutritional balance, e.g. N/P ratios, thereby impacting root colonisation (Miller et al. 2002). The idea that the availability of nutrients or improved efficiency of use, due to the existence of biochar, needs a critical review to determine the function of soil microorganisms.

6.4 Application of Biochar

Biochar has a vast number of applications in agricultural and environment field, some of are described below:

6.4.1 Increased Soil Fertility

Biochar soil additions cause pH, electrical conductivity (EC), cation exchange capacity (CEC) and nutrient levels to change (Gundale and DeLuca 2006). The increase in soil pH induced by biochar application is not surprising given the well-known use of materials such as wood ash for pH modification and the availability of nutrients, especially P and K. Elevated CECs are caused by a rise in the load density per unit surface of organic matter, which is equivalent to a higher degree of oxidation, or an increase in the surface area for cation adsorption, or a combination of both.

Liang et al. (2006) reported an increase in organic matter adsorption and load density (CEC per unit surface area) in anthrosols due to black carbon particle surface oxidation. Ammonium leaching, although from greenhouse biochar experiments, has been reduced (by 60%) (Lehmann et al. 2003a, 2003b), although in some cases N2O emissions may be reduced (Spokas and Reicosky 2009). Other research using soils in Amazonian field studies have reported that biochar can act as an adsorber to minimise N leaching and increase N use quality (Steiner et al. 2008). A large area of research will appear to be needed to ensure that the pyrolysis process and the feedstock used have the potential to optimise soil N for plant availability while reducing leaching. Efficiency of N use will be an absolute requirement to sustain future population growth. In order to do this, much more needs to be considered with regard to the mechanistic effect of biochar (direct and indirect) on nitrification and Na-validity. Expectation of enhanced soil fertility benefits emerges from terra preta studies that involve high proportions of black carbon (Haumaier and Zech 1995; Glaser et al. 2002; Lehmann et al. 2003a, 2003b). The apparent fertility of the *Terra preta* is usually due to the high content of soil organic matter—organic matter contributes to the preservation of water, soil solution and cation and the retention ability of the aged biochar itself for nutrients and water. Black carbon present in terra preta is thought to come from partially-combusted biomass residues resulting from a variety of anthropogenic activities, including cooking and field fires. An especially striking aspect is the stronger relationship between soil carbon content and soil CEC in these soils compared to neighbouring soils, suggesting that biochar represents a higher proportion of soil carbon (Liang et al. 2006). Since CEC is representative of the ability to maintain key soil nutrient cations in a plant-available form and mitigate leaching losses, this is cited as a key factor where variations in crop productivity are observed. High levels of biocharging in the tropical environment have been correlated with increased plant uptake of P, K, Ca, Zn and Cu (Lehmann and Rondon 2006a, 2006b). In comparison to mainstream chemical fertilisers, biochar often contains bioavailable elements, such as selenium, which have the potential to help increase crop development.

Much speculation has been made about the possible effects of biochar on soil microbial activity, which Steiner has studied in depth in the sense of *Terra preta* (Steiner et al. 2003). Assuming that plant inputs and thus microbial substrate remain unchanged, enhanced microbial activity alone will minimise soil organic matter. This is, however, contrary to the finding in *Terra preta*, where soil organic matter is usually higher than in comparable surrounding soils (Liang et al. 2006). However, a shift in the balance of microbial activity between different functional groups could gain crop nutrition, specifically the enhancement of mycorrhizal fungi (Ishii and Kadoya 1994), which could lead to higher net primary productivity and carbon production. Relatively comprehensive literature records the stimulation of indigenous arbuscular mycorrhizal fungi by biochar, which has been expressed in plant growth Rondon et al. (2007), Nishio (1996). Warnock et al. (2007) analysed this literature in some detail, suggesting four mechanistic hypotheses, the most plausible of which were mixed nutrient, water and CEC effects.

6.4.2 Water Retention in Soil

Both the mineral and organic components of the soil contribute to the soil water holding capacity, but only the latter can be actively controlled. Water is kept tighter in small pores, so clay soils hold more water. Lower soil bulk density commonly correlated with higher soil organic matter is a partial indicator of how organic matter changes soil structure and pore size distribution. Many researches, in which the impact of biochar on crop yield was assessed, cited moisture retention as a key factor in the results. Given that the pore size of the biochar is relatively constant, while that of the mineral soil is determined primarily by its texture, it can be predicted that the available moisture in sandy soil will increase, will have a neutral effect on medium textured soils and will decrease the available moisture in clay soils. Any impact of the size of the biochar particle could be short-lived, as it appears to be relatively rapidly disintegrating into fine fractions. Experimentally, the usual technique for determining pore size characteristics is the moisture-release curve, which shows how easily soil moisture is drawn from soil under increasing stress.

The method is well adapted for distinguishing differences between soils of contrasting texture, but its sensitivity may be less adequate for discriminating against the effect of contrasting management at a particular location: high levels of replication may be required to demonstrate a significant impact of a management intervention of a realistic magnitude. In a more recent study (Gaskin et al. 2007), moisture-release curves were determined using samples of loamy sand soil from field experiments where biochar was applied at rates of up to 88 t ha-1. In soils where biochar was applied at concentrations of up to 22 t ha-1, there was no difference from non-modified soil, although at the maximum rate the difference was substantial at water potentials in the range of 0.01–0.20 MPa. The mean volumetric effect of the water content was doubled by the addition of biochar at the maximum potential. Soil temperature, soil cover, evaporation and evapotranspiration influence the available soil water. The comparison of the actual volumetric water content between biochar-modified and control soils in field experiments can, therefore, be confounded by any indirect impact of biochar on plant growth and soil thermal properties. Soil organic matter increases soil water holding capacity and in the biochar-enriched terra preta with their associated higher levels of soil organic matter, Glaser (2002) reported a water retention capacity that was 18% higher than in adjacent soils in which charcoal was low or absent. This was likely a combined effect of the char itself and the higher levels of organic matter that this promotes.

The effect of biochar (BC) and hydrochar (HTC) on water retention characteristics (WRC) as well as on the wettability of sandy soils has been documented using lab and field studies. Sandy soils with varying quantities of organic matter were mixed with BCz (feedstock maize) and HTC. The total added was 1, 2,5 and 5 wt percent, respectively. The mixtures were packed in 100 cm3 of soil columns. In the field campaign, the same quantities of BCf (feedstock beech wood) were applied to the soil. Samples of undisturbed soil were taken 6 months after incorporation. Accessible water capacity (AWC) was calculated for these field samples. The WRC was measured in the pressure head range from saturation to wilting point (15,848 cm) for the packed soil columns. The amount of water repellency was calculated for all samples using the water drop penetration time test.

6.4.3 Increased Crop Yield

Glaser et al. (2002) analysed a number of early studies performed in the 1980s and 1990s, which appeared to demonstrate major impacts of low-carbon additions (0.5 t ha-1) on different plant species. Higher rates appeared to inhibit plant growth. In later studies, the combination of higher biochar application rates together with NPK fertiliser increased crop yields on tropical Amazonian soils (Steiner et al. 2008) and semi-arid soils in Australia (Ogawa 2006). Due to year-on-year variance in climate and its effect on short-term dynamics, the results of a number of recent field experiments have not yet been reported though generating data. The essence and mechanistic basis for interactions between crops, soil type, biochar feed stock and production method and application rate will have to be understood in order to increase the predictive potential for soil biochar output and to open up the possibility for large-scale deployment.

6.4.4 Restoring the Soil Properties

6.4.4.1 Effects of Biochar on Soil Physical Properties

Biochar is a long-term adaptation technique that enhances soil physical and chemical properties that contribute to soil fertility. Possible mechanism for improving yield by increasing porosity and water storage space, as well as reducing bulk density (Jeffery et al. 2011; Lu et al. 2014; Nelissen et al. 2015). For e.g., ash content in biochars ranged from 0.35 to 59.05%, which was rich in available nutrients, in particular cationic elements such as K (0–560 mmol kg − 1), Ca (3–1,210 mmol kg − 1), Mg (0–325 mmol kg − 1) and Na (0–413 mmol kg − 1) (Rajkovich et al. 2012).

Soil Structure

Biochar incorporation into soil can alter physical properties such as structure, pore size, bulk density, soil aeration, water holding ability, plant growth and soil fertility. The introduction of biochar into soil will alter the physical properties of soils such as structure, pore size distribution and density with logical consequences for soil aeration, water holding capability, plant growth and soil workability. Sohi et al. (2009) suggested an analogy between the effect of the incorporation of biochar and the observed increase in soil water repellency due to burning. The re-arrangement of amphiphilic molecules by fire heat, as suggested by Doerr et al. (2000), does not affect the soil, but may have an effect on the biochar itself during pyrolysis. In addition, soil hydrology can be influenced by the partial or complete blockage of soil pores by the smallest fraction of the particle size of the biochar, thus reducing the rate of water infiltration. Liu et al. (2014) stated that when 40 t ha-1 biochar was applied, the soil water stable aggregate (>0.25 mm) in the 0–15 cm soil layer increased considerably compared to other treatments.

6.4.4.2 Porosity, Aggregate Stability, Soil Surface, Bulk Density, Penetration Resistance Porosity

The application of biochar may boost the physical properties of the soil, in particular its high porosity and wide inner surface area. Porosity depends on the carbon temperature and pyrolysis activation of up to 750 °C and the parent feed stock forms. Pore sizes in biochar have been reported to range from <2 nm to >50 nm, with an increase in the small diameter pore fraction as the pyrolysis temperature increases. However, the high porosity of carbon dioxide particles does not inherently increase the amount of plant-available water in the soil, since pore sizes <200 nm tend to retain water at a higher water potential than those produced by plants (Lal and Shukla 2004). Herath et al. 2013 concluded that biochar increased macroporosity in the soil of Tokomaru and mesoporosity in the soil of Egmont.

Soil Density

Biochar has a much lower mass density than mineral soils; therefore, the application of biochar will reduce the overall mass density of the soil. Tensile strength of hard soil under investigation also decreased with a rising rate of biochar application. Jein and Wang (2013) stated that the application of 5% biochar decreased the bulk density (1.08 mg m-3) from 1.42 mg m-3.

Surface Area

The specific surface area of biochar, which is typically higher than sand and equal to/or higher than clay, would result in a net increase in the total soil-specific surface when applied as an amendment. Evidence suggests that biochar application to soil can increase the overall net surface area of the soil and, as a result, improve soil water retention and soil aeration. The direct effect is related to the broad inner surface of the biochar. Increased soil-specific surface area and physical conditions can also support native microbial communities.

Soil Water

Influence of biochar on soil physical properties will affect soil response to water, aggregation and workability, shrinking dynamics, permeability and soil water retention. This change may be attributable to physical changes in soil where small carbohydrate particles block soil pores and reduce water penetration rates. Glaser et al. (2002) found that Amazonian char rich anthrosols had a field water retention potential of 18%, which is higher than the non-charcoal surrounding soil. The hydrophobic polyaromatic backbone reduces the flow of water into the aggregate pores, resulting in improved aggregate stability and availability of water. The results of this study also suggest that the application of cow manure biochar to sandy soil is not only beneficial for crop growth, but also significantly improved the physico-chemical properties of the coarse soil. Uzoma et al. (2011) reported that the lower bulk density and porous nature of added biochar increased water use efficiency consequent to improvement in field capacity and hydraulic conductivity. Granatstein et al. (2009) also reported an improvement in water holding ability of both sandy and silt loam soil due to the use of biochar.

6.4.4.3 Liming Effect in Soil/Reduced Toxicity and pH

Biochar can act as a liming agent, resulting in increased pH and availability of nutrients for a variety of soil types (Glaser et al. 2002; Lehmann and Rondon 2006a, 2006b). Biochar carbonate concentration promotes liming in soils and can increase the pH of neutral or acidic soils (Van Zwieten et al. 2009). Mbagwu and Piccolo (1997) reported increases in pH of different soils and textures by up to 1.2 pH units from pH 5.4 to 6.6. Tryon (1948) recorded a higher increase in pH in sandy and loamy soils than in clay soils. The pH of the various soils increased more after the use of hardwood charcoals (pH 6.15) than coniferous charcoals (pH 5.15), possibly due to different ash contents of 6.38% and 1.48% respectively (Glaser et al. 2002).

The biochar-associated liming effect may not be suitable for all soil types and plant communities. Increased soil pHs associated with biocharging have resulted in micronutrient deficiencies in agricultural crops (Kishimoto and Sugiura 1985) and forest vegetation (Mikan and Abrams 1995); thus, prior to application, it is necessary to recognise the existence of calcifuge vegetation. In addition, many forest plants, fungi and bacteria thrive in lower pH soils (Meurisse 1976, 1985); thus, altering the pH of the forest soil by adding biochar can result in unfavourable shifts in the upper and lower soil flora. Understanding the interactions between biochar production and application conditions and soil texture, organic matter and pH will be crucial in evaluating the long-term effects of biochar application on forest soils.

Soil pH is an essential factor influencing the supply of biochar nutrients (Silber et al. 2010). The release of PO4− and NH4+ was pH-dependent while the release of K+ and NO3− was not pH-dependent (Zheng et al. 2013). In addition, at pH 2–7, the content of PO4− and NH4+ released from biochars would have decreased with an increase in pH values, whereas that of K+ remained relatively constant (Zheng et al. 2013). Similarly, the initial release of Ca and Mg from corn straw biochar was also pH-dependent, with a rise in releases as the pH decreased from 8.9 to 4.5 (Silber et al. 2010). As a result, pH and lower temperature pyrolysis can increase the availability of N and P, whereas higher temperature pyrolysis can increase the availability of K.

6.4.5 Improve Soil Organic Carbon (SOC)

The soil carbon reservoir, composed of organic and inorganic carbon, is the largest carbon pool in the terrestrial ecosystem and has a gross reserve of approximately 3.3 times that of the atmospheric carbon pool (Wang et al. 1999). Soil organic carbon content (SOC) is also used as an important index for determining potential soil fertility (Spaccini et al. 2001; Dalal et al. 2003) and its complex balance has a direct effect on soil fertility and crop yields.

Human activities, such as high fertilisation rates and intensive crop rotation systems, have resulted in decreases in soil organic matter and carbon/nitrogen ratios and global imbalances in soil carbon pools (Lemenih et al. 2005; Collard and Zammit 2006). Overall decreases in soil microbial communities and microbial imbalances (Bell et al. 1998; Oldeman et al. 2017) have severely reduced the supply of nutrients and soil transformation capability. These changes have resulted in major nutrient imbalances in the soil environment (Parton et al. 1987), which have influenced crop production.

Biochar usually refers to the highly aromatic organic matter derived from the pyrolysis of any solid biomass. It can persist in the environment and plays an important role in global biogeochemical cycling, climate change and environmental systems as a part of the SOC pool (Marin-Spiotta et al. 2014; Brodowski et al. 2006). Biochar is generally referred to as highly aromatic organic matter derived from the pyrolysis of any solid biomass. It can stay in the ecosystem and plays an important role in global biogeochemical cycles, climate change and environmental processes as part of the SOC pool (Marin-Spiotta et al. 2014; Brodowski et al. 2006). Biochar is also known to be a significant source of atmospheric CO2 (Forbes et al. 2006). Furthermore, as a potential source of the highly aromatic portion of soil humus, biochar plays a key role in sustaining and growing the SOC pool and maintaining soil nutrients, improving soil fertility and maintaining the balance of the soil ecosystem (Hart and Luckai 2013; Chan et al. 2008). Microbial soil populations are considered to be acutely sensitive to changes in soil environment (Zhang et al. 2005).

Biochar is alkaline and porous, has a high specific surface area and multiple negative surface loads, and contains high-charge dense materials (Steiner et al. 2008). Increased soil biochar content can alter the soil ecosystem and microbial habitat, altering the biogeochemical cycle of soil carbon. As a material with a high carbon content, the addition of biochar to the soil would directly complement the organic carbon sources required for soil carbon cycling (Yin et al. 2014; Wu et al. 2012). The high stability of the biochar is due to its complex aromatic structure and its physical and chemical defensive effects. As the biochar enters the soil, its stable group significantly enriches the SOC pool and is retained in the soil for a long time, thereby enriching the total amount of organic matter in the soil (Woolf and Lehmann 2012). Labile components of biochar, such as aliphatic carbon speciation, can complement the soil carbon pool in the form of soluble organic matter (Brewer et al. 2009; Schmidt and Noack 2000). Biochar can boost soil water holding capacity (Laird et al. 2010), reduce soil bulk density, promote soil ECE (Cation Exchange Capacity) and pH (Van Zwieten et al. 2009), alter soil biochemical reactions and stimulate soil enzyme activity (Acosta-Martinez and Tabatabai 2000) and promote soil microbial reproduction.

6.4.6 Role of Biochar in Climate Change

6.4.6.1 N_2O and CH_4 Emissions

The main greenhouse gases associated with agriculture are nitrous oxide (N2O) and methane (CH4). Cropland and grassland are an important agricultural source of N2O emissions, whereas paddy fields, livestock waste and enteric fermentation are the major sources of CH4 emissions. When added to soil, biochar will minimise greenhouse gas emissions by dramatically reducing N2O emissions. Emissions of N2O, a greenhouse gas that is nearly 300 times higher than CO2 in terms of global warming potential, have been decreased by 40%. Laboratory studies show that the reduction of N2O emissions from biochar-treated soil depends on soil moisture and soil aeration (Yanai et al. 2007). Greenhouse gas emission reductions could be 12% to 84% higher if biochar is used instead of combusted for energy purposes (Lehman et al. 2006).

The retention of nutrients by biochar can depend on the temperature of the biochar pyrolysis, soil types, fertiliser doses and soil water content. Some studies have shown that the inclusion of biochar in soil effectively decreases N2Oemission from different soils. For example, Rondon et al. (2005) stated that 50% reductions in N2Oemissions were found in soybean systems, while 80% reductions in N2Oemissions were found in grass systems. Similarly, treatment with biochars could minimise N2Oemissions from 1768 to 45–699 μgN2O-N m − 2 h − 1 (Wang et al. 2013) and suppress N2Oemissions between 21.3% and 91.6% (Stewart et al. 2012). However, several studies have confirmed that there is no impact (Cheng et al. 2012) or even increase (Clough et al. 2010) was detected on N_2Oemissions after the application of biochar.

The global warming potential (GWP) of the gas represents two aspects: the efficiency of the molecule in the absorption of incoming solar radiation and its rate of chemical breakdown in the atmosphere. By definition, the global warming potential (GWP) of CO2 is1.0; by contrast, the nitrous oxide GWP is 310. Under anaerobic conditions, N2O is released from soil by means of denitrification, a method in which specialised microbes that obtain energy from nitrate reduction (NO3-) or intermediate gases are dinitrogen-based. However, it appears that nitrifying bacteria, which are generally involved in the conversion of N2 to ammonium (NH4+), i.e. nitrification, can simultaneously denitrify (Bateman and Baggs 2005).

The availability of NH4+ is usually regulated by the climate-driven mineralisation of organic matter, but its concentration is greatly enhanced by the use of nitrogen fertiliser or by the use of dung or slurry in livestock and grassland systems. Regardless of the environment or source, the majority of soil nitrogen is in organic form and N2O emanates through the use of a relatively small and dynamic nitrogen reservoir. Life cycle evaluations quantifying the benefits of biochar-based energy strategies are very heavily dependent on a decrease in N2O emissions that often accompanies the addition of mineral nitrogen fertilisers. Accounting for this effect makes a great difference to the overall analysis of how a biochar to soil strategy impacts on net greenhouse gas balance (Gaunt and Lehmann 2008).

Yanai et al. (2007) showed that 'bio-waste' carbon dioxide was used during the re-wetting of a former grassland soil, high in organic matter, in laboratory incubation (25 °C). Nine-tenths of N2O was suppressed in five-day soil-wetting emission episodes to 73% and 78% water-filled pore space. At a marginally higher pore-filled water region (83%), carbon dioxide had the opposite effect, increasing N2O emissions. The rate of biochar applied used in the study was equivalent to a relatively high application rate of 180 t ha-1 in topsoil.

In arable soil with a much lower C content (2.2% C), Sohi et al. (2009) studied the impact of willow charcoal at a much lower rate of 10 t C ha-1, which was assessed during 20 °C incubation of wet (70% water holding capacity) and re-wetted (20% water holding capacity) soils, with and without simultaneous addition of small amounts of inorganic N (equivalent 75 kg N ha-1). More moderate suppression of 15% was proportionally equivalent for all interventions where there was some response at all (the already-wet soil did not emit significant N2O). After 6 months, the available soil N would have been mostly consumed and the soil would have been well balanced. A second inorganic N addition (without new charcoal) at this time showed no difference in N_2O emissions between amended and control soils.

It is currently estimated that 1325% of N in N of corn fertiliser is converted to N2O emissions (Wang 2008). Biochar reportedly decreases N2O soil emissions resulting from the use of N fertilisers (Yanai et al. 2007; Rondon et al. 2006; Van Zwieten et al. 2009). A laboratory research in Japan (Yanai et al. 2007) showed that soils changed with 10 wt percent of soil as biochar suppressed 89% of N2O emissions. In the meantime, laboratory incubation studies (Van Zwieten et al. 2009) have shown that soils modified with biochar from poultry litter emit approximately 40%–80% less N2O than control. However, the same study found that yard waste biochar generated at lower temperatures increased N_2O emissions by 100%. These findings show that not all biochars can minimise N2O emissions equally (Van Zwieten et al. 2009). For this study, the baseline scenario assumes that biochar processing takes place under conditions such that soil N2O emissions from N fertiliser applications are reduced by 50%. Therefore, 0.394 kg of N2O emissions to the air would be avoided with each tonne of biochar added. Sensitivity analysis also looks at the effects of varying soil N2O emissions.

Despite some claims that good management of compost piles results in minimal or zero CH4 or N2O emissions, numerous studies have found that even well maintained compost piles (turning, aeration, proper moisture content) emit quantifiable quantities of CH4 and N2O. The CH4 is produced by microorganisms when the biomass is processed under anaerobic conditions. N2O is produced during the decomposition of biomass by denitrification and nitrification processes. Studies of yard waste compost (grass clippings, green leaves and brush) have reported that 0.030 kg and 2.79 kg of N2O and CH4 are released, respectively, per tonne of dry organic material (Lehmann et al. 2006) assuming moisture content.

6.4.6.2 Carbon Sequestering

Carbon sequestration is the capture and preservation of carbon to prevent its release into the atmosphere. Studies indicate that biochar sequestration of approximately

50% of the available carbon in the biomass feedstock is pyrolysed, depending on the feedstock type (Lehmann et al. 2006). Carbon sequestration in agricultural soils is also promoted as a realistic solution to slow down the rate of CO2 rise in the atmosphere (Mekuria and Noble 2013). Over the last two to three decades, a variety of land and crop management activities have been promoted to recover organic soil carbon and reduce the net CO2 emissions from agricultural systems in the tropics (Smith et al. 2008; Sohi and Shackley 2009; Woodfine 2009). Practices for restoring organic soil carbon and minimising net CO2 emissions include, but are not limited to, crop rotation, avoiding the use of bare fallow, conservation of tillage, management of organic inputs such as manure and crop residues, restoration of degraded agricultural land, water management and agroforestry (Banger et al. 2010; Batlle-Bayer et al. 2010; Shafi et al. 2010; Wang et al. 2010; Bangroo et al. 2011; Fallahzade and Hajabbasi et al. 2012).

Studies have shown that smallholder farmers can minimise greenhouse gas emissions and preserve carbon stocks in soil and vegetation at relatively low cost by adopting crop and land management practices (Nair et al. 2007). However, the study by Giller et al. (2009) and Sanchez (2000) established a number of constraints, including a low degree of mechanisation within the smallholder system, a lack of suitable implements, an issue of weed control under the no-till system and a lack of adequate technical knowledge that hinders the large-scale adoption of practices by smallholder farmers. Woodfine (2009) added that the availability of funds to catalyse the initial transition is a crucial obstacle for realising the implementation of several mitigation activities. Operational improved crop and land management practices can require more manual labour than traditional agricultural practices (Suprayogo et al. 2010a, 2010b). Optimising these advantages and disadvantages can be a difficult job, which in itself is a drawback as there is a lack of skilled staff and extension workers to provide information and advice to farmers. In addition, the time pattern of influence to reduce the rise in CO2 varies between practices and, in most situations, the decrease in CO2 emissions as a result of the recommended practices is temporary (Smith et al. 2008). For example, a study in Kenya showed that the residual effect of manure applied for 4 years only lasted another 7 or 8 years when assessed by yield, SOC and OlsenP (Suprayogo et al. 2010a, 2010b). The consequences of no-till practices are often quickly reversed and contribute to the release of CO2 into the environment as soon as the system has begun to be interrupted.

6.4.7 Bioenergy from Agricultural and Forestry Residues

Ogawa (2006) proposed a scheme for carbon sequestration by forestation and carbonisation. The scheme concerned fast-growing plantation tree species fixing atmospheric CO2, with items consisting not only of traditional wood, wood chips and pulp, but also of the transfer of waste and residues to the carbonisation process and the re-application of this stabilised carbon back to the plantation soil. This method has been formally suggested under the Clean Development Framework in Sumatra, Indonesia. In Minas Gerais, Brazil, a current commercial project says, under CDM, a carbon credit for the replacement of coal-derived coke for the smelting of iron by pyrolysed eucalyptus plantation. The project produces 300,000 t-1 of charcoal. The 'fines' of carbon dioxide, which account for about 5% of the substance, are used for the manufacture of briquettes rather than for use in soils. In Australia, the potential for integration of oil production from mallee trees with the processing of wood waste for the production of biochar for use in crop production has been investigated examined (McHenry 2009).

Seifritz (1993) measured the size and expense of the carbon gain that could be realised by converting plantation forest biomass directly to biochar. The scenario included no energy capture in the conversion, demonstrating instead the net primary efficiency that is retained by the cultivation process and the long life of charcoal relative to the nature and fate of conventional timber products. In the tropical sense, 'slash and char' scenarios have been addressed, where one-off biochar inputs are made during the conversion of land from forest to agriculture (Steiner 2006) or maybe 'crop and char' with a positive feedback loop between one-off, periodic or rotational biochar inputs and an increase in the productivity of biomass and feedstock. However, as examples of viable, village-scale bioenergy based on gasification technology are growing in developing countries, it is conceivable that technological progress in tandem with increased income from 'crop and charcoal' practices may eventually lead to combined biochar production and energy capture on the same scale.

In the absence of sufficient technological growth, charcoal processing and production could not deliver the same benefits to human health as, for example, replacing current biomass burning practices with basic yet cleaner and more efficient combustion technology. Submicron soot particles formed by condensation reactions in combustion gas streams are the most recalcitrant types of black carbon but, despite the relatively small amount of carbon involved, can have a major effect on the albedo of both the global atmosphere and the ice caps, alter the radiative balance and intensify climate change (Ramanathan and Carmichael 2008). At present, global emissions of soot are expected to decline as rural biomass consumers in developing countries turn to clean fossil fuel combustion (Streets et al. 2004). Charcoal processing produces less soot than open burning, but despite the potential size of future biochar production, its future contributions to the global soot inventory have not been formally evaluated.

6.5 Conclusion

In the light of the available literature relating to each and every phase of biochar development and its application, we have concluded that there is an urgent need for low-cost biochar-based waste material for the sustainability of agriculture and environmental remediation, as well as for the daily erosion and infertility of agricultural land.

Acknowledgements The authors acknowledge Indian Institute of Technology Guwahati for providing facilities to do the literature review.

References

Acosta-Martinez V, Tabatabai M (2000) Enzyme activities in a limed agricultural soil. Biol Fertil Soils 31:85–91

Ahmad M, Rajapaksha AU, Lim JE, Zhang M, Bolan N, Mohan D et al (2014) Biochar as a sorbent for contaminant management in soil and water: a review. Chemosphere 99:19–33

Banger K, Toor GS, Biswas A, Sidhu SS, Sudhir K (2010) Soil organic carbon fractions after 16-years of applications of fertilizers and organic manure in a Typic Rhodalfs in semiarid tropics. Nutr Cycl Agroecosyst 86(3):391–399

Bangroo SA, Kirmani NK, Ali T, Wani MA, Bhat MA, Bhat MI (2011) Adapting agriculture for enhancing ecoefficiency through soil carbon sequestration in agroecosystem. Res J Agric Sci 2:164–169

Barrow CJ (2012) Biochar: potential for countering land degradation and for improving agriculture. Appl Geogr 34:21–28

Bateman EJ, Baggs EM (2005) Contributions of nitrification and denitrification to N2O emissions from soils at different water-filled pore space. Biol Fert Soils 41:379–388

Batlle-Bayer L, Batjes NH, Bindraban PS (2010) Changes in organic carbon stocks upon land use conversion in the Brazilian Cerrado: a review. Agric Ecosyst Environ 137(1–2):47–58

Bell M, Moody P, Connolly R, Bridge B (1998) The role of active fractions of soil organic matter in physical and chemical fertility of Ferrosols. Soil Res 36:809–820

Berglund LM, DeLuca TH, Zackrisson TH (2004) Activated carbon amendments of soil alters nitrification rates in scots pine forests. Soil Biol Biochem 36:2067–2073

Bonanomi G, Ippolito F, Scala F (2015) A" black" future for plant pathology? Biochar as a new soil amendment for controlling plant diseases. J Plant Pathol 1:97(2)

Brewer CE, Schmidt-Rohr K, Satrio JA, Brown RC (2009) Characterization of biochar from fast pyrolysis and gasification systems. Environ Progr Sustain Energy 28:386–396

Brodowski S, John B, Flessa H, Amelung W (2006) Aggregate-occluded black carbon in soil. Eur J Soil Sci 57:539–546

Brown D, Gassner M, Fuchino T, Maréchal F (2009) Thermo-economic analysis for the optimal conceptual design of biomass gasification energy conversion systems. Appl Therm Eng 29:2137–2152

Cantrell KB, Hunt PG, Uchimiya M, Novak JM, Ro KS (2012) Impact of pyrolysis temperature and manure source on physicochemical characteristics of biochar. Bioresour Technol 107:419–428

Cao X, Ma L, Gao B, Harris W (2009) Dairy-manure derived biochar effectively sorbs lead and atrazine. Environ Sci Technol 43:3285–3291

Chan K, Van Zwieten L, Meszaros I, Downie A, Joseph S (2008) Using poultry litter biochars as soil amendments. Soil Res 46:437–444

Chen B, Chen Z, Lv S (2011) A novel magnetic biochar efficiently sorbs organic pollutants and phosphate. Bioresour Technol 102:716–723

Chen WH, Peng JH, Bi XTT (2015) A state-of-the-art review of biomass torrefaction, densification and applications. Renew Sust Energ Rev 44:847–866

Cheng Y, Cai Z, Chang SX, Wang J, Zhang J (2012) Wheat straw and its biochar have contrasting effects on inorganic N retention and N2O production in a cultivated black Chernozem. Biol Fertil Soils 48:941–946

Clough TJ, Bertram JE, Ray JL, Condron LM, O'Callaghan M, Sherlock RR, Wells NS (2010) Unweathered wood biochar impact on nitrous oxide emissions from a bovine-urine-amended pasture soil. Soil Sci Soc Am J 74:852–860

Collard S, Zammit C (2006) Effects of land-use intensification on soil carbon and ecosystem services in Brigalow (Acacia harpophylla) landscapes of Southeast Queensland, Australia. Agric Ecosyst Environ 117:185–194

Dalal R, Eberhard R, Grantham T, Mayer D (2003) Application of sustainability indicators, soil organic matter and electrical conductivity, to resource management in the northern grains region. Aust J Exp Agric 43:253–259

DeLuca TH, MacKenzie MD, Gundale MJ, Holben WE (2006) Wildfire-produced charcoal directly influences nitrogen cycling in ponderosa pine forests. Soil Sci Soc Am J 70:448–453

Demirbas A, Arin G (2002) An overview of biomass pyrolysis. Energy Sources 24:471–482

Doerr SH, Shakesby RA, Walsh RPD (2000) Soil water repellency: its causes, characteristics and hydro-geomorphological significance. Earth-Sci Rev 51:33–65

Elmer W, White JC, Pignatello JJ (2010) Impact of biochar addition to soil on the bioavailability of chemicals important in agriculture. Report, University of Connecticut, New Haven

Fallahzade J, Hajabbasi MA (2012) The effects of irrigation and cultivation on the quality of desert soil in central Iran. Land Degrad Dev 23(1):53–61

Forbes M, Raison R, Skjemstad J (2006) Formation, transformation and transport of black carbon (charcoal) in terrestrial and aquatic ecosystems. Sci Total Environ 370:190–206

Gaskin JW, Speir A, Morris LM, Ogden L, Harris K, Lee D, Das KC (2007) Potential for pyrolysis char to affect soil moisture and nutrient status of a loamy sand soil. Georgia Institute of Technology

Gaunt JL, Lehmann J (2008) Energy balance and emissions associated with biochar sequestration and pyrolysis bioenergy production. Environ Sci Technol 42:4152–4158

Giller KE, Witter E, Corbeels M, Tittonell P (2009) Conservation agriculture and smallholder farming in Africa: the heretics view. Field Crops Res 114:23–34

Glaser B, Lehmann J, Zech W (2002) Ameliorating physical and chemical properties of highly weathered soils in the tropics with charcoal–a review. Biol Fertil Soils 35(4):219–230

Granatstein D, Kruger C, Collins H, Garcia-Perez M, Yoder J (2009) Use of biochar from the pyrolysis of waste organic material as a soil amendment. Center for Sustaining Agric. Nat. Res. Washington State University, Wenatchee, WA. WSDA Interagency Agreement. C.;800248

Gundale MJ, DeLuca TH (2006) Temperature and source material influence ecological attributes of ponderosa pine and Douglas-fir charcoal. For Ecol Manag 231:86–93

Hart S, Luckai N (2013) Review: charcoal function and management in boreal ecosystems. J Appl R Soc Ecol 6:181499. 50:1197–1206

Haumaier L, Zech W (1995) Black carbon—possible source of highly aromatic components of soil humic acids. Org Geochem 23(3):191–196

Herath HMSK, Camps-Arbestain M, Hedley M (2013) Effect of biochar on soil physical properties in two contrasting soils: an Alfisol and an Andisol. Geoderma 209–210:188–197

Higashikawa FS, Conz RF, Colzato M, Cerri CEP, Alleoni LRF (2016) Effects of feedstock type and slow pyrolysis temperature in the production of biochars on the removal of cadmium and nickel from water. J Clean Prod 137:965–972

Ishii T, Kadoya K (1994) Effect of charcoal as a soil conditioner on citrus growth and vesicular-arbuscular mycorrhizal development. J Jpn Soc Hort Sci 65:529–535

Jeffery S, Verheijen FGA, Van Der Velde M, Bastos AC (2011) A quantitative review of the effects of biochar application to soils on crop productivity using meta-analysis. Agric Ecosyst Environ 144:175–187

Jein SH, Wang CS (2013) Effects of biochar on soil properties and erosion potential in a highly weathered soil. Catena 110:225–233

Kambo HS, Dutta A (2015) Comparative evaluation of torrefaction and hydrothermal carbonization of lignocellulosic biomass for the production of solid biofuel. Energy Convers Manag 105:746–755

Kastner JR, Miller J, Geller DP, Locklin J, Keith LH, Johnson T (2012) Catalytic esterification of fatty acids using solid acid catalysts generated from biochar and activated carbon. Catal Today 190:122–132

Kishimoto S, Sugiura G (1985) Charcoal as soil conditioner. Int Arch Future 5:12–23

Kumar G, Shobana S, Chen WH, Bach QV, Sang HK, Atabani AE et al (2016) A review of thermo chemical conversion of microalgal biomass for biofuels: chemistry and processes. Green Chem 19:44–67

Laird DA, Fleming P, Davis DD, Horton R, Wang B, Karlen DL (2010) Impact of biochar amendments on the quality of a typical Midwestern agricultural soil. Geoderma 158:443–449

Lal R, Shukla MR (2004) Principles of soil physics. Marcel Dekker, New York

Lee XJ, Lee LY, Gan S, Thangalazhy-Gopakumar S, Ng HK (2017) Biochar potential evaluation of palm oil wastes through slow pyrolysis: thermo chemical characterization and pyrolytic kinetic studies. Bioresour Technol 236:155–163

Lehmann J, Rondon M (2006a) Bio-char soil management on highly weathered soils in the humid tropics. Biol Approaches Sustain Soil Syst 113(517):530

Lehmann J, Rondon M (2006b) Bio-char soil management on highly weathered soils in the humid tropics. In: Uphoff N (ed) Biological approaches to sustainable soil systems. CRC Press, Boca Raton, pp 517–530

Lehmann J, Kern D, German L, McCann J, Martins GC, Moreira L (2003a) Soil fertility and production potential. Chapter 6. In: Lehmann J, Kern DC, Glaser B, Woods WI (eds) Amazonian dark earths: origin, properties, management. Kluwer Academic, Dordrecht, pp 105–124

Lehmann J, Kern DC, Glaser B, Woods WI (2003b) Amazonian dark earths: origin, properties and management. Kluwer Academic Publishers, Dordrecht

Lehmann J, Gaunt J, Rondon M (2006) Bio-char sequestration in terrestrial ecosystems—a review. Mitigat Adaptat Strateg Glob Chan 11:403–427

Lemenih M, Karltun E, Olsson M (2005) Assessing soil chemical and physical property responses to deforestation and subsequent cultivation in smallholders farming system in Ethiopia. Agric Ecosyst Environ 105:373–386

Li J, Li Y, Wu M, Zhang Z, Lü J (2013) Effectiveness of low-temperature biochar in controlling the release and leaching of herbicides in soil. Plant Soil 370:333–344

Lian F, Xing B (2017) Black carbon (biochar) in water/soil environments: molecular structure, sorption, stability, and potential risk. Environ Sci Technol 51:13517–13532

Liang B, Lehmann J, Solomon D, Kinyangi J, Grossman J, O'Neill B, Skjemstad JO, Thies J, Luizão FJ, Petersen J, Neves EG (2006) Black carbon increases cation exchange capacity in soils. Soil Sci Soc Am J 70:1719–1730

Liu Z, Chen X, Jing Y, Li Q, Zhang J, Huang Q (2014) Effects of biochar amendment on rapeseed and sweet potato yields and water stable aggregate in upland red soil. Catena 123:45–51

Lu SG, Sun FF, Zong YT (2014) Effect of rice husk biochar and coal fly ash on some physical properties of expansive clayey soil (Vertisol). Catena 114:37–44. https://doi.org/10.1016/j.catena.2013.10.014

Mackie KA, Marhan S, Ditterich F, Schmidt HP, Kandeler E (2015) The effects of biochar and compost amendments on copper immobilization and soil microorganisms in a temperate vineyard. Agric Ecosyst Environ 201:58–69

Marin-Spiotta E, Chaopricha NT, Plante AF, Diefendorf AF, Mueller CW, Grandy AS, Mason JA (2014) Long-term stabilization of deep soil carbon by fire and burial during early Holocene climate change. Nat Geosci 7:428–432

Matsubara Y-I, Hasegawa N, Fukui H (2002) Incidence of Fusarium root rot in asparagus seedlings infected with arbuscular mycorrhizal fungus as affected by several soil amendments. J Jpn Soc Hortic Sci 71:371–374

Mbagwu JSC, Piccolo A (1997) Effects of humic substances from oxidized coal on soil chemical properties and maize yield. In: Drozd J, Gonet SS, Senesi N, Weber J (eds.) The role of humic substances in the ecosystems and in environmental protection. IHSS, Polish Society of Humic Substances, Wroclaw, pp 921–925

McHenry MP (2009) Agricultural bio-char production, renewable energy generation and farm carbon sequestration in Western Australia: certainty, uncertainty and risk. Agric Ecosyst Environ 129(1-3):1–7

Mekuria W, Noble A (2013) The role of biochar in ameliorating disturbed soils and sequestering soil carbon in tropical agricultural production systems. Appl Environ Soil Sci:354965

Meurisse RT (1976) Soil chemical and other properties of western hemlock soils in Oregon—their relationship to productivity. In: Atkinson WA, Zazoski RT (eds.) Western hemlock management: proceedings of a conference; May 1976; Alderbrook, WA. Contrib. No. 34. University of Washington, College of Forest Resources, Institute of Forest Products: Seattle, WA. 317 p

Meurisse RT (1985) Properties of Andisols important to forestry. In: Beinroth FH, Lusio WL, Maldonado FP, Eswaran H (eds.) Proceedings of the sixth international soil classification workshop, Chile and Ecuador. Part 1: papers. U.S. Department of Agriculture, Soil Management Support Services, Washington, pp 53–67

Mikan CJ, Abrams MD (1995) Altered forest composition and soil properties of historic charcoal hearths in southeastern Pennsylvania. Can J For Res 25:687–696

Miller RM, Miller SP, Jastrow JD, Rivetta CB (2002) Mycorrhizal mediated feedbacks influence net carbon gain and nutrient uptake in Andropogon gerardii. New Phytol 155:149–162

Mohan D, Pittman CU, Steele PH (2006) Pyrolysis of wood/biomass for bio-oil: a critical review. Energy Fuel 20:848–889

Nair PKR (2007) The coming of age of agroforestry. J Sci Food Agric 87(9):1613–1619

Nelissen V, Ruysschaert G, Manka'Abusi D, D'Hose T, De Beuf K, Al-Barri B, Cornelis W, Boeckx P (2015) Impact of a woody biochar on properties of a sandy loam soil and spring barley during a two-year field experiment. Eur J Agron 62:65–78. https://doi.org/10.1016/j.eja.2014.09.006

Nerome M, Toyota K, Islam TM, Nishimima T, Matsuoka T, Sato K et al (2005) Suppression of bacterial wilt of tomato by incorporation of municipal biowaste charcoal into soil. Soil Micro 59:9–14

Nishio M (1996) Microbial fertilizers in Japan. National Institute of agro-environmental sciences, Ibaraki, Japan

Ogawa M (2006) Charcoal use in agriculture in Japan. In: Keynote address, 1st Asia Pacific biochar conference, May 17–20, Gold Coast, Australia

Oldeman L, Hakkeling R, Sombroek WG (2017) World map of the status of human-induced soil degradation: an explanatory note. In: International soil reference and information centre

Park J, Hung I, Gan Z, Rojas OJ, Lim KH, Park S (2013) Activated carbon from biochar: influence of its physicochemical properties on the sorption characteristics of phenanthrene. Bioresour Technol 149:383–389

Parton W, Schimel DS, Cole C, Ojima D (1987) Analysis of factors controlling soil organic matter levels in Great Plains Grasslands1. Soil Sci Soc Am J 51:1173–1179. https://doi.org/10.2136/sssaj1987.03615995005100050015x

Rajkovich S, Enders A, Hanley K, Hyland C, Zimmerman AR, Lehmann J (2012) Corn growth and nitrogen nutrition after additions of biochars with varying properties to a temperate soil. Biol Fertil Soils 48:271–284. https://doi.org/10.1007/s00374-011-0624-7

Ramanathan V, Carmichael G (2008) Global and regional climate changes due to black carbon. Nat Geosci 1(4):221

Rondon M, Ramirez J, Lehmann J (2005) Charcoal additions reduce net emissions of greenhouse gases to the atmosphere. In: Proceedings of the 3rd USDA symposium on greenhouse gases and carbon sequestration in agriculture and forestry. Baltimore, MD, p 208

Rondon MA, Molina D, Hurtado M, Ramirez J, Lehmann J, Major J, Amezquita E (2006) In Enhancing the productivity of crops and grasses while reducing greenhouse gas emissions through bio-char amendments to unfertile tropical soils, 18th world congress of soil science, 9–15, Philadelphia, PA

Rondon MA, Lehmann J, Ramirez J, Hurtado M (2007) Biological nitrogen fixation by common beans (Phaseolus vulgaris L) increases with bio-char additions. Biol Fertil Soils 43:699–708

Safaei Khorram M, Zhang Q, Lin D, Zheng Y, Fang H, Yu Y (2016) Biochar: a review of its impact on pesticide behavior in soil environments and its potential applications. J Environ Sci 44:269–279

Saifullah DS, Naeem A, Rengel Z, Naidu R (2018) Biochar application for the remediation of salt-affected soils: challenges and opportunities. Sci Total Environ 625:320–335

Sanchez PA (2000) Linking climate change research with food security and poverty reduction in the tropics. Agric Ecosyst Environ 82(1–3):371–383

Schmidt MW, Noack AG (2000) Black carbon in soils and sediments: analysis, distribution, implications, and current challenges. Global Biogeochem Cycles 14:777–793

Seifritz W (1993) Should we store carbon in charcoal? Int J Hydrog Energy 18:405–407

Sevilla M, Mokaya R (2014) Energy storage applications of activated carbons: supercapacitors and hydrogen storage. Energy Environ Sci 7:1250–1280

Shafi M, Bakht J, Attaullah A, Khan MA (2010) Effect of crop sequence and crop residues on soil C, soil N and yield of maize. Pak J Bot 42(3):1651–1664

Silber A, Levkovitch I, Graber ER (2010) PH-dependent mineral release and surface properties of corn straw biochar: agronomic implications. Environ Sci Technol 44:9318–9323. https://doi.org/10.1021/es101283d

Smith P, Martino D, Cai Z et al (2008) Greenhouse gas mitigation in agriculture. Philos Trans R Soc B 363:789–813

Sohi SP, Shackley S (2009) Biochar: carbon sequestration potential. Copenhagen, Denmark

Sohi S, Lopez-Capel E, Krull E, Bol R (2009) Biochar, climate change and soil: a review to guide future research. In: CSIRO land and water science report

Spaccini R, Zena A, Igwe C, Mbagwu J, Piccolo A (2001) Carbohydrates in water-stable aggregates and particle size fractions of forested and cultivated soils in two contrasting tropical ecosystems. Biogeochem 53:1–22. https://doi.org/10.1023/A:1010714919306

Spokas KA, Reicosky DC (2009) Impacts of sixteen different biochars on soil greenhouse gas production. Ann Environ Sci

Steiner C (2006) Slash and char as alternative to slash and burn: soil charcoal amendments maintain soil fertility soil charcoal amendments maintain soil fertility and establish a carbon sink. Doctoral dissertation, Dissertation Faculty of Biology, Chemistry and Geosciences University of Bayreuth, Germany

Steiner C, Teixeira W, Lehmann J, Zech W (2003) Microbial response to charcoal amendments of highly weathered soils and Amazonian dark earths in Central Amazonia—preliminary results. In: Lehmann J, Kern DC, Glaser B, Woods WI (eds) Amazonian dark earths: origin, properties, management. Kluwer Academic Publishers, Dordrecht

Steiner C, Das KC, Garcia M, Förster B, Zech W (2008) Charcoal and smoke extract stimulate the soil microbial community in a highly weathered xanthic Ferralsol. Pedobiologia 51:359–366. https://doi.org/10.1016/j.pedobi.2007.08.002

Stelt MJCVD, Gerhauser H, Kiel JHA, Ptasinski KJ (2011) Biomass upgrading by torrefaction for the production of biofuels: a review. Biomass Bioenergy 35:3748–3762

Stewart CE, Zheng J, Botte J, Cotrufo MF (2012) Co-generated fast pyrolysis biochar mitigates greenhouse gas emissions and increases carbon sequestration in temperate soils. GCB Bioenergy 5:153–164. https://doi.org/10.1111/gcbb.12001

Streets DG, Bond TC, Lee T, Jang C (2004) On the future of carbonaceous aerosol emissions. J Geophys Res Atmosph 27(D24):109

Suprayogo D, Hairiah K, Noordwijk MV, Cadisch G (2010a) Agroforestry interactions in rainfed agriculture: can hedgerow intercropping systems sustain crop yield on an ultisol in Lampung (Indonesia)? Agri 32:3

Suprayogo D, Hairiah K, Van Noordwijk M, Cadisch G (2010b) Agroforestry interactions in rainfed agriculture: can hedgerow intercropping systems sustain crop yield on an ultisol in Lampung (Indonesia)?AGRIVITA. J Agric Sci 32(3):205–215

Tang J, Zhu W, Kookana R, Katayama A (2013) Characteristics of biochar and its application in remediation of contaminated soil. J Biosci Bioeng 116:653–659

Thies JE, Rillig MC (2009) Characteristics of biochar: biological properties. Chapter 6. In: Lehmann J, Joseph S (eds) Biochar for environmental management science and technology. Earthscan, London, pp 85–105

Tryon EH (1948) Effect of charcoal on certain physical, chemical, and biological properties of forest soils. Ecol Monogr 18:82–115

Uzoma KC, Inoue M, Andry H, Fujimaki H, Zahoor A, Nishihara E (2011) Effect of cow manure biochar on maize productivity under sandy soil condition. Soil Use Manag 27:205–212

Van Zwieten L, Singh B, Joseph S, Kimber S, Cowie A, Chan KY (2009) Biochar and emissions of non-CO2 greenhouse gases from soil. In: Lehmann J, Joseph S (eds) Biochar for environmental management: science and technology 1. Earthscan, London, pp 227–250

Wang M (2008) Greenhouse gases, regulated emissions, and energy use in transportation (GREET) model, 1.8b; U Chicago Argonne, LLC

Wang Y, Zhao S, Niu D (1999) Research state of soil carbon cycling in terrestrial ecosystem. Chin J Ecol 18:29–35

Wang Q, Li Y, Alva A (2010) Cropping systems to improve carbon sequestration for mitigation of climate change. J Environ Protec 1:207–215

Wang Z, Zheng H, Luo Y, Deng X, Herbert S, Xing B (2013) Characterization and influence of biochars on nitrous oxide emission from agricultural soil. Environ Pollut 174:289–296. https://doi.org/10.1016/j.envpol.2012.12.003

Warnock DD, Lehmann J, Kuyper TW, Rillig MC (2007) Mycorrhizal responses to biochar in soil—concepts and mechanisms. Plant Soil 300:9–20

Wilk M, Magdziarz A (2017) Hydrothermal carbonization, torrefaction and slow pyrolysis of Miscanthus giganteus. Energy 140:1292–1304

Windeatt JH, Ross AB, Williams PT, Forster PM, Nahil MA, Singh (2014) Characteristics of biochars from crop residues: potential for carbon sequestration and soil amendment. J Environ Manag 146:189–197

Woodfine A (2009) Using sustainable land management practices to adapt to and mitigate climate change in sub-Saharan Africa, resource guide version 1.0. TERR AFRICA. http://www.terrafrica.org/

Woolf D, Lehmann J (2012) Modelling the long term response to positive and negative priming of soil organic carbon by black carbon. Biogeochemistry 111:83–95

Wu W, Yang M, Feng Q, McGrouther K, Wang H, Lu H, Chen Y (2012) Chemical characterization of rice straw-derived biochar for soil amendment. Biomass Bioenergy 47:268–276

Yanai Y, Toyota K, Okazaki M (2007) Effects of charcoal addition on N2O emissions from soil resulting from rewetting air-dried soil in short-term laboratory experiments. Soil Sci Plant Nutr 53:181–188

Yin Y-F, He X-H, Ren G, Yang YS (2014) Effects of rice straw and its biochar addition on soil labile carbon and soil organic carbon. J Integr Agr 13:491–498

Yu KL, Lau BF, Show PL, Ong HC, Ling TC, Chen WH et al (2017) Recent developments on algal biochar production and characterization. Bioresour Technol 246:2–11

Zhang W, Wei H, Gao H, Hu Y (2005) Advances of studies on soil microbial diversity and environmental impact factors. Chin J Ecol 24:48–52

Zheng H, Wang Z, Deng X, Zhao J, Luo Y, Novak J, Herbert S, Xing B (2013) Characteristics and nutrient values of biochars produced from giant reed at different temperatures. Bioresour Technol 130:463–471

Life Cycle Analysis to Estimate Environmental Impact of the Food Industry 7

Anulipt Chandan, V. R. Sankar Cheela, and Brajesh Dubey

7.1 Introduction

Food is essential for human survival, and access to ample amount of nutritious food throughout human life influences their health and well-being. A food system is a complex interconnection of physical, social and political system to provide nutrition (Heller et al. 2013). Inconsistency inadequate nutrition supply is also responsible for paradoxical coexistence of malnutrition and obesity within the same population. Fulfil the demand for nutrition to current seven billion people and for future generation brings challenges for finite available natural resources. Nonetheless, food production and consumption also have the highest environmental impact (Bruno et al. 2019). With a growing population and proportionately increasing food demand put further stress on the food production system, resulting in additional environmental challenges.

Although food production is hotspot in the food supply chain, it is also essential for food security and safety (Roy et al. 2009). Therefore, adoption of sustainable practices in food production and consumption is vital for food security. Sustainability concern in the food sector varies from environment to socio-economic. It includes excessive use of natural resources, greenhouse gas emissions,

A. Chandan
Discipline of Information System, School of Management, Curtin University, Perth, WA, Australia

V. R. S. Cheela
Department of Civil Engineering, Indian Institute of Technology Kharagpur, Kharagpur, West Bengal, India

Sustainable Engineering Group, Curtin University, Perth, WA, Australia

B. Dubey (✉)
Department of Civil Engineering, Indian Institute of Technology Kharagpur, Kharagpur, West Bengal, India
e-mail: bkdubey@civil.iitkgp.ac.in

I. Haq, A. S. Kalamdhad (eds.), *Emerging Treatment Technologies for Waste Management*, https://doi.org/10.1007/978-981-16-2015-7_7

loss of biodiversity, impact on ecosystem, health issue, trade and finance, equality and nutrition value. Food consumption is responsible for 15–25% of the total greenhouse gas emissions in developed countries. Along with greenhouse gas emissions, food production has a major impact on water resources, biodiversity and aquatic ecosystem. Use of chemical fertilizer in intensive farming disrupts the nitrogen and phosphorus cycle, leaching of chemical fertilizer deteriorates the water quality and threats the marine life. Due to the growing concern of health and nutrition diet, consumers are looking for organic food (Lee and Yun 2015). With the growing scandal in food industries like the horsemeat scandal, baby formula powder scandal, consumers are becoming more concerned about food they consumed (Barnett et al. 2016). Therefore, demand for a transparent food supply chain is arising and thus pressure on a food processor to provide more information like provenance, process used during manufacturing is increasing (Bánáti 2011).

The demand for sustainability in the food sector and demand for sustainable products are increasing rapidly. However, assessing sustainability of a product is a major concern. There are various methods available to access sustainability of products like Life Cycle Assessment (LCA), indicators based methods and eco-footprints (Manning and Soon 2016). However, LCA is the most popular method to find the environmental sustainability of a product during its life cycle (Roy et al. 2009). LCA analysis can be used for various purposes like making environmental policies, innovation to develop new sustainable method and technology to reduce environmental impact. This chapter will give a detailed overview of recent advancement in LCA studied in the food sector and challenges in LCA. Rest of the book chapters organized are as follows: Sect. 7.2 will describe the LCA analysis steps in brief, Sect. 7.3 will review LCA study in various food products at product level and diet level, Sect. 7.4 discussed the challenges in LCA study and then Sect. 7.5 presented a discussion and future direction and finally a conclusion presented in the last section.

7.2 LCA in Food Industry

Life Cycle Assessment (LCA) is one of the most applied methods in food industry to assess the potential environmental impacts of a food product during its whole lifetime. It accounts and uses various resources and its impact on environment during the entire life cycle of a product. International standard organization (ISO) has standardized the process of LCA analysis, yet it has a lot of flexibility in selection of parameter and boundaries under study (Klöpffer 2012). Due to this flexibility, it is adopted across different Industry. In general, LCA involves objective and boundary definition, data collection, analysis and interpretation of result (Arvanitoyannis 2008). According to ISO 14040 (Finkbeiner et al. 2006), an LCA comprises four main stages: goal definition, scope definition, life cycle inventory, life cycle impact assessment and interpretation of the results (Fig. 7.1). LCA enables the quantification of emissions during lifetime of a product for comparison with other product, and

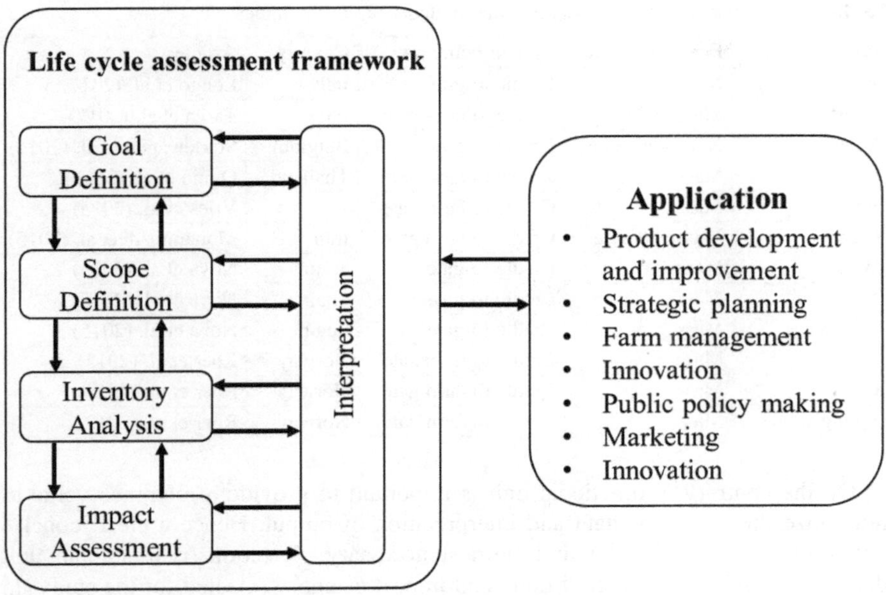

Fig. 7.1 Different stages in LCA analysis (Source: Roy et al.(2009) and Laurent et al.(2014))

the identification of hotspots to maximize efficiency and/or minimize the environmental impact (Castellani et al. 2017).

7.2.1 Goal Definition

Goal Definition is the first and one of the crucial stages in LCA analysis. It is important in a sense that it decides and steers the selection among various available option in next phase of analysis. It aimed to identify the objectives of the studies.

7.2.2 Scope Definition

In the next phase of LCA analysis, defining scope of studies is important to carry out modelling in next phase. Selection of different parameters like functional unit, system boundaries, data sources, cut-off limit is determined in this phase. These parameters are important factors for both life cycle inventory development and result interpretation.

System boundaries define what stage of the food supply chain should be included and what should be excluded, it is generally described by input and output flow map. Most frequently used system boundaries in LCA study in the food system is cradle to farm gate, whereas cradle to grave approach is not preferred because of unavailability of food consumption data.

Table 7.1 Application of functional unit and boundary in LCA studies

Food produce	Functional unit	System boundary	Country	Article
Apple	Mass	Cradle to grave	Italy	Longo et al. (2017)
Endive	Mass and area	Cradle to farm gate	Italy	Tasca et al. (2017)
Meal	Nutrition value	Cradle to grave	Belgium	Sturtewagen et al. (2016)
Beef	Mass	Cradle to farm gate	Thailand	Ogino et al. (2016)
Beef	Mass	Cradle to farm gate	–	Vries et al. (2015)
Rice	Mass and area	Cradle to farm gate	Iran	Mohammadi et al. (2015)
Wheat	Mass	Cradle-to-gate	Spain	Noya et al. (2015)
Maize	Mass	Cradle-to-gate	Spain	Noya et al. (2015)
Triticale	Mass	Cradle-to-gate	Spain	Noya et al. (2015)
Barley	Mass	Cradle to farm gate	Norway	Roer et al. (2012)
Oat	Mass	Cradle to farm gate	Norway	Roer et al. (2012)
Spring wheat	Mass	Cradle to farm gate	Norway	Roer et al. (2012)

On the contrary, a functional unit is important to provide a reference value to normalize the inventory data and interpretation of output. Hence a clear, concise definition of a functional unit is utmost necessary. Selection of functional unit depends on objective and environmental impact category selected for the study. In the food sector, mass and area based often selected as functional unit (Roy et al. 2009). However, functional unit based on nutritional value also being used in the study (Sturtewagen et al. 2016). Therefore, the functional unit shall be clearly defined and measurable (Table 7.1).

7.2.3 Life Cycle Inventory

Life Cycle Inventory (LCI) involved collection of all input data like material, energy data and output data such as emissions data at every stage on analysis. Data collection phase in case of a food supply chain is challenging due to involvement of various actors. As compared to other phases of life cycle analysis LCI phase is utmost work-intensive and time-consuming process due to data collection and compilation. It can be easier to collect data in case a good database is available respective to the food produce understudied and all actors in the supply chain are willing to help. The data required for LCA analysis in food sector can be divided into two parts one is site-specific data that can be collected by interviewing supply chain actors or by visiting the site. On the other hand, generic data such as electricity production, mineral extraction, transportation data and production of packaging material and their disposal can be collected from the database.

7.2.4 Impact Assessment

According to ISO 14004 (2006) Life Cycle Impact Assessment (LCIA) "*aimed at understanding and evaluating the magnitude and significance of the potential environment impact for a product system throughout the life cycle of the period*". LCIA impact assessment results in relative environmental impact categories of a product based on functional unit. The general framework LCIA has three mandatory elements: selection and classification of impact categories and characterization. The first one is selection of impact category selection that is very crucial in the interpretation phase. Food system used chemical fertilizer, fossil fuel, energy and water in large quantity. Therefore, it is important to select all relevant category impacted by these inputs. The optional element of impact assessment category is to find a single index from different impact category using normalization and weighting method. However, a single index provides concise information about environmental sustainability yet not provide the exact picture. The primary reason behind this is the selection of weighting factor, which is selected based on expert knowledge that is ambiguous.

In addition to finding the environmental impact of a product, impact assessment also includes the uncertainty analysis and sensitivity analysis. The uncertainty analysis helps in finding the ambiguity in available data and understanding reliability of the outcome. However, sensitivity analysis is used to understand the impact of change in data and methodology, which helps in finding a way to reduce the environmental impact of a food product.

7.2.5 Interpretation

In the final phase of LCA analysis, interpretation of data analysis from LCIA is involved. Interpretation helps in finding the hotspot in environmental performance and comparing different alternative cases. Interpretation of the LCA result can be made in two ways in one is based on LCIA in which inventory analysis, along with impact assessment output is presented. On the other side, the interpretation of result can be based on LCI analysis in which only inventory assessment is presented.

The interpretation of assessment involves the identification of significant environmental issues and also recommendation of the corrective action to reduce impact. A recommendation can be a qualitative or quantitative or combination of both measures to improve environmental performance of the food system. The improvement measure can be optimal fertilizer use, alternative feed, better waste management, better co-product handling and change in industrial processing. Figure 7.2 depicted a detail life cycle assessment structure.

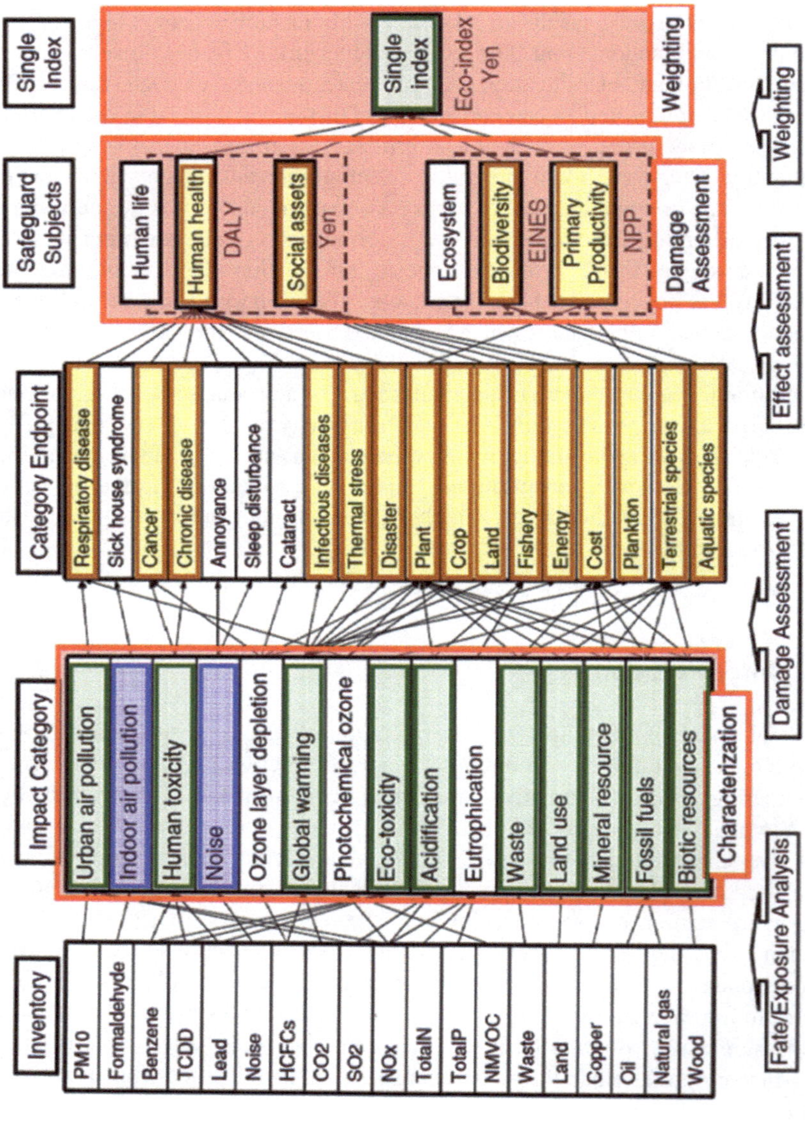

Fig. 7.2 LCA structure based on endpoint modelling (Source: Roy et al. (2009))

7.3 LCA Studies on Food Products

LCA is a widely utilized analysis toll to evaluate potential environmental impacts of product and services during its life period. It accounts emission from resources use, production process, distribution, consumption and disposal. LCA studies are used in understanding the environmental performance of a product and provide an opportunity to improve it. It also helps policymaker to formulate and design an environmentally friendly policy. There are numerous LCA studies available in the food system, which are varied in terms of selection of input data based on different region and environmental condition. Studies are varied in their boundary condition like cradle to grave and cradle to farm gate. Comparative studies utilized LCA analysis to compare the environmental impact of food products using a different method (homemade vs canteen vs industrial production). Another type of study analyses food product made of different component like a meal from canteen, tomato ketchup. LCA study is also utilized to find the environmental hotspot stage.

7.3.1 LCA Approach

In general, a food product can be viewed as individual food (Rice, meat) or specific to dietary requirement (meal) food. LCA analysis can be used to find the environmental impact in both cases. Therefore, LCA can be approached in two ways: one is produced base approach, and another is diet-based approach. Selection of approach totally depends on the objective and scope of a study. However, each approach has its own pro and cons. Such as in case of dietary approach, complexities arise due to variation in diet pattern from person to person, whereas product level approach lacks in integration of nutrition value. Another major complexity is the collection of data, in diet level approach, multi-ingredient food product brings challenges in data collection (Poore and Nemecek 2018).

7.3.1.1 Product Approach

In produce level approach, the goal was to find the environmental impact of food product without consideration of its nutrient value. In this approach, most of the studies didnot include the consumption in their study due to lack of consumption data. The product approach LCA can be applied from Agri product to dairy product to meat. This section will highlight the recent study in this field (Fig. 7.3).

Agri-Food Product

Rice and wheat are the most produced agriculture commodities in the world, which are used as food. The life cycle assessment of these two major kinds of cereal helps in understanding the environmental performance of its production. Mohammadi et al. (2015) applied LCA to evaluate environmental performance of the paddy field in two growing seasons, namely summer and spring in northern Iran. Studied showed that emission from the paddy field has greater impacts on GWP, followed by energy used for irrigation. Irrigation was also having a major impact on cumulative

Fig. 7.3 LCA approach

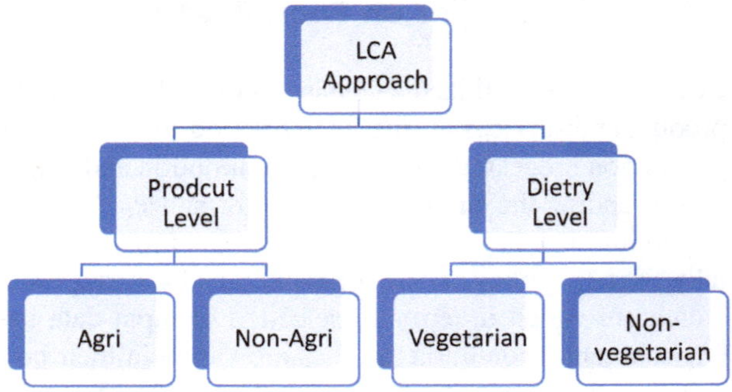

energy demand (CED) followed by using chemical fertilizer. Studied also showed that all impact categories considered in this study like GWP, (Water Depletion) WD, CED have more impact in summer farming as compared to spring farming due to increased demand water thus required more resource for an irrigation system. However, the difference in environmental performance in these two farming cases was varied according to the selection of function unit (FU). In case of area as FU, the difference was lower, whereas in mass as FU has considerably higher differences.

Zhang et al. (2017) utilized LCA method to estimate the carbon footprint of grain production in China. This study showed that the carbon footprint was higher in China as compared to other countries like Canada, USA due to change in irrigation method and different amount of fertilization used. They reported that key factors for GHG emission in grain production were production and application of nitrogen fertilizer, straw burning, fossil fuel used in machinery, energy used for irrigation and CH4 emissions from rice paddies, whereas Wang et al.(2017) studied environmental impact of wheat-maize double-cropping grain production on different farm size in China employing LCA method. Their study showed that a larger farm showed slightly better environmental performance of double grain production as compare to small farm due to efficiently used machinery in large farm reduced demand for labour, thus reducing energy demand. In case of chemical fertilization impact on environment was reduced in the large farm due to reduced emission, biogas utilization and efficient use of fertilization to increase total production. They reported that the irrigation was another major concern in grain farming due to use of freshwater and require electrical energy to pump groundwater to farm (Boone et al. 2016). They suggested that an improved sprinkle irrigation system can reduce the water requirement and lessen the demand for electricity.

In contrast, Noya et al. (2015) assessed environmental profile of three different kinds of cereal, namely wheat, maize and triticale to identify best environmentally sustainable cereals production in Po valley, Italy. They also studied the different varieties of maize, like class 300, 400, 500, 600, 700. The finding of this study showed that agrochemical production contributed noticeably in almost all impact categories considered in the study. Hence use of fertilizer in different varieties of wheat was an important deciding factor for environmental profile. Maize class 300 had the highest environmental impact, whereas class 600 and 700 have the least environmental impact due to lower requirement of chemical fertilizer in class 600 and 700 production. On the contrary, in terms of farming activities of maize production, irrigation has the greatest environmental impacts due to use of diesel fuel to extract water from a source. Therefore, irrigation activity has a major impact on water and fossil depletion, and GWP. However, in the case of wheat and triticale, which required less amount of water activities like cultivation, harvesting, harrowing and ploughing were environmental hotspot. In terms of fertilizer in wheat and triticale production, herbicide production has the greatest impact on environment.

Roer et al. (2012) did life cycle assessment of barley, oat and spring wheat and their finding showed that the CO2 emission from added lime, humus mineralization, direct N2O emissions and indirect N2O emissions was a prime cause for the climate change. During various stage in lifecycle assessment, the fertilization production stage was most crucial and had the highest impact in environment. In case of barley, fertilizer production has highest impact compare to oat and wheat due to higher use of fertilizer per kg yield. Field emission caused due to runoff of applied mineral fertilizer has highest impact on marine ecotoxicity, whereas use of pesticide has highest impact on terrestrial ecotoxicity. The machinery used in farming has two components one is fuel used, and other is machinery production itself. Diesel was used as fuel source that has a major impact on fossil depletion and ozone depletion, whereas the machinery production impacted human toxicity, freshwater ecotoxicity and marine ecotoxicity. Johnson et al. (2016) studied greenhouse gas emission of wheat and rice in the USA. This study showed that the GHG emission varied based on different factors like regions due to soil fertility, farm size due to farm efficiency, and cropping system.

In the case of fruit and vegetable, there are various literatures available that showed environmental performance of their production. Tasca et al. (2017) studied the environmental impact of endive production in Italy. They consider two farming method, namely organic and integrated endive farming to compare the environmental performance in these two farming methods. A cradle to farm gate LCA approach was selected to determine environmental performance. The study showed that based on per hectare, organic farming performed better in climate change, oxidation potential, human toxicity cancer effect, photochemical ozone formation and land use, then integrated farming methods. In integrated farming primary reason for higher climate change impact was due to use of chemical fertilizer. In fruit production, Longo et al. (2017) applied LCA method in organic and conventional apple production in northern Italy. They reported that organic farm has lower productivity compared to conventional farming. However, it has less environmental impact. In case of conventional farming, use of fertilizer, pesticide and fossil fuel consumption were contribution to the major impacts on environment.

Meat

Beef, chicken and pork are the most consumed meat product around the world. There are several LCA studies available focused on meat industries to determine its environmental impact. A study by Vries et al. (2015) on suckler-based and dairy-based beef production showed that the dairy-based beef production system has a lesser effect in all environmental impact categories as compare to suckler-based production. In contrast, concentrate-based diet to fattening calves beef production has lesser GWP, energy use and land use as compare to roughage-based fattening diet. Similarly, Tichenor et al. (2017) compared the environmental impact of grass-fed intensive grazing farm to confine dairy beef production system. They reported that the GHG potential was 33.7 and 12.7 kg CO_2 eq. per kg carcass weight in case of grass-fed dairy production and dairy-based beef production farm, respectively. Enteric methane was contributed 57% and 37% of total GHG emissions in grass-fed and dairy-based farm, respectively.

Considering the calve fattening system, there is generally two types of beef production system one is extensive and other is an intensive beef production system. Huerta et al. (2016) studied the extensive and intensive beef production system in Mexico. Grazing cattle feeders considered into extensive system, whereas feedlot feeders production includes into the intensive system. The result showed that the extensive system performed better in most of the categories except in climate change, freshwater eutrophication and agriculture land occupation. Climate change impact was higher in the extensive system as compare to intensive beef production due to enteric fermentation and manure management. They reported that the enteric fermentation contributed around 67% of total GHG emissions. Likewise, Ogino et al. (2016) assessed the environmental performance of an extensive and intensive beef production system in northeastern Thailand. The study reported that intensive system contributes 25% less GHG than extensive system. The enteric fermentation was the highest contributor in GHG emissions in extensive system accounts 77%, whereas in intensive farming it was 65% of total GHG emissions in respective system. Manure management had a significant contribution to GHG emissions. The extensive system has lesser energy consumption compare to intensive system by 69% due to energy consumption in purchased feed in intensive farming. Which showed the intensification of beef production system has a great impact on environmental performance. However, in both systems, a balanced approach between production and environment needs to adopt in the intensive system.

In contrast to beef, pork production has a lesser environmental impact (Caro et al. 2017). Winkler et al. (2016) studied environmental impacts of fresh Austrian pork from farm to fork using LCA analysis. They reported that the farming stage was a major contribution to GWP, which was 92.3% of total emissions. The farming stage also has a major contribution to acidification and eutrophication potential, whereas transportation and slaughterhouse have lesser contribute to GWP 3.53% and 2.99%, respectively. In the farming stage, the feed has a large share in GWP compare to enteric fermentation, manure management and energy used. Environmental impact of pork production can be reduced by applying a better management system. Sonesson et al. (2016) studied the pork supply chain environmental impact using the LCA analysis in Sweden. They suggested that there were possibilities to reduce GWP by 21–54% by taking certain measure at farm level. These measures were improved reduction management, crop rotation, better manure management, improving animal health and increasing nitrogen efficiency.

In contrast to beef and pork production, chicken meat production is more efficient and environmental-friendly (Caro et al. 2017; McCarthy et al. 2015). As compared to beef, chicken is most widely used in a different meal around the globe. Kalhor et al. (2016) performed LCA analysis to determine the environmental impact of chicken meat production in Iran. They studied the broiler chicken production in summer and winter season. The system boundaries were considered from input to chicken farm to slaughterhouse, excluding the infrastructure and capital goods. They reported that feed production and transportation were the key contributor in all impact categories. Electricity produced from natural gas contributes a significant amount in GWP, whereas farm emission was a major contributor in acidification and eutrophication impact category. Environmental impact of chicken meat production in winter was higher in all impact categories compare to summer chicken meat production due to heating application in the winter season that was provided by combustion of fossil fuel. In terms of energy used in various phase of meat production, the study showed that the production stage was consumed 56% of total energy followed by slaughterhouse (31%) and transportation (13%) which showed that distance travelled or food mile does not provide a clear picture of sustainability. Similarly, Lopez-Andres et al.(2018) studied the environmental impact of chicken meat production in Mexico. They utilized the LCA analysis to determine environmental impact from cradle to slaughterhouse. They reported that the chicken farms are a major contributor to environment in all impact categories. In contrast, McCarthy et al. (2015) extended the system boundary to grave to assess the environmental performance of chicken meat production. They considered two different scenarios in their study that includes a whole bird and portions operation. The result showed that the feed has a significant impact on global warming potential in both scenarios. They reported that packaging material used in portion scenario has greater impact in marine aquatic and ecotoxicity due to landfill disposal of polypropylene available in packaging material. Comparing the two different scenarios it was reported that portions have larger impact in all end point category, i.e. resources, human health and ecosystem in LCA analysis. The differences in environmental impact between these two scenarios were due to extra raw material used in case of portions production to produce the same quantity of meat at retail.

7.3.1.2 Dietary Approach

On every day, large quantities and varieties of food are produced, processed, delivered and consumed. The choice of diet changes from region to region and that brings changes in ingredients of diet and thus environmental impact varying from one diet to another. Dietary choices like vegetarian, non-vegetarian and vegan have a different impact on environment. Heller et al.(2013) presented a framework to integrate nutrition approach into environmental impact analysis using LCA methods (Fig. 7.4). LCA analysis of food-based dietary choice is varied from choices of functional unit. There are various studies that showed that environmental impact is associated with the choice of diet.

A shift of dietary requirement and reducing food loss improve the environmental profile of the food system. Bene et al. (2019) presented a model (Fig. 7.5) to show

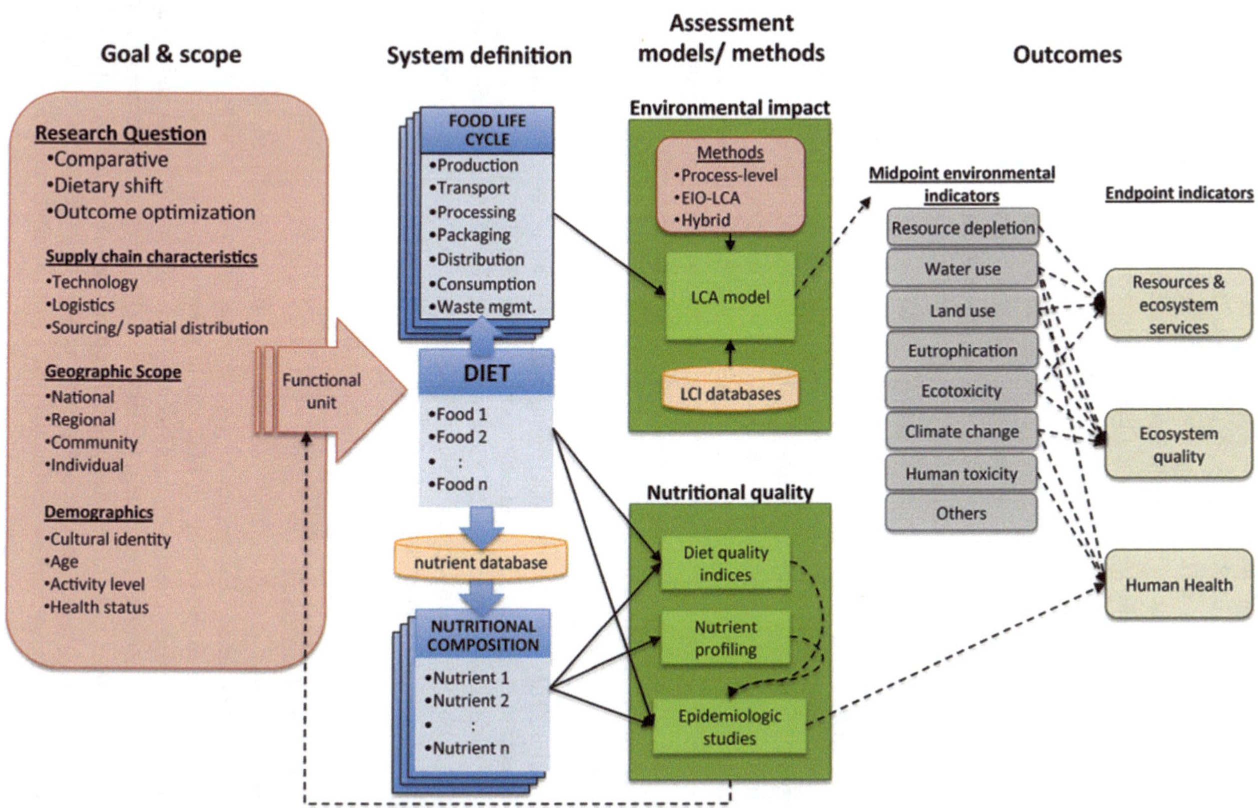

Fig. 7.4 Frame work for diet level integration to LCA (Source: Heller et al.(2013))

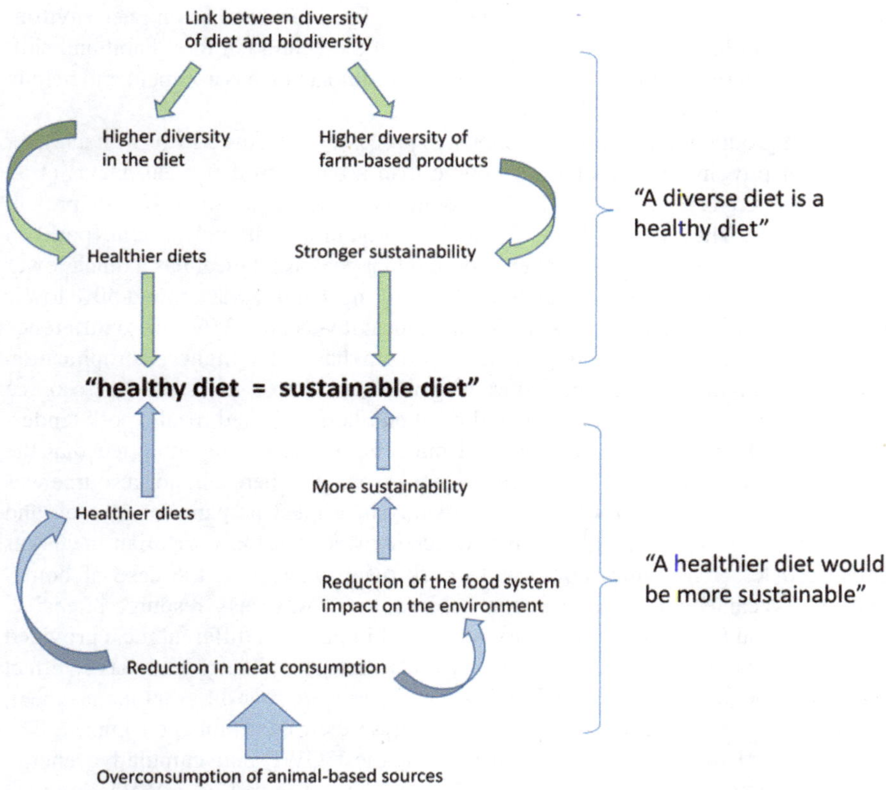

Fig. 7.5 Two pathway for a sustainable and healthy diet (Source: (Béné et al. 2019))

trade-off between a healthy diet and a sustainable diet. Meier and Christen (2013) analysed the environmental impact of nutrition in Germany considering the energy uptake as functional unit. They studied various dietary requirements that include vegan, vegetarian, and two other dietary recommendations, one from German Nutrition Society (D-A-CH) and other is from Federation for Independent Health Consultation (UGB) with more vegetables and less meat. The studied showed that vegan dietary has lesser GHG emissions and land use, whereas diet with higher meat content has the highest CO2 contribution and land use per person. However, blue water use was higher in the vegan, and vegetarian diet due to these diets contains water-intensive products group like nuts and seed. Carlsson-Kanyama et al. (2003) studied environmental impact of different meal with equivalence protein and energy content. This study was focused on analysing the global warming potential of a meal with ingredient meat and vegetable. Similarly, Baroni et al.(2007) compared environmental impact of three types of diets, namely conventional, vegetarian and vegan. In this study, they also studied the impact of organically and conventionally produced raw material used in three different diets. They reported that the conventional diet, which includes consumption of meat and related products has higher environmental impact on all categories. They stated that the change in food habit and shift towards vegetarians and vegans will have a lesser impact on environment and help in preserving the environmental resources.

Protein production is a resource-intensive process. Therefore a meal with a higher amount of protein demands more resource than a balanced diet meal. Davis et al. (2008) applied LCA to study the environmental impact of two different protein source meals. They study the meal with pea protein and animal protein (pork) in Spanish and Swedish scenario. They reported that pea-based meal has a much lower contribution to the GWP compare to pork-based meal and it was around 50% lower in Swedish meal, whereas in case of Spanish meal it was only 35% due to difference in energy production. Similarly, animal protein has a lot higher eutrophication potential than the pea protein. Sturtewagen et al. (2016) assessed the resource consumption of typical Belgian catered meal and home-cooked meals, pork tenderloin with potato croquettes, carrots and mushroom sauce. They reported that the production, processing and packaging were a key stage, where a major resource was consumed. They also did a comparative study of a meal prepared at canteen and home and their vegetarian alternative, which showed that the vegetarian meal was consumed lesser resources compare to pork meal whereas in the case of home-cooked vs canteen catered, canteen catered meal was less resource-intensive. Raimondi et al.(2011) studied the environmental impact of a different meal provided to a worker using LCA. The result showed that a change in dietary habit has an effect on environmental sustainability. In this study, there were four different menus, beef, poultry, pork and vegetarian, included to analyse their environmental impact. The result showed that the beef menu has the highest GWP and cumulative energy demand, whereas vegetarian menu has the least impact with 11% GWP lesser than beef and 53% lesser demand for energy. This showed that the vegetarian menu has minimum environmental impact compare to other.

Nutrition and diet-based LCA analysis provides an answer to an important environmental question that what would be environmental impact if everybody shift to one particular diet. However, a simple answer can get out from studies due to variation in diet ingredient and in some studies it was created hypothetically from a secondary data source (Van Kernebeek et al. 2014) which did not provide a real picture. In studies, it was observed that the vegetarian meal was more environmentally sound compared to a non-vegetarian meal, yet it needs to be cautious in comparison due to hypothetical consideration of diet composition and nutrition value (Hallström et al. 2015; Vieux et al. 2013). Jones et al. (2016) argued that consumer behaviours need to include in diet analysis. Even though it is hard to do a specific survey on consumer diet behaviours come to standard diet for analysis, it would be better to rely on consumer preference on case to case basis. Despite the differences in diet level approach among different studies, it was unambiguously reported that reducing meat consumption in the diet will improve the food system environmental performance (González-García et al. 2018).

7.4 Challenges in LCA Studies

The international LCA community is still battling with various challenges in quality LCA database, inventory and field data collection. Specific to the food sector, the challenges are varied from the variability in food production, system boundaries and appropriate function unit (Notarnicola et al. 2012).

7.4.1 Database and Its Quality

LCA analysis effective implementation directly depends on the availability of inventory data. A reliable and latest inventory data plays a crucial role in LCA impact assessment. With increasing interest from industry and academia database in food supply chain evolving continuously. However, with a wide range of agriculture methods, the material used in farming and food products, it is a great challenge to find the exact data required for the LCA analysis in context to specific region and product (Notarnicola et al. 2012). In this scenario, an alternative dataset is used, which is near to the required dataset, which impacts the outcome of the analysis. Therefore, it is challenging to a developed database of food system relevant to a specific region and food item-wise, which should be transparent and of good quality.

7.4.2 Consumer Requirement

With increasing consumer awareness towards sustainable, they started demanding more sustainable food products. Policymaker in a different country also looking various ways to integrate the sustainable use policy in the food supply chain. LCA is among the most used tools to implement for product sustainable policy. European union's Integrated Production Policy is fundamentally based on LCA analysis. This policy helps in defining the sustainability label like eco-label that helps a consumer to buy sustainable produce item (Basset-Mens et al. 2010). However, with limited knowledge of LCA consumer believed on the label on the package. With too many labelling systems, it becomes difficult for the consumer to compare sustainability of two product.

Therefore, there need to develop LCA guidelines to product specific that can provide reliable, concise and in comprehend way. Eco-labelling is generally providing sustainability information of a product or two stages of product processing during its lifecycle. This was a case of limited system boundaries. Therefore, to meet the consumer demand LCA analysis needs to extend the boundaries and provide more environmental impact information to the consumer (Basset-Mens et al. 2010).

7.4.3 Divergence of Interpretations

Interpretation of LCA analysis outcome is varied on various factors such as system boundaries, functional unit selection, data uncertainty and modelling approaches. Selection of system boundary greatly influences the outcome and thus misinterpretation can be possible. When an impact is assessed with respect to different functional unit interpretation can be changed. This can be easily seen in the LCA analysis that compares extensive versus intensive farming. Intensive farming was better environmental performer when mass was selected as functional unit; however, in case of area-based function unit intensive unit is better option. Similarly, in LCA using diet approach if the function unit selected mass compared to nutrition value, the interpretation is different (Notarnicola et al. 2012). Another interpretation that raised questions about the LCA analysis is relative versus absolute value. Relatively environmental impact (impact per unit produce) can be lesser; while the absolute impact may be higher. Therefore, it will be better if each impact categories have an absolute benchmark, and LCA result will be compared to those impact categories.

7.4.4 Impact Categories Selection

A wide range of impact category is selected in LCA studies. However, there is inconsistency in the selection of the impact category. There are 18 impact categories, however, only a few studied selected among all of them in their study. Only a few impact categories are common in studies like climate change, eutrophication and acidification potential. However, larger number of impact categories will provide a wider environmental impact of food product. In contrast to selection of impact categories, there are different methods to select the impact categories like ReCiPe, CML that impact the comparison between two studies, and it is challenging to find which products or methods are good for environment. The impact categories currently used are developed for the industrial system, which is not providing a better understanding of the food system (Roy et al. 2009). A collection of impact categories specific to food system require to interpret the impact better (Roy et al. 2009).

7.5 Discussion and Future Research Direction

LCA study in the food system is varied from the system boundary to function unit to impact category selection. The variation in selection of parameters in LCA analysis effect the understating of the actual environmental impact. LCA analysis gives the relative impact analysis based on a functional unit that can be misinterpreted in some case where an absolute value was very high. In the food system, there are multiple function unit used in studies. The most common functional unit selected in LCA study is mass, area, protein content and energy content. The mass-based functional unit also varied like in case of beef, some study selected the Caracas weight as function weight, whereas some selected only meat weight. The change in functional unit changes the relative outcome from impact assessment method. A wide range of functional unit helps in understanding environmental impact of food product from different perspective thus helps in decision and policy-making to reduce the environmental product. For example, LCA analysis of beef and vegetarian alternative with the same nutritious value showed that beef meal has higher environmental impact this information can be utilized in decision making to reduce the beef consumption. Multiple function unit also helps in better understanding the environmental impact of various farming method. Such as in conventional vs organic, where the production difference is largely due to high amount of fertilizer and pesticide that has an adverse impact on environment. Whereas organic farming produces less emission, however it requires more land.

On the other side, system boundary definition also varied from one study to another. However, in food LCA studied the study generally considered the cradle to farm gate as boundary yet in this some study does not include the various component like exclusion of infrastructure, capital goods, machinery production and transportation. Since the definition of system boundaries varies from one study to another, that also get partial information of environmental product in their life cycle. Some impact categories like aquatic toxicity selected in food LCA may lead to wrong interpretation. Current impact categories used in study are industrial based on which some categories can be used in the food system directly, whereas some new categories need to include specific to food sector.

7.6 Conclusion

LCA method is effective in determining and understanding the environment of food products. This study discussed LCA analysis based on product approach and dietary approach. Product level approaches are good to understand impact of specific food product at the farm level and processing level. However, the food consumption depends on diet level that is important to understand the environmental impact at this level which informs the consumer. The consumer is one who can have more possibility to the environmental sustainability due to availability of alternative food product. Like, it was understood from the literature that meat production has a higher environmental impact compared to their alternative vegetarian diet. With this information, a consumer has a choice to choose the alternative product. However, the LCA analysis at diet level is complex due to various ingredient of the meal. Hence, it requires in-depth knowledge of food products processing and quality data to evaluate the environmental soundness of a food product using LCA. In the food system, production stage is crucial to environment and land use and water use are primary concern along with fertilizer use. Hence, new impact categories which showed the impact on land water pollution need to be integrated into LCA.

References

Arvanitoyannis IS (2008) ISO 14040: life cycle assessment (LCA)–principles and guidelines. In: Waste management for the food industries. Elsevier, Amsterdam, pp 97–132

Bánáti D (2011) Consumer response to food scandals and scares. Trends Food Sci Technol 22 (2–3):56–60

Barnett J et al (2016) Consumers' confidence, reflections and response strategies following the horsemeat incident. Food Control 59:721–730

Baroni L, Cenci L, Tettamanti M, Berati M (2007) Evaluating the environmental impact of various dietary patterns combined with different food production systems. Eur J Clin Nutr 61 (2):279–286

Basset-Mens C et al. (2010) Is LCA-based eco-labelling reasonable? the issue of tropical food products. Proceedings of the 7th International Conference on Life Cycle Assessment in the Agri-Food sector, pp 461–466

Béné C et al (2019) When food systems meet sustainability – current narratives and implications for actions. World Dev 113:116–130

Boone L et al (2016) Environmental life cycle assessment of grain maize production: an analysis of factors causing variability. Sci Total Environ 553:551–564

Bruno M et al (2019) The carbon footprint of Danish diets. Clim Chang 156(4):489–507

Carlsson-Kanyama A, Ekström MP, Shanahan H (2003) Food and life cycle energy inputs: consequences of diet and ways to increase efficiency. Ecol Econ 44(2–3):293–307

Caro D, Davis SJ, Bastianoni S, Caldeira K (2017) Greenhouse gas emissions due to meat production in the last fifty years. In: Ahmed M, Stockle CO (eds) Quantification of climate variability, adaptation and mitigation for agricultural sustainability. Springer, Cham, pp 27–37. https://doi.org/10.1007/978-3-319-32059-5_2

Castellani V, Sala S, Benini L (2017) Hotspots analysis and critical interpretation of food life cycle assessment studies for selecting eco-innovation options and for policy support. J Clean Prod 140:556–568

Davis J, Sonesson U, Baumgartner D, Nemecek T (2008) Veggie versus meat–environmental analysis of meals in Spain and Sweden. In: Life cycle assessment in the agri-food sector. Springer, Cham

Finkbeiner M et al (2006) The new international standards for life cycle assessment: ISO 14040 and ISO 14044. Int J Life Cycle Assess 11(2):80–85

González-García S, Esteve-Llorens X, Moreira MT, Feijoo G (2018) Carbon footprint and nutritional quality of different human dietary choices. Sci Total Environ 644:77–94

Hallström E, Carlsson-Kanyama A, Börjesson P (2015) Environmental impact of dietary change: a systematic review. J Clean Prod 91:1–11

Heller MC, Keoleian GA, Willett WC (2013) Toward a life cycle-based, diet-level framework for food environmental impact and nutritional quality assessment: a critical review. Environ Sci Technol 47(22):12632–12647

Huerta AR, Güereca LP, Lozanoa MSR (2016) Environmental impact of beef production in Mexico through life cycle assessment. Resour Conserv Recycl 109:44–53

ISO (International Organization for Standardization) (2006) ISO 14040:2006(E) environmental management–life cycle assessment – principles and framework

Johnson MD et al (2016) Greenhouse gas Emissions from US grain farms. J Crop Improv 30 (4):447–477

Jones AD et al (2016) A systematic review of the measurement of sustainable diets. Adv Nutr 7 (4):641–664

Kalhor T, Rajabipour A, Akram A, Sharifi M (2016) Environmental impact assessment of chicken meat production using life cycle assessment. Inform Process Agric 3(4):262–271

Klöpffer W (2012) The critical review of life cycle assessment studies according to ISO 14040 and 14044. Int J Life Cycle Assess 17(9):1087–1093

Laurent A et al (2014) Review of LCA studies of solid waste management systems–part II: methodological guidance for a better practice. Waste Manag 34(3):589–606

Lee H-J, Yun Z-S (2015) Consumers' perceptions of organic food attributes and cognitive and affective attitudes as determinants of their purchase intentions toward organic food. Food Qual Prefer 39:259–267

Longo S, Mistretta M, Guarino F, Cellura M (2017) Life cycle assessment of organic and conventional apple supply chains in the north of Italy. J Clean Prod 140:654–663

López-Andrés JJ et al (2018) Environmental impact assessment of chicken meat production via an integrated methodology based on LCA, simulation and genetic algorithms. J Clean Prod 174:477–491

Manning L, Soon JM (2016) Development of sustainability Indicator scoring (SIS) for the food supply chain. Br Food J 118(9):2097–2125

McCarthy D, Matopoulos A, Davies P (2015) Life cycle assessment in the food supply chain: a case study. Int J Log Res Appl 18(2):140–154

Meier T, Christen O (2013) Environmental impacts of dietary recommendations and dietary styles: Germany as an example. Environ Sci Technol 47(2):877–888

Mohammadi A et al (2015) Joint life cycle assessment and data envelopment analysis for the benchmarking of environmental impacts in Rice Paddy production. J Clean Prod 106:521–532

Notarnicola B, Hayashi K, Curran MA, Huisingh D (2012) Progress in working towards a more sustainable Agri-food industry. J Clean Prod 28:1–8

Noya I et al (2015) Comparative life cycle assessment of three representative feed cereals production in the Po Valley (Italy). J Clean Prod 99:250–265

Ogino A et al (2016) Environmental impacts of extensive and intensive beef production Systems in Thailand Evaluated by life cycle assessment. J Clean Prod 112:22–31

Poore J, Nemecek T (2018) Reducing food's environmental impacts through producers and consumers. Science 360:987–992

Raimondi A, Sanfilippo S, Fino D, Ruggeri B (2011) Towards the Assessment of an Ecological Index for Quantifying Sustainability of Day Life. A Case Study of the Environmental Consequences of Dietary and Transport in a Standard Workday. In ELETTRONICO, pp 1–10

Roer A-G et al (2012) The influence of system boundaries on life cycle assessment of grain production in central Southeast Norway. Agric Syst 111:75–84

Roy P et al (2009) A review of life cycle assessment (LCA) on some food products. J Food Eng 90 (1):1–10

Sonesson UG et al (2016) Paths to a sustainable food sector: integrated design and LCA of future food supply chains: the case of pork production in Sweden. Int J Life Cycle Assess 21 (5):664–676

Sturtewagen L et al (2016) Resource use profile and nutritional value assessment of a typical Belgian meal, catered or home cooked, with pork or Quorn™ as protein source. J Clean Prod 112:196–204

Tasca AL, Nessi S, Rigamonti L (2017) Environmental sustainability of Agri-food supply chains: an LCA comparison between two alternative forms of production and distribution of endive in northern Italy. J Clean Prod 140:725–741

Tichenor NE et al (2017) Life cycle environmental consequences of grass-fed and dairy beef production Systems in the Northeastern United States. J Clean Prod 142:1619–1628

Van Kernebeek HRJ et al (2014) The effect of nutritional quality on comparing environmental impacts of human diets. J Clean Prod 73:88–99

Vieux F, Soler L-G, Touazi D, Darmon N (2013) High nutritional quality is not associated with low greenhouse gas emissions in self-selected diets of French adults. Am J Clin Nutr 97(3):569–583

Vries M, Middelaar CE, Boer IJM (2015) Comparing environmental impacts of beef production systems: a review of life cycle assessments. Livestock Science 178: 279–288

Wang X et al (2017) Preliminary analysis on economic and environmental consequences of grain production on different farm sizes in North China plain. Agric Syst 153:181–189

Winkler T, Schopf K, Aschemann R, Winiwarter W (2016) From farm to fork–a life cycle assessment of fresh Austrian pork. J Clean Prod 116:80–89

Zhang D et al (2017) Carbon footprint of grain production in China. Sci Rep 7(1):4126

Food Wastes: Perceptions, Impacts and Management

8

John Onolame Unuofin, Oluwatosin Joseph Aladekoyi, and Olubusola Ayoola Odeniyi

8.1 Introduction

According to the Food and Agriculture Organization of the United Nations, approximately 795 million people were chronically undernourished between 2014 and 2016, worldwide, with an estimated 780 million from developing countries and 14.7 million people from developed countries (FAO 2015). Despite the soaring prevalence of undernourishment, food wastage remains correspondingly high (Kings 2018), and has become a substantial constituent of municipal wastes, which are constantly attracting major environmental concerns, worldwide (Ren et al. 2017).

In 2016, the amount of municipal waste generated was estimated to be 2.01 billion tons; this statistic is foreshadowed to reach an estimated 3.4 billion by 2050 (World Bank 2019) due to urbanization and population growth, which had also been estimated to reach 9.5 billion (FAO 2009) in the same year. Approximately 1.3 billion tons of food produced globally are either lost or wasted before being consumed by humans; this waste was projected to be one-third of what is intentionally produced for human consumption annually (FAO 2011). Moreover, out of the 1.5 quadrillion kcal (energy content) of food produced globally for human consumption, about 24% ends up becoming wasted. This essentially means that approximately one out of every four calories produced for human consumption ends up not being consumed by humans due to food waste.

J. O. Unuofin (✉)
Department of Microbial, Biochemical and Food Biotechnology, University of the Free State, Bloemfontein, South Africa

O. J. Aladekoyi
Environmental Applied Science and Management, Ryerson University, Toronto, ON, Canada

O. A. Odeniyi
Department of Microbiology, University of Ibadan, Ibadan, Nigeria

I. Haq, A. S. Kalamdhad (eds.), *Emerging Treatment Technologies for Waste Management*, https://doi.org/10.1007/978-981-16-2015-7_8

Ultimately, the effects of food waste can be assessed by studying how it affects the environment, society and the economy in both developed and developing nations of the world (Garcia-garcia et al. 2015). Vandermeersch et al. (2014) explained these effects by considering the strain they place on sustainability, ranging from crop losses during harvest or storage (which indicates economic impact) to hunger in developing countries (social), and the depletion of natural resources without accomplishing its intended purpose (environmental).

To our knowledge, the most habitually espoused approach to solving world hunger crisis and the constantly growing human population, till date, is the increase in food production. However, while this approach has proven to be seemingly advantageous, over the years, it does not solve the problem of food insecurity and food wastage, if its lifecycle is ratiocinated. During the different stages involved in transformation of food through the supply chain (initial agricultural production, processing and final household consumption), a very high amount of waste is generated. In Europe and the USA, about 89 million tonnes and 40 million tonnes of food are wasted, respectively. It has also been reported that twice the amount of food needed for actual nourishment is available to the consumers while half is wasted (Moorcroft 2012). These amounts of food wasted could cost sufficiently enough to feed the millions of hungry people globally (Food and Agriculture Organization, International Fund for Agricultural Development, UNICEF, World Food Programme and World Health Organization 2017); this implies that increasing food production might not be a sustainable solution. Therefore, a more effective approach to the management and sustainability of the food supply chain will be reduction of food lost or wasted and the adoption of effective treatment options that encourages the generation of value-added products.

There are several conventional methods for food waste management; however, the most common method employed worldwide is landfilling, which is non-beneficial, highly detrimental to the environment and poses health risk to humans. According to Eurostat (2017), 31% of municipal solid waste eventually end up in landfills while others are composted, recycled or incinerated (Fig. 8.1). Food waste has become the largest single component of municipal solid waste reaching landfills where it is spontaneously bio-converted to methane, a greenhouse gas (GHG) which contributes at least 25 times more in global warming outputs than carbon dioxide (CO_2) (Gunders 2012).

A report by The Waste and Resources Action Program (WRAP), United Kingdom (UK), estimated that food waste generates 17 million tonnes of CO_2 (Quested and Parry 2011), which is equivalent to the emissions released by 1 in 5 cars on UK roads. Greenhouse gases, such as methane, CO_2, when present in excess amount in the environment, absorb and re-emit infrared radiation, heating up the earth's atmosphere, thereby causing global warming and climate change.

Apart from these environmental factors, food waste also leads to waste in the resources used in the production, processing and manufacturing of food; which include capital, water, land, fertilizers and labour (Fig. 8.2).

Consequently, they encourage land, air and water contamination, such as eutrophication, acidification and photooxidation (Monier et al. 2010). According to a

Fig. 8.1 Classification of Municipal waste treatment in EU-28 according to their mode of treatment, (kg per capita) between 1995 and 2017. Source: Eurostat (2017)

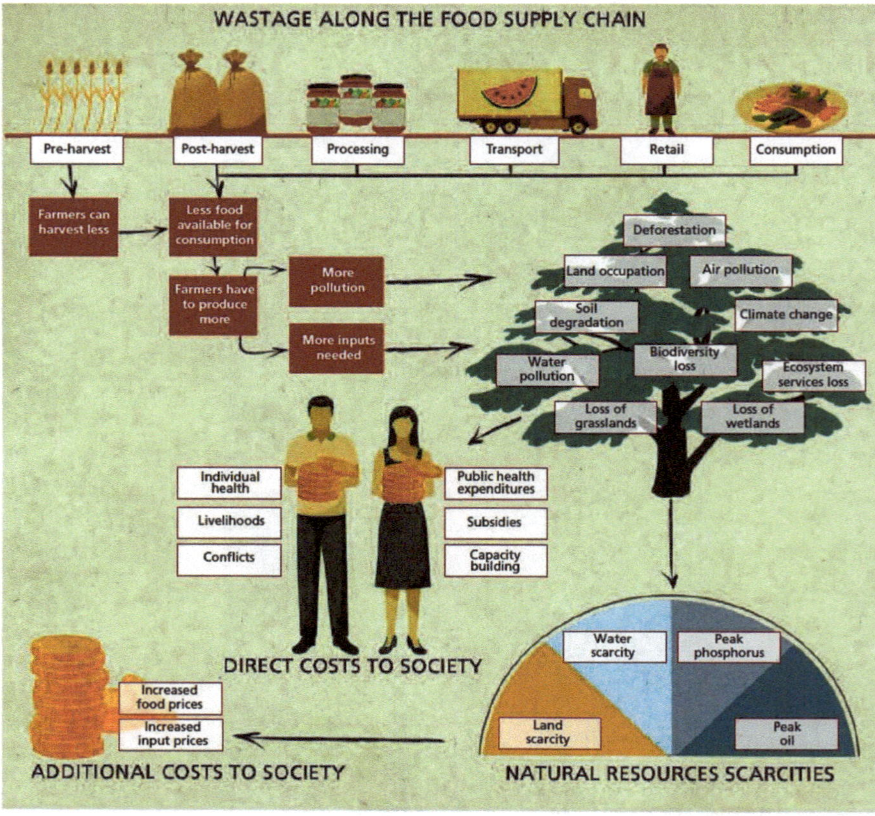

Fig. 8.2 The impacts of food wastage along the food supply chain. Source: FAO (2014)

report published by FAO (2014), the cost of food wastes as it specifically relates to the environment, economy and society could reach an estimated 700 Billion, 1 trillion and 900 billion U.S. Dollars, respectively, each year. Therefore, there is a need for the development of an effective management approach that will reduce the amount of food wastes generated across the food supply chain.

The Sustainable Development goal 12 (SDG 12) established by the United Nations in 2015 is a very good approach. It is aimed to cut the global food waste in half at the retail and consumer levels by 2030 through prevention, collection and treatment of food waste within the confines of cities (Newman et al. 2018). Though this approach has received so much support and collaborations from organizations, ranging from non-profit establishments to food campaigners and researchers in different nations and cities of the world, there is still much work to be done in dealing holistically with the global impacts that food wastes have on the environment, economy and society.

This study, therefore, gives a historian understanding on food wastes, their associated impacts, the different management options for food wastes and how they are being adopted around the world for the development of a sustainable future.

8.1.1 Food Waste/Loss Perception

According to Newman et al. 2018, food is defined as any substance, whether processed, semi-processed or raw, with parts intended (edible) and not intended (inedible) for human consumption. A good example is orange, which is a food, but its skin is inedible. Waste, according to the European Parliament and the Council of the European Union (2008), is any substance or object which the holder discards, or intends to, or is required to discard. Although, food loss and waste appear throughout this text, they are different in terms of how and where they occur across/within the food chain.

Food loss is a phenomenon whereby food undergoes an unintentional reduction in quality or quantity as a result of bruising, wilting, food spills, spoils and other damages, or gets lost before reaching the consumer. Food loss practically occurs due to agricultural or technical limitations at the production, storage, processing and distribution stages of the food value chain (Lipinski et al. 2013). Food waste refers to food that does not get consumed even though it is of good quality and fit for human consumption. In essence, it is food discarded, whether before or after it spoils, and this typically occurs at the retail and consumption stages in the food chain. It is majorly the result of negligence or intentional decision to throw food away (FAO 2013a, b, c).

However, throughout the food supply chain, food wastes can either be avoidable or unavoidable. Avoidable food waste is discarded food, even though it was intended for human consumption. Examples are meat, slices of bread, fruits, cake, etc. Unavoidable food waste is food that is discarded because it is not intended for human consumption. Examples are bones, peelings, eggshells, pineapple skin, corn cobs, etc. (WRAP 2009). Food loss and waste describe food in the supply chain right from the production stage, when the crops are ripe in the field, animals are matured in the farm or fishes are matured and have been harvested in nets. The food value chain ends when food produced have been consumed by humans, or discarded, or removed from the value chain before human consumption, even if it is used as animal forage, or for bioenergy production (FAO 2013a, b, c).

While clear distinctions have been highlighted between loss and waste by various researchers, the definition of food waste as rendered by European Project FUSIONS has been adopted as the most convenient and complete, as it includes both the edible and the inedible parts of food and does not discriminate between food loss and waste. This definition will also be adopted throughout this text. According to FUSIONS, food waste is edible and inedible parts of food that is discarded from the food supply chain which can be recovered or completely disposed. The food wastes in this category include those that are meant to be composted, unharvested crops, anaerobically digested for bioenergy production, incinerated, disposed to sewer, landfilled or discarded to sea (Östergren et al. 2014).

8.1.2 Food Wastes Categorization

Pfaltzgraff et al. (2013) distinguished the food waste generated in the food supply chain into two categories: the pre-consumer and post-consumer wastes. Pre-consumer wastes are waste generated during the stages of food production (wastes generated due to agricultural practices), food processing (wastes from food processing residues such as tomato pomace) and food distribution. Post-consumer wastes are wastes generated from meal preparation and consumption. Since food can be lost at each stage of the food supply chain, FAO (2013a, b, c) classified food waste in the food supply chain using five segments and explained how they can be lost within those segments using the following examples:

Agricultural Production: Food loss as a result of mechanical damage or spillage during harvest operation. This also covers grains left behind due to poor harvesting equipment, discarded fish, fruits discarded as a result of failing to meet quality standards and reduction in milk production due to dairy cow sickness.

Post-harvest handling and storage: This refers to wastages that occur due to spillage or degradation during handling, storage and transportation between farm and distribution. This includes loss caused by pests, fungus and disease; like fruit rot as noticed in tomatoes, the death and condemnation of an animal before arriving at the slaughter house, fish degradation during storage and milk spillage and degradation during transportation between farm and distribution.

Processing: Processed foods may be lost due to poor order forecasting and inefficient factory process. This includes milk spillage during industrial processing (such as yoghurt production), damaged fishes which are unsuitable for processing (such as canning or smoking), fruits which are unsuitable for juice production as a result of damage and meat loss during processing and packaging.

Distribution: This refers to edibles which are discarded or removed as a result of non-compliance with regulatory standards or the nonsalable status they assume after expiry dates. This includes fast moving consumer goods which are wasted in the market system, especially the non-durable goods like fruits and vegetables, breakfast cereals, yoghurt, margarine, snacks, etc.

Consumption: This includes waste generated during consumption at the household level. Advertisements and promotional offers have been linked to some of the factors which stimulate consumers to buy more than they need before the expiration date of a product. On or close to expiration, consumers dispose products due to misinformation that expired products have become stale and unfit for human consumption.

8.1.3 Quantification of Food Wastes

Any attempt to understand how to effectively manage food waste should begin with understanding where it occurs regionally. According to FAO (2011), data from a 2009 survey revealed about 56% of total food loss and waste occurred in the developed world (Industrialized Asia, Oceania, Europe and North America) while about 44% of total food loss occurs in the developing nations of the world (Fig. 8.3); however, the amount of food wasted varies significantly across different regions of the world.

Figure 8.4 shows that 280–300 kg of food loss (from production to retailing) is generated in Europe and North America per capita per year; in South/Southeast Asia and sub-Saharan Africa, per capita food loss is 120–170 kg/year. The amount of food waste (at the consumer level) generated per capita in Europe and North America is 95–115 kg per year whereas in sub-Saharan Africa and South/Southeast, it is 6-11 kg per year.

Food waste can also be quantified by observing where they occur in the food supply chain, using the different stages of food production discussed earlier. About 24% of total food loss occurs at the stage of production while during handling and storage, about 24% food loss is also recorded, and 35% during consumption (Fig. 8.5). Total food waste in developed countries is as high as in developing nations of the world. In developing countries, over 40% of food wastes occur at the production and processing stages, while in developed countries, over 40% of food wastes occur at the distribution and consumption stages of the food supply

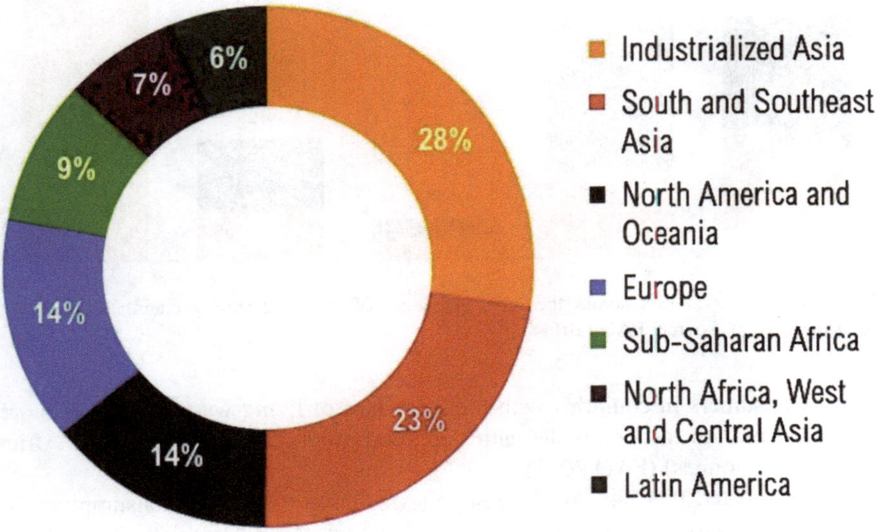

Fig. 8.3 Quantification of Food Loss and Waste by Region in 2009. Source: The World Resources Institute (WRI) analysis from data generated by FAO (2011)

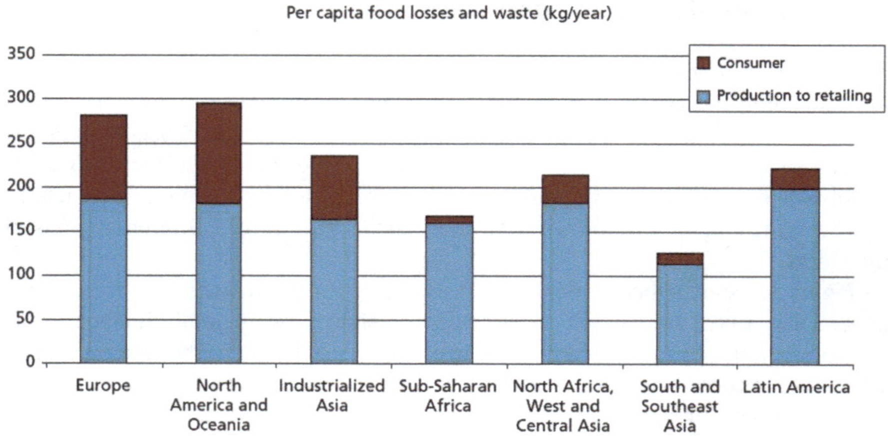

Fig. 8.4 Per capita food losses and waste (kg/year), at production and consumption stages, in different regions. Source: FAO (2011)

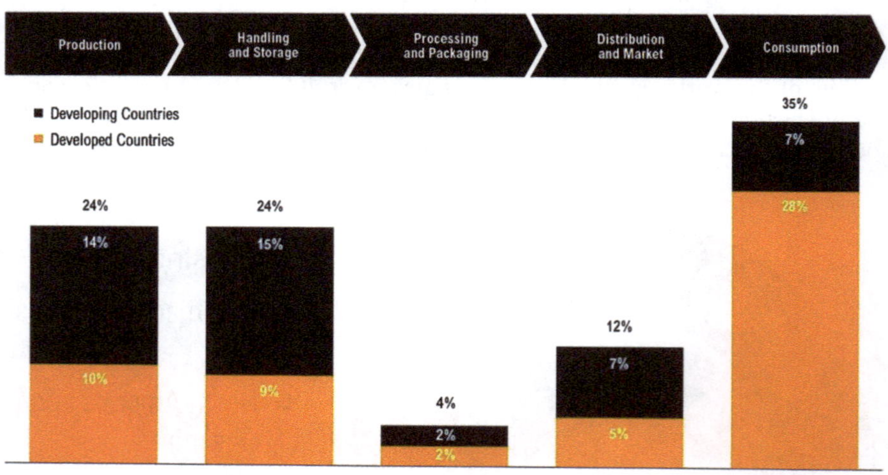

Fig. 8.5 Total food loss across the different stages of the food supply chain (100% = 1.5 quadrillion kcal). Source: FAO (2011)

chain. Consumers in countries with high standard of living waste almost as much food (222 million tonnes) as the entire net food production of sub-Saharan Africa (230 million tonnes) (FAO 2011).

It is estimated that 34% of food supplies suitable for human consumption are being lost or wasted in Near East and North Africa (NENA) region (FAO 2011). According to World Wildlife Fund for Nature (2017), food waste in South Africa was estimated to reach 33% per annum. About 40% of food is lost and wasted in the USA during processing and distribution by retailers, restaurants and consumers (Gunders 2012). Also, approximately 30–40% of food produced in the United

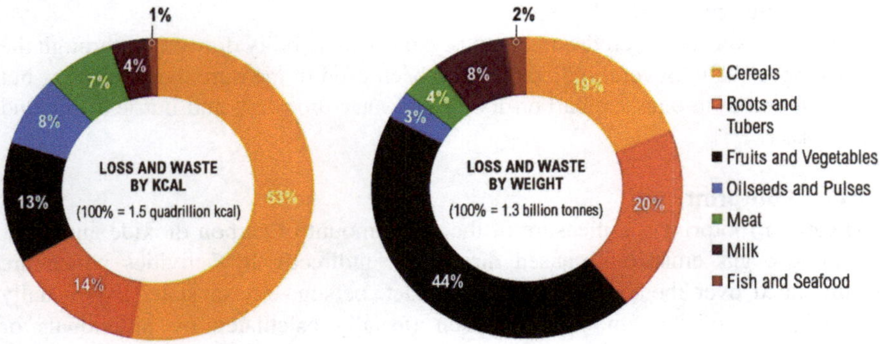

Fig. 8.6 Categorization of food loss and waste based on caloric content and weight. Source: WRI analysis based on data generated by FAO (2011)

Kingdom and Japan are removed from the food supply chain every year due to food waste (Kosseva 2009).

Quantification of food waste has also been achieved by observing the loss in weight and caloric content of different types of agricultural products. Fig. 8.6 shows that 44% of food removed from the supply chain are fruits and vegetables; this value represents the largest source of food waste by weight, followed by root and tubers (20%) and cereals (19%). This is primarily because of variation in water content; much of which is lost and wasted in fruits and vegetables. However, observing food waste by caloric content loss highlights a totally different agricultural product. While fruits and vegetables are the largest source of food loss based on weight, cereals constitute the most common source of food loss based on caloric content.

Whether by weight or caloric content, the reduction of food waste cannot be overemphasized. Although meat represents a small amount (4% by weight and 7% by caloric content) of food lost or wasted in the food supply chain, its significant environmental impact in relation to land use, water consumption, eutrophication and the release of greenhouse gases coupled with its economic impact suggests that it is worthy of as much attention as other agricultural products (Searchinger et al. 2013).

Since a significant portion of fruits and vegetables is composed of water, reducing their removal from the food supply chain will also reduce the loss of water, vitamins and minerals needed for healthy living (Gebhardt and Thomas 2002).

8.1.4 Impacts of Food Waste

Food waste is both an environmental and socioeconomic problem that affects food security for the impoverished, food quality and safety, the environment, economy and the society. It represents a waste of resources (like energy, land and water) used in the production of food (FAO 2011).

8.1.4.1 Environmental Impact

Food waste exposes the environment to several sustainability drawbacks through the food supply chain. Several indicators have been used to measure these impacts, but the most common ones are carbon footprint, water footprint, and nutrient loss and land loss.

Carbon Footprint

The carbon footprint is a measure of the total amount of carbon dioxide and other greenhouse gas emissions caused directly or indirectly by activities which are accumulated over the life stages of a product, person, city or state. It is usually the reflection of the amount of carbon (usually calculated by individuals or organizations) that has been emitted over a period of time or during the life cycle of a project (Goodier 2010). Carbon dioxide (CO_2), methane (CH_4) and nitrous oxide (N_2O) are greenhouse gases that contribute the most to global warming and climate change (Garnett 2011). They are emitted at all the stages of the food supply chain, including the release of carbon stored in cleared biomass due to change in land use from forests to agriculture, emissions from livestock (e.g. cattle), slurries and manures, production of energy from fossil fuels and decomposition of food waste in landfill sites (Newman et al. 2018). The global carbon footprint of food waste was estimated to be 3.3 CO_2Gtonnes equivalent per year with Asia being the largest contributor majorly because of the generation of cereal waste (FAO 2013a, b, c). Globally, cereal is the largest contributor to carbon footprint (34%) followed by meat (21%) and vegetables (21%). It is worthy of note that 15% of the total food waste is of animal origin, but it contributes 33% of the total carbon footprint.

Water Footprint

Animals and plants rely essentially on water to produce food that is consumed by humans. According to Hoekstra et al. (2011), the water footprint of a product is defined as the total volume of fresh water that is used directly or indirectly to produce a product. About 92% of the water footprint by humans is as a result of the consumption of agricultural products (Hoekstra and Mekonnen 2012). How water is being used can be measured by distinguishing the water footprint into three sub-categories: blue, green and grey water. Of all the three sub-categories, blue water footprint is the most relevant to food waste because it measures the total amount of surface water and ground water consumed for agricultural purposes. The utilization of blue water in irrigated agriculture can lead to serious environmental issues such as salinization, water logging for soil degradation and water depletion (Aldaya et al. 2010).

In 2007, the global blue water footprint of food wastage for agriculturally produced food is about 250km^3 which is more than 38 times the blue water footprint of the USA households (FAO 2013a, b, c). The water footprint varies widely between different products. The major contributor to the blue water footprint is cereal, representing 52% of total blue water footprint, followed by fruits with 18%. Animal feed crops have a higher water footprint per kg of food product due to the essentially large amounts of water that is needed to grow them. Therefore, a significant reduction and proper management of food waste will greatly improve water sustainability, especially through appreciable reduction in the formation of leachates, which might serves as niches for pathogenic microorganisms, and contamination of groundwater. Improving the irrigation system also can result in up to 33% efficiency in water use (Roka 2019).

Nutrient Loss

The United Nations Sustainable Development Goal 2, Target 2.2 is aimed to end all forms of malnutrition by 2030. Malnutrition may address an individual's state of health, marked by the deficiencies in macronutrients (carbohydrate, proteins and fats) or micronutrients (minerals and vitamins). This may in turn directly impact the population of a region throughout the lifecycle—from childhood into adulthood—leading to acute and chronic cases of malnutrition, and even death (Food and Agriculture Organization, International Fund for Agricultural Development, UNICEF, World Food Programme and World Health Organization 2017).

About 95% of all food consumed by humans are grown in soil (FAO 2015). The main constituents of plants are carbon and water; nitrogen (N), phosphorus (P) and potassium (K), amongst other nutrients, are also essential for their growth. Plants utilize carbon through photosynthesis, whereas these nutrients are obtained from soil, as well as organic or synthetic fertilizers which are applied by farmers. Conversely, over 133 billion tonnes of carbon have been lost from soils, globally, since the dawn of agriculture (Ecowatch 2017). This lost carbon ends up in the atmosphere as greenhouse gas emission, thereby, reducing the quality of air, and ultimately, leading to climate change. Moreover, with the rapid increase in global population and urbanization, the desperate demand on food production would become intense, thereby subjecting already fertile lands to exhaustive unsustainable agricultural practices. These practices have resulted in the depletion of the organic matter and nutrients found in soil and provide little or no room for agricultural lands to recover lost nutrients. Therefore, prevention of food waste will reduce the pressure on land for the higher yields needed to meet with the growing population concern. If food waste is alleviated, the resultant less aggressive food production would afford agricultural land the opportunity to replenish, thereby, reducing land degradation (Newman et al. 2018), and ultimately, impede occurrences of synthetic fertilizers and nutrient run-off to surface water bodies which causes eutrophication and algal bloom; a great threat to water potability.

Land Use

In the past 50 years, modern technologies and farming techniques have helped in the expansion of agricultural land by 12%, and 4.9 Global hectares (Gha) out of the 10 Gha of usable land surface is currently being used for food production (Roka 2019). Regrettably, an estimated 30% of agricultural land (about 1.4 billion hectares), worldwide, is used to produce food that is eventually removed from the food supply chain without human consumption. Meat and milk are the major contributors of land use, having the highest impact per kg of product, occupying mainly non-arable land. Meat and milk represent 78% of total surface although their share of total food wastage is only 11% (FAO 2013a, b, c). The management of produced food can help reduce the pressure to increase land use.

8.1.4.2 Socioeconomic Impact of Food Waste

The impact of food waste goes beyond the excessive use of natural resources and greenhouse emissions; how it affects the society and economy is also considered. The total socioeconomic costs of food waste to the global economy are in the order of 1.9 trillion US Dollars (FAO 2014). The socioeconomic problem of food waste is characterized by the loss of money and the resources like land, water, labour, energy and other inputs, used in producing the food; thereby leading to food insecurity and inequality of food availability for the future population. Food insecurity is best captured as a scenario where the amount of food consumed by one or members of a household is reduced and dietary patterns are disrupted due to lack of money to make food purchase (Roka 2019). According to the Food and Agriculture organization of the United Nations (FAO 2013a, b, c), vegetables are the major contributors to the economic cost of food waste, representing 23% of total cost, followed by meat (21%), fruits (19%) and cereals (18%). Meat is also a high contributor to the economic cost of food lost and wasted; although it represents 4% of total food wasted, the total economic cost of meat wastage is about 20%. Regionally, the major contributors to food wastage volumes and the economic cost of food lost and wasted are Industrialized Asia, with 31% of total cost and South/Southeast Asia accounting for 18% of total cost. A product's contribution to the total cost of food wastage is driven by its cost of production per kilogram.

In summary, feeding the world's growing population by proportionally increasing food production will also lead to more food wastage. Globally, a 70% increase in food production is estimated to feed the world's growing population, but preventing food waste can help in making the increase in food production sustainable while at the same time alleviate poverty and food insecurity in developing and developed countries, worldwide (Roka 2019).

8.1.5 Effective Waste Management Options

Different researchers all over the world have suggested different approaches to food waste management and a few of them are going to be discussed in this section. Memon (2010) proposed a 3 R's approach, which is "Reducing" the amount of waste generated, "Reusing" the waste, and "Recycling" them. This approach does not only aim to minimize the amount of waste generated but also emphasizes the use of waste to generate value-added products. Furthermore, to effectively manage the food waste that occurs throughout the food supply chain, there is a pressing need for an accurate waste categorization that does not only consider the environmental, social and economic impacts of food waste but also employ these impacts to implement a more effective plan for waste management.

Categorization is very important because it encourages the selection of the best method in managing food waste. Garcia-Garcia et al. (2017) developed a nine-step categorization for food waste using different indicators at every step to select the best management options for different kinds of food waste. Table 8.1 gives a technical summary to this approach, where the nine-staged design restricts the wastes found in each category to two or three types, thereby allowing treatment using the most effective management option. This approach is better than the less effective approaches that simply group waste and treat them generically.

Out of the five catalogues that were designed for managing and minimizing waste, as prepared by Riemer and Kristoffersen (1999) in a technical report for the European Topic Centre on Waste, catalogue 5 and catalogue 2 were aimed at addressing preventive approaches to waste generation (waste minimization) and management of generated waste, respectively. While the former encourages the reduction of waste from the manufacturing step and addresses the recycling and re-use of a product internally, the latter addresses the re-use of the product or part of the product for any other purpose (externally) or for energy recovery.

In 1975, the European Parliament Council introduced a waste management policy using waste hierarchy which was later legislated by the waste management strategy team in 1989 (European Parliament Council 1989). Since its legislation, it has been adopted globally by different waste management networks such as The Waste and Resources Action Programme (WRAP (Waste and Resources Action Programme) 2011) and Department for Environment, Food and Rural Affairs (DEFRA, Environment Agency, WRAP 2011) as the primary module for waste management. The purpose of having a waste hierarchy is to identify the most effective waste management option that will be least harmful to the environment. The waste hierarchy, when used for the management of food waste, is a very important tool for selecting alternative management options that will encourage sustainability.

Figure 8.7 represents the food and drink material hierarchy that was developed by the United Nations Environment Program (UNEP (United Nations Environment Programme) 2014), which presents a guide on the preferable options of dealing with waste in order to minimize their environmental and social impact. Garcia-garcia et al. (2015) described a more detailed approach to categorize food wastes and their ideal management option by using the common waste hierarchy (reduce, re-use, recycle, recover and disposal), showing the most preferred option at the top of the hierarchy and the least preferred option at the bottom (Fig. 8.8). While this approach highlights the environmental impact of each treatment, it also takes into consideration its socioeconomic benefits.

While prevention of food removal (food waste) from the food supply chain is imperative, it is also sagacious to have a secondary plan to optimize the waste generated, by putting them into alternative uses, either to be redistributed to people or sent to animal feed. If a manufactured food (which is still considered edible and fit to be consumed) cannot be used by the manufacturer, it should be redistributed to people who can utilize it. But if it is considered unsafe for human consumption, it should be used to feed animals. However, in this regard, some products have been considered unsafe for animal consumption. The European Commission laws enshrined in the Animal by-product regulations ban the use of some food wastes to feed animals (European Union 2017).

Table 8.1 Indicators to categorize food waste for effective waste management

	Indicators	Description
1	Edibility -Edible -Inedible	Edible waste is the part of food that is expected to be consumed by humans, e.g. meat, bread, etc. Inedible wastes are by-products obtained during production, e.g. bones peels, etc. However, wastes that possess edible parts are considered edible (e.g. egg, even when it has an inedible shell)
2	State -Eatable -Uneatable -Uneatable for humans -Eatable for humans	A product is eatable if it is still fit for human consumption at the point of disposing as waste. Uneatable waste can be a badly processed food, rotten or expired food (e.g. burnt food). At this stage, the preferred management option for a waste that is both edible and eatable is redistribution to people. Also, products considered uneatable for humans can be considered for animal feeding
3	Origin -Animal-based -Plant-based	A waste is considered animal-based if it was part of an animal or produced by an animal (e.g. meat, dairy, honey, etc.); otherwise, the waste is plant-based
4	Complexity -Single product -Mixed product	Waste is single if it is the product of only one type of ingredient and it has not been in contact with other food material. Waste is mixed if it is the product of different types of foods
5	Animal Product Presence -Meat/Animal product/ By-product from animal bodies -In contact with animal-based product -Not in contact with animal-based product	This option is in two stages. If the product is animal-based, it must be classified as meat (including fish), animal product (e.g. eggs, honey, dairy, etc.) or by-product from animal bodies not intended for human consumption (e.g. by-products from slaughterhouses like hides, skins, horns, etc.). Compounds of interest can also be recovered through the treatment of by-products (e.g. fat obtained through rendering can be used to make soaps). According to The European Union (2009), waste should be classified into Category 1, 2 or 3. These rules are used in preventing disease outbreak in plant-based that has been in contact with animal-based product. Animal-based products should be composted (WRAP (Waste and Resources Action Programme) 2011)
6	Treatment -Processed -Unprocessed	Food is considered processed when it has completed the production process, or if it requires no processing before being distributed (e.g. vegetables and fresh fruits). Unprocessed food is food that still requires treatment at the point of its management as waste
7	Packaging -Packaged -Unpackaged (or separable from packaging)	Unpackaged product is a product that is not contained in a packaging material or that is contained in an easily separable package. Otherwise, the product is packaged
8	Packaging biodegradability -Biodegradable -Non-biodegradable	Biodegradable materials are those that be degraded by microorganisms in the absence of oxygen (anaerobic digestion) or in the presence of oxygen (composting). Biodegradable packaging includes paper, bioplastics, wood, etc. Non-biodegradable packaging materials include plastic, metal or glass. If the packaging is not biodegradable, the product must not be subjected to anaerobic digestion, composting or landspreading

(continued)

Table 8.1 (continued)

	Indicators	Description
9	Stage of the supply chain -Catering waste -Non-catering waste	Catering waste includes domestic waste and waste from food services (e.g. hotels, schools, restaurants, etc.). Waste generated in the earlier stages of the food supply chain (i.e. manufacturing, distribution or retailing) is known as the non-catering waste. According to law, catering waste must not be used for animal feeding (DEFRA. Department for Environment, Food and Rural Affairs 2014)

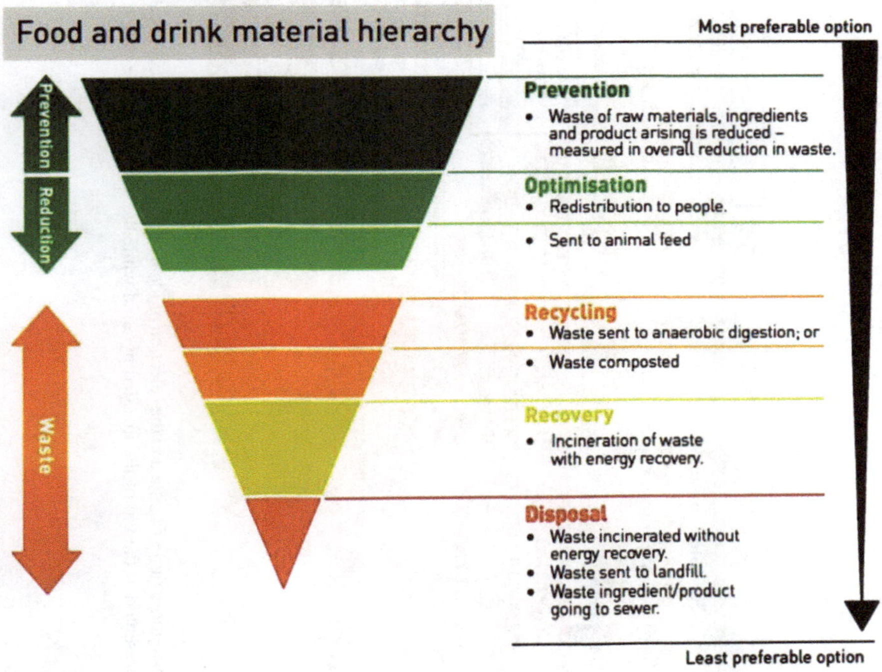

Fig. 8.7 Food and drink material hierarchy for effective management of food waste. Source UNEP (United Nations Environment Programme) (2014)

Notwithstanding, disposal is the least preferred option because it leads to loss of energy and nutrients, and also poses serious environmental threats such as the release of environmentally noxious gases that incite climate change, generation of offensive odours and the contamination of water bodies (Newman et al. 2018). In this regard, food wastes could be repurposed thus: (1) once food has been considered unsafe for both human and animal consumption, it can be treated for the extraction of compound of interest. Essential oils can be extracted from meats, fruits and vegetables for the production of fuel (bioethanol) and soaps. A review by Osorio et al. (2021) and references therein have epitomized our optimism regarding the inexhaustibility of derivatives of the agri-food industry, and their adaptability as sustainable

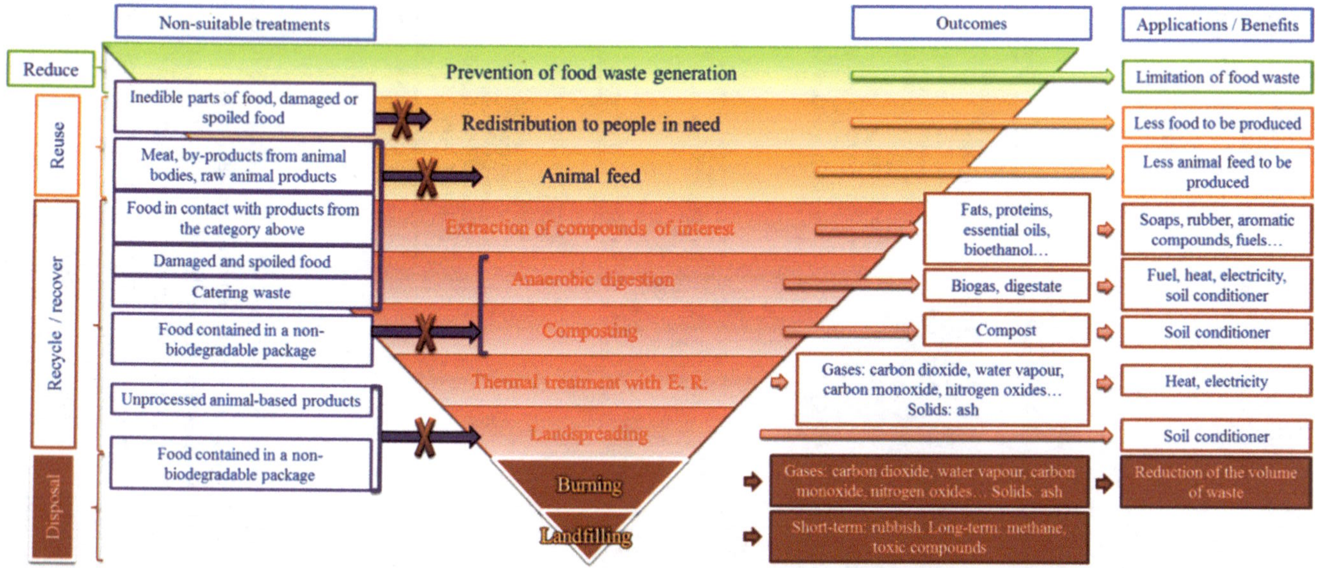

Fig. 8.8 Food waste hierarchy for the effective categorization of waste and the selection of ideal management options. Source: Garcia-garcia et al. (2015). *The most preferable option, which is prevention, is presented at the top of the inverted pyramid while the least preferable option (disposal) is at the bottom

feedstock for some much-needed products, spanning the food, agriculture and pharmaceutical sectors (2) Where the extraction of compound of interest is not possible, food waste can be recycled via anaerobic digestion for the generation of biogas (bioenergy), which can be used to generate heat, fuel or electricity. In this regard, a recent review by Kavathia and Prajapati (2021, in press) deliberated on the circular economy-oriented use of fruit and vegetable waste for a biomass-fired combined heat and power system (CHP), intended for an organic Rankine cycle (ORC) operation. Here, it was proposed that inferred that the method is more sustainable, with regard to energy, capital and maintenance costs (3) The digestate from the anaerobic digester can be composted for nutrient recovery, especially if combined with treated wastewater colloids and particulates as a mulch mix (in the case of lignocellulose fibrous residues) (Unuofin et al. 2019a) (4) When food waste cannot be recycled, incineration is used for energy recovery; when energy cannot be recovered from wastes, the residues from combustion can be used as inert materials for cement composites (Roychand et al. 2021), which signifies an innovative bid to ensure a circular economy of agri-food industry derivatives. Alternatively, non-edible food composites, such as agroresidues as well as aquaculture and animal husbandry disposables could be adopted as substrates for the microbial-enabled production of vital fine chemicals of importance in industrial, food and environmental biotechnology. This is evident from a series of studies where some agroresidues showed potentials for high yield of novel biomolecules (Odeniyi et al. 2017; Unuofin et al. 2019b, c, d; Unuofin 2020a, b).

Worldwide, different functional agencies have been set up in many countries and cities, with the enormous task of providing unique management plans to generated wastes. Table 8.2 shows the examples of food waste management activities that are employed in different countries and the involved agencies. The management activities range from food waste collection to how the wastes are recycled to obtain value-added products.

8.1.6 Conclusion and Future Anticipations

Removal of food from the supply chain, either as loss at the manufacturing stage or as waste at the consumption stage, has become a primary concern both locally and globally. Food waste has socioeconomic impacts throughout the supply chain, right from the production cost by the producer to purchasing and waste disposal costs. These impacts can be clearly observed in how they influence the total socioeconomic cost of wasted food, the amount of food wasted and a rapid rise in food insecurity within several households worldwide, especially in developing countries. Nutrient loss, depletion of natural resources, greenhouse gas emission and climate change are the predominant impacts of food waste on the environment. This text addressed the need for an effective and sustainable management of food waste. While there is a need for increase in food production due to expected increase in population, there is also an importunate need for a sustainable management option for the amount of food produced, as well as the amount of food consumed, due to the increasing

Table 8.2 Food waste collection and treatment activities in selected countries

Country and agency involved	Collection and treatment activities
Hartberg, Austria—private waste management company (Saubermacher)	Food waste from catering and other businesses are collected and delivered to a biogas plant which is about 2 km away from Hartberg's town Centre Annually, about 5450 tonnes of food waste are collected from catering services, 530 tonnes from beverage production industry as well as waste from slaughterhouses, dairy farms, fruits and vegetables Treatment of food waste is via anaerobic digestion. The biogas produced from this process is used to generate heat and electricity through a combine heat power (CHP) unit. Digestate is used by farmers as soil amendment for crops
New York City, USA—Department of Sanitation of New York (DSNY)	In 2006, DSNY released a comprehensive solid waste management plan that collects yard waste and food scraps, and also created a Compost Facility Siting Task Force Food scraps such as meat fruit, vegetable and dairy food waste are collected and treated via composting which is used as compost for parks, community gardens street trees, etc.
Copenhagen, Denmark—private waste management company	Private companies are hired by the municipality. The city launched a mandatory food waste management scheme that covers a population of about 600,000 inhabitants, including 300,000 households (in multifamily houses and 20,000 in single family houses) About 10,000 tonnes of food waste is expected to be collected as breakfast products, fruits and vegetables, eggs and eggshells, nuts and nutshells, etc. The waste is treated by anaerobic digestion. During the pretreatment stage, the non-biodegradable waste is carefully sorted out while the biodegradable waste is converted into a bio-pulp that can be pumped into a biogas reactor tank along other wastes from slaughterhouses, fishing industries and breweries to produce biogas with quantity reaching 7,500,000 m^3 per year. The gas produced is used to produce electricity and generate heat for houses in local districts. The digestate from the tank is used by local farmers as fertilizers for the fields
Cajica, Colombia—Empresa de Servicios Públicos de Cajicá (EPC) and IBICOL	EPC and IBICOL run a source-segregated organic waste collection programme since 2008. This programme serves 25,000 and 88,000 inhabitants About 480 tonnes of organic matter is collected from homes and schools per month Organic matter is composted using Bokashi EM (effective microorganisms), a rice/wheat bran-based material which has been fermented with

(continued)

Table 8.2 (continued)

Country and agency involved	Collection and treatment activities
	mixed microbial cultures and dried. The composting process takes place at an IBICOL facility using kitchen waste by crushing and spraying with activated EM. The compost is used for growing vegetables
Auckland, New Zealand—Auckland Council	Food waste accounts for 40% of the waste stream in the highly urbanized city. Auckland council aims to achieve zero waste by 2040 by turning its waste into resources The feed stock collected is food waste with volume expected to reach 50,000 tonnes per annum from 2500 tonnes in 2018 Treatment is by composting using a combination of aerated static pile and Gore-Tex cover system
Seoul, South Korea	Volume-based bags, weight-based radio frequency identification (RFID) and waste containers are used to collect household wastes. Collected food wastes are converted to animal feed, anaerobically digested, composted and also used directly on farmlands. Over 500 tons of waste water is converted to biogas

Adapted from the Global Food Waste Management Full Report by Newman et al. (2018)

concerns about the quantity and impacts of food removed from the food chain annually without being consumed by humans.

Different approaches have been introduced for the management of food waste, and these approaches vary and have been applied in different regions of the world. Categorization is necessary for identifying the most effective options for the management of food waste. Using the waste hierarchy method, the impacts of food waste to the environment, economy and society can be addressed. The food waste hierarchy can be used as a framework for the prevention (or the minimization) of food waste (by improving the agricultural practices used in food production), optimization of waste (by redistributing food to humans, and animals when food is unsafe for human consumption) and recycling of waste to produce value-added products for energy and power. Disposal of waste through landfilling and thermal treatments without energy recovery should be discouraged as much as possible due to their environmental impact and effect on human health; instead, different avenues for incorporating the ideology of circular economy into food waste management must be sought after.

Additional research is needed to engender transition from less sustainable approaches on waste management to more sustainable ones. Also, effective waste management policies should be introduced, integrated and implemented by relevant policy makers across the food supply chain. Efforts should also be made to ensure that the currently inefficient policies are restructured and amended so that food waste can be efficiently minimized and prevented. Since some of the data presented by regulatory bodies concerning food waste are based on estimated values and projections, more accurate findings should be made to address the amount of food waste that occurs regionally and where it occurs along the food chain. In developing countries, improvement of harvesting equipment, harvesting and storage techniques will greatly reduce food waste. For developed countries, more efficient approaches should be employed at the processing and consumption stage to reduce food waste. The major concerns in the conversion of food waste to generate value-added products are the production cost of these products and the quality of the products generated. Novel methods that will improve the quality of end products while putting the environmental and economic impacts into consideration should be developed.

References

Aldaya MM, Allan JA, Hoekstra AY (2010) Strategic importance of green water in international crop trade. Ecol Econ 69(4):887–894. catering%20waste%20leaflet.pdf

DEFRA (Department for Environment, Food and Rural Affairs) (2014) Feeding catering waste to farmed animals is illegal. http://www.chippingnortonvets.co.uk/modules/downloadspl2/files/DEFRA%20

DEFRA, Environment Agency, WRAP (2011) Applying the waste hierarchy: evidence summary. London

EcoWatch (2017) World's soils have lost 133bn tonnes of carbon since the dawn of agriculture. https://www.ecowatch.com/soil-carbon-loss-2478725457.html

European Parliament and the Council of the European Union (2008) Directive 2008/98/EC of the European Parliament and of the council of 19 November 2008 on waste and repealing certain directives. Official Journal of the European Union. 312: 3–30.

European Parliament Council (1989) A community strategy for waste management, SEC/89/934 (final). Brussels

European Union (2017) Commission regulation (EU) 2017/893 of 24 May 2017 amending annexes I and IV to regulation (EC) no 999/2001 of the European Parliament and of the council and annexes X, XIV and XV to commission regulation (EU) no 142/2011 as regards the provisions on processed animal protein (text with EEA relevance). OJ L. Vol. 138

Eurostat (2017) Municipal waste treatment in EU-27 by type of treatment, (kg per capita), 1995–2017

FAO (Food and Agriculture Organization of the United Nations) (2011) Global food losses and food waste: extent, causes and prevention. Rome

FAO (Food and Agriculture Organization of the United Nations) (2013a) Food waste footprint: Impacts on natural resources. http://www.fao.org/docrep/018/i3347e/i3347e.pdf

FAO (Food and Agriculture Organization of the United Nations) (2013b) Food loss and waste: definition and scope

FAO (Food and Agriculture Organization of the United Nations) (2013c) Food wastage footprint. Impacts on natural resources. In: Presentation of the FAO FWF study. Food Wastage Footprint project

FAO (Food and Agriculture Organization of the United Nations) (2015) Healthy soils are the basis for healthy food production. http://www.fao.org/soils-2015/News/news-detail/en/c/277682/

FAO (Food and Agriculture Organization) (2009) How to feed the world in 2050. Food and Agriculture Organization

FAO (Food and Agriculture Organization) (2015) The state of food insecurity in the world meeting the 2015 international hunger targets: taking stock of uneven progress. Food and Agriculture Organization of the United Nations, Rome

Food and Agriculture Organization, International Fund for Agricultural Development, UNICEF, World Food Programme and World Health Organization (2017) The state of food security and nutrition in the world 2017: building resilience for peace and food security

Food and Agriculture Organization of the United Nations, FAO. 2014. Food wastage footprint. Full-cost accounting. Final report. Rome, Italy

Garcia-garcia G, Woolley E, Rahimifard S (2015) A framework for a more efficient approach to food waste management. Int J Food Eng 1(1):65–72

Garcia-garcia G, Woolley E, Rahimifard S, Colwill J, White R, Needham L (2017) A methodology for sustainable management of food waste. Waste Biomass Valor 8:2209–2227

Garnett T (2011) Where are the best opportunities for reducing greenhouse gas emissions in the food system (including the food chain)? Food Policy 36:S23–S32

Gebhardt S, Thomas R (2002) Nutritive value of foods. U.S. Department of Agriculture, Agricultural Research Service, Home and Garden Bulletin 72

Goodier C (2010) Carbon footprint. In: Cohen N, Robbins P (eds) Green cities: an A-to-Z guide. SAGE Publications, London, pp 49–53

Gunders D (2012) Wasted: how America is losing up to 40 percent of its food from farm to fork to landfill. In: Natural resources defense council issue paper, IP: 12-06-B

Hoekstra AY, Mekonnen MM (2012) The water footprint of humanity. Proc Natl Acad Sci U S A 109(9):3232–3237

Hoekstra AY, Chapagain AK, Aldaya MM, Mekonnen MM (2011) The water footprint assessment manual. Setting the global standard. Water Footprint Network, London

Kings A (2018) Food waste: a growing global concern with dire consequences. Biotech. https://born2invest.com/articles/food-waste-growing-global-concern/

Kosseva MR (2009) Processing of food wastes. In: Taylor SL (ed.) Advances in food and nutrition research, vol 58. pp 57–136

Lipinski B, Hanson C, Lomax J, Kitinoja L, Waite R, Searchinger T (2013) Reducing food loss and waste. In: Working paper, installment 2 of creating a sustainable food future. World Resources Institute, Washington. http://www.worldresourcesreport.org

Memon MA (2010) Integrated solid waste management based on the 3R approach. J Mater Cycles Waste 12(1):30–40

Monier V, Mudgal S, Escalon V, O'Connor C, Gibon T (2010) Preparatory study on food waste across EU 27. European Commission (DG ENV), Directorate C – Industry

Moorcroft S (2012) A coming food waste revolution? https://innovationmanagement.se/2012/05/16/a-coming-food-waste-revolution/

Newman D, Cepeda-Márquez R, Zeller K (2018) Global food waste management: an implementation guide for cities. World Biogas Assoc:20

Odeniyi OA, Unuofin JO, Adebayo-Tayo BC, Wakil SM, Onilude AA (2017) Production characteristics, activity patterns and biodecolourisation applications of thermostable laccases from Corynebacterium efficiens and Enterobacter ludwigii. J Sci Ind Res 76:562–569

Osorio LLDR, Flórez-López E, Grande-Tovar CD (2021) The potential of selected Agri-food loss and waste to contribute to a circular economy: applications in the food, cosmetic and pharmaceutical industries. Molecules 26(2):515

Östergren K, Gustavsson J, Bos-Brouwers H, Timmermans T, Hansen O (2014) FUSIONS definitional framework for food waste. Full report, Food Use for Social Innovation by Optimising Waste Prevention Strategies (FUSIONS)

Pfaltzgraff LA, De Bruyn M, Cooper EC, Budarin V, Clark JH (2013) Food waste biomass: a resource for high-value chemicals. Green Chem 15:307–314

Quested T, Parry A (2011) New estimates for household food and drink waste in the UK. Waste and Resources Action Program (WRAP)

Ren Y, Yu M, Wu C, Wang Q, Gao M, Huang Q, Liu Y (2017) A comprehensive review on food waste anaerobic digestion: research updates and tendencies. Bioresour Technol 247:1069–1076

Riemer J, Kristoffersen M (1999) Information on waste management practices. A proposed electronic framework. European Environmental Agency, Copenhagen

Roka K (2019) Environmental and social impacts of food waste. In: Leal Filho W, Azul A, Brandli L, Özuyar P, Wall T (eds) Responsible consumption and production. Encyclopedia of the UN sustainable development goals. Springer, Cham

Roychand R, Patel S, Halder P, Kundu S, Hampton J, Bergmann D, Surapaneni A, Shah K, Pramanik BK (2021) Recycling biosolids as cement composites in raw, pyrolysed and ashed forms: a waste utilisation approach to support circular economy. J Build Eng:102199

Searchinger T, Hanson C, Ranganathan J, Lipinski B, Waite R, Winterbottom R, Dinshaw A, Heimlich R (2013) The great balancing act. In: Working paper, installment 1 of creating a sustainable FoodFuture. World Resources Institute, Washington. http://www.worldresourcesreport.org

The European Union (2009) The European Parliament and the Council of the European Union: regulation (EC) No 1069/2009 of the European Parliament and of the Council of 21 October 2009. Off J Eur Union 300: 1–33

UNEP (United Nations Environment Programme) (2014) Prevention and reduction of food and drink waste in businesses and households – Guidance for governments, local authorities, businesses and other organisations, Version 1.0

Unuofin JO (2020a) Treasure from dross: application of agroindustrial wastes-derived thermo-halotolerant laccases in the simultaneous bioscouring of denim fabric and decolorization of dye bath effluents. Ind Crop Prod 147:112251

Unuofin JO (2020b) Sustainability potentials of novel laccase tinctures from Stenotrophomonas maltophilia BIJ16 and Bordetella bronchiseptica HSO16: from dye decolourization to denim bioscouring. Biotechnol Rep 25:e00409

Unuofin JO, Okoh AI, Nwodo UU (2019a) Aptitude of oxidative enzymes for treatment of wastewater pollutants: a laccase perspective. Molecules 24(11):2064

Unuofin JO, Okoh AI, Nwodo UU (2019b) Utilization of agroindustrial wastes for the production of laccase by Achromobacter xylosoxidans HWN16 and Bordetella bronchiseptica HSO16. J Environ Manag 231:222–231

Unuofin JO, Okoh AI, Nwodo UU (2019c) Maize Stover as a feedstock for enhanced laccase production by two gammaproteobacteria: a solution to agroindustrial waste stockpiling. Ind Crop Prod 129:611–623

Unuofin JO, Moubasher HA, Okoh AI, Nwodo UU (2019d) Production of polyextremotolerant laccase by Achromobacter xylosoxidans HWN16 and Citrobacter freundii LLJ16. Biotechnol Rep 22:e00337

Vandermeersch T, Alvarenga RAF, Ragaert P, Dewulf J (2014) Environmental sustainability assessment of food waste valorization options. Resour Conserv Recycl 87:57–64

World Bank (2019) Solid waste management. http://www.worldbank.org/en/topic/urbandevelopment/brief/solid-waste-management

WRAP (2009) Household food and drink waste in the UK. http://www.wrap.org.uk/sites/files/wrap/Household_food_and_ drink_waste_in_the_UK_-_report.pdf

WRAP (Waste and Resources Action Programme) (2011) Applying the waste hierarchy. A guide to business. Banbury

WWF (World Wildlife Fund for Nature) (2017) Food loss and waste: facts and futures taking steps towards a more sustainable food future. http://awsassets.wwf.org.za/downloads/WWF_Food_Loss_and_Waste_WEB.pdf

Hydrothermal Carbonization of Organic Fraction of Municipal Solid Waste: Advantage, Disadvantage, and Different Application of Hydrochar

9

Saikrishna Venna, Hari Bhakta Sharma, V. R. Sankar Cheela, and Brajesh Dubey

9.1 Introduction

In recent times, the mutation of complex municipal solid waste (MSW) biodegradable portion to an energy recovery resource became an extraordinary innovation as it conserves our precious natural resources and maintains the ecological balance. The various thermo-chemical and biological treatment approaches also succeeded in synthesizing efficient energy recovery resources like solid biofuel (char) and biogas (Table 9.1). Despite, the process instability due to precursor heterogeneity, environmentally non-benign, and non-viable processes have shown their inefficiency in on-site executions (Liu and Balasubramanian 2012). Nevertheless, on the other side intercontinental MSW generation is almost 1.3×10^3 million tons and reaches 2.2×10^3 million tons by 2025 (Xu 2013). Therefore, a promising futuristic viable and sustainable treatment technology for producing active energy recovery resource is the need of the hour. Likewise, hydrothermal carbonization (HTC) has recently gained well-built recognition as key enabling mechanization for converting the MSW biodegradable portion to desirable commodities (Wang et al. 2018; Zhao et al. 2014). The HTC process operates under lower temperatures, making use of distilled/de-ionized water as a catalyst or medium within a residence time of minutes to several hours (Ramke et al. 2009). The process finally yields into an energy-dense

S. Venna
Department of Civil Engineering, National Institute of Technology, Warangal, Andhra Pradesh, India

Present Address: Department of Civil Engineering, Vardhaman College of Engineering, Hyderabad, Telangana, India

H. B. Sharma · V. R. S. Cheela · B. Dubey (✉)
Department of Civil Engineering, Indian Institute of Technology - Kharagpur, Kharagpur, West Bengal, India
e-mail: bkdubey@civil.iitkgp.ac.in

I. Haq, A. S. Kalamdhad (eds.), *Emerging Treatment Technologies for Waste Management*, https://doi.org/10.1007/978-981-16-2015-7_9

Table 9.1 Thermal conversion technologies for the treatment of municipal solid waste

Option	Pyrolysis	Gasification	Hydrothermal carbonization
Pre-treatment			
Waste sorting required	Yes	Yes	–
Shredding	Yes	Yes	Yes
Drying	Yes	Yes	Yes
Addition of water	No	No	Yes
Operating conditions			
Reaction temperature (°C)	400–900	500–1300	180–350
Moisture content	<20%	<15%	–
Residence time	Seconds to weeks	10–20 s	Few seconds to weeks
Catalyst	Optional	Optional	Optional
End products and application			
End products	High syngas, oxygenated oils char	Syngas Char	Energy-dense hydrochar Organic rich process water Gases
Material utilization	Engine fuels Soil conditioners	Transport fuel Power and heat Fertilizer and catalyst	Soil remediation Fuel Energy source
Product yield			
Liquid (weight %)	30–80	–	60–75
Solid (weight %)	12–60	–	08–20
Gas (weight %)	06–35	1–2.6 (m^3 gas/kg biomass	15–20
Reference	Malav et al. (2020), Munir et al. (2019), Patel et al. (2016)		

product named "hydrochar", can be the replacement by lignite/bituminous coal. The usage of HTC as a waste conversion technology can notably minimize the greenhouse gas emissions (GHG's) via compost centres and municipal solid waste (MSW) landfill sites (Berge et al. 2013). Also, the synthesized energy-dense hydrochar can be practiced as an economical replacement to different environmental applications to deal with the imminent universal energy shortage and environmental challenges (Wang et al. 2020). However, HTC is manifesting as a robust technology, along with the process advantages and hydrochars wide variety applications in environmental and energy sectors, the process disadvantages are also discussed in this chapter for its better practical scale execution.

9.2 Introduction to Hydrothermal Carbonization (HTC)

"Hydrothermal Carbonization", a solely futuristic route in terms of efficient biomass yield through various HTC mechanisms such as hydrolysis, dehydration, decarboxylation, condensation-polymerization, and aromatization (Kent Hoekman et al. 2018). The process will take place in an aqueous phase under high temperatures (180–350 °C) and autogenous pressure, oxygen-limited environment (Ramke et al. 2009). However, to balance the water in its liquid state, the pressure will be increased above 1 ATM mechanically (Wyczalkowska and Sengers 2002). The process uses the water as solvent and catalyst to break the complex material in an easy way (Ramke et al. 2009). The ultimate HTC yields are energy-dense hydrochar, water, gaseous. HTC of solid waste is so advantageous for the synthesis of valuable carbon product excluding toxic chemicals usage (Titirici and Antonietti 2010). The process synthesized products have gained significant applications in energy sectors, such as solid biofuels, carbon electrode in supercapacitors and batteries, electro-catalysis, adsorption, and CO_2 sequestration (Saqib et al. 2019). The abundant surface functional groups, thermo-chemical steadiness, and the readiness of molding their functional and structural behaviour consummate the prerequisite applications (Jain et al. 2016). The presence of cationic and anionic ions on the surface of the hydrochar also used to remove specific pollutants.

9.3 Mechanism During Hydrothermal Carbonization

The mechanism during HTC of wet waste is very complex and varies according to the precursor type and operating situations (Kruse et al. 2013). Therefore, it is highly difficult to explain the proper chemical process for HTC. But many researchers concluded that HTC undergoes a series of simultaneous chemical processes: (a) hydrolysis; (b) dehydration; (c) decarboxylation; (d) re-condensation &polymerization; and (e) aromatization (Libra et al. 2011). Hydrolysis is the initial chemical reaction occurs during the HTC process where water reacts with larger constituents of wet waste to break them into smaller constituents. Later, these smaller fragments will interact with other chemical reactions to produce hydrochar-a solid biofuel. Generally, the biomass constituents like hemicellulose, carbohydrates, glucose, lipids, and fats are less stable and are easily hydrolysed under lower HTC temperatures (180 °C). Due to the uniform and linear structure, cellulose requires elevated temperatures to hydrolyse (230 °C). Even lignin restricts more to hydrolyse under temperatures at 230 °C, remains unreacted with any of the chemical reactions in synthesizing the hydrochar and it is almost absent in the hydrochar. Further reaction of smaller fragments with successive chemical processes develops water-soluble products such as furans, acetic acid, furfurals, and others. These intermediates further react under a series of reaction like dehydration, decarboxylation, re-condensation, polymerization, and aromatization to form carbon-rich hydrochar (Sharma et al. 2020a).

9.4 Effect of Different Process Matter

The crucial parameter of the HTC process is known as temperature. The degradation of the biomass bond using heat is the primary criteria of the temperature. The hydrochars quality increases with the rise in the temperature (Akhtar et al. 2012). In general, the hydrochars mass efficiency will be higher at 150–250 °C and the higher liquid yield is observed at 250–350 °C, whereas exceeding 350 °C, the formation of gases will be higher. The efficiency of the liquefaction productfor different operating temperatures of the paper industry is shown in Fig. 9.1. It indicated that maximum char yields at lower temperatures. Likewise, the experimental analysis of Sun et al. and Liu et al. showed, higher char yields were observed under 200 °C (Sun et al. 2010; Liu et al. 2012). The results also demonstrated that rise in temperature to 200–250 °C and 280 °C favoured the char efficiency to reduce. However, along with the temperature increase, char yield gradually reduces, while by-products yield improves. However, the elevated operating conditions during the HTC are observed to gear the aromatization quality that can be used as an alternate solid fuel.

The surrounding environment and time of growth vary the feedstock characteristics in a different zone of areas. The leading biomass components are hemicellulose, cellulose, and lignin. As discussed, the temperature is crucial HTC parameter, therefore due to different temperature variations each component will behave uniquely based on their distinguishable characteristics. The studies exclaimed that a higher percentage of hemicellulose and cellulose increase the oil

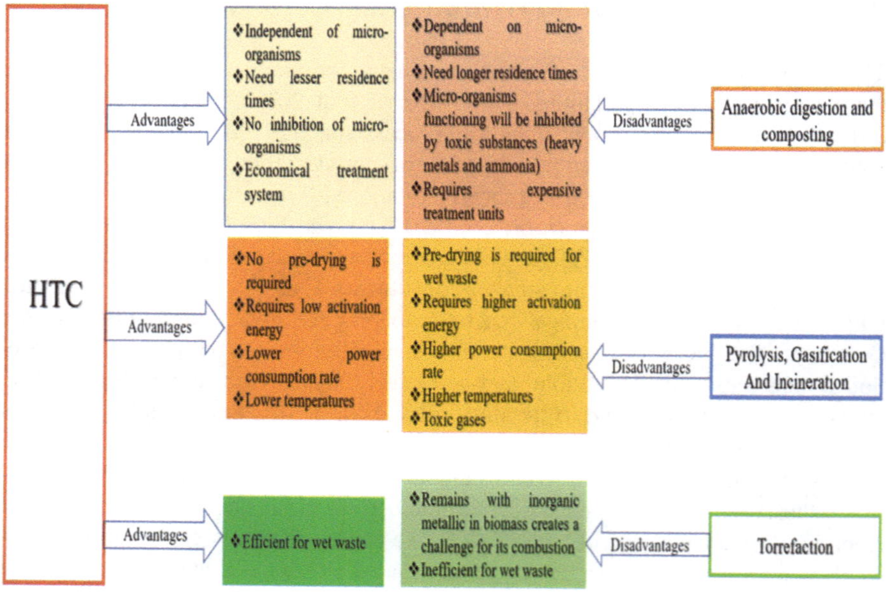

Fig. 9.1 Advantages and disadvantages of thermo-chemical processes

yield. Similarly, the higher percentage of lignin in biomass lefts greater amount of hydrochar as a residue due to its complex branching (Lachance et al. 2008). The studies of yang et al. also showed that 94.5% loss of cellulose @ 400 °C, 80% loss of hemicellulose @ 800 °C, and 54.3% loss of lignin @ 900 °C, respectively (Yang 2007). Therefore, char production will be higher for higher lignin content feedstock. Finally, this concluded that the calorific value (HHV) of hydrochar is strongly connected with the initial feedstock composition. Also, understanding the behaviour of HTC mechanisms depends upon feedstock type and the level of hydrochar quality depends upon the feed composition.

HTC has been recognized as a slow thermo-chemical reaction process. The residence time for HTC varies from several minutes to hours. However, residence time create an impact on the hydrolysis mechanism until a certain time range, it does not have any particular impact on the process post to residence time. Commonly, the researchers noticed that a higher amount of char is yielded at longer retention times. However, at higher operating conditions, the rate of hydrolysis and breaking down of feedstock are relatively fast. Therefore, relatively shorter reaction times are necessary for the effective degradation of biomass. The longer residence times favour mild bio-oil efficiency (Yan et al. 2021). On the other hand, the analysis perceived that hydrochar and gaseous yields are dominant concerning water products (Zhang et al. 2009). Also, the quantity of hydrochar components depends upon the HTC retention times. In opposite, the shorter running times of HTC enhances the hydrochars calorific value as same as or higher than lignin could be due to the elimination of oxygen from the feedstock.

The pressure is the other important parameter that affects the feedstock breakdown during hydrolysis mechanism. Both the feedstock degradation and rate of hydrolysis can be managed by balancing the pressure ahead of the medium's critical pressure. During the HTC, density of the solvent increases with the rise in pressure. The high-density solvent media can degrade the feedstock at larger proportions (Kersten et al. 2006). The reason could be that the high-density solvent medium infiltrates into the feedstock structure which will improve the degradation rate. It is also concluded that the influence of pressure on a solvent medium is insignificant (Akhtar et al. 2011). However, the pressure can be increased by increasing the reaction temperature or via purging the inert gases such as argon, nitrogen, etc. The hydrochar quality also highly relies upon the pressure conditions. The hydrochar obtained from high temperatures results with extreme quality.

The utilization of catalysts significantly improves the rate of hydrolysis. Acid catalysts support higher hydrochar formation compared to alkaline catalysts. The catalyst blending immediately transforms NO_x to N_2 and H_2O that reduces the NO_x emissions. The studies of karagoz et al. revelled that the influence of CsOH and RbOH on pinewood increased liquid yield compared to solid yield (Karago et al. 2006). Some of the catalysts chosen by international scientists for synthesizing qualified biofuel were Citric acid, K_2CO_3, NaOH, KOH, C_3H_8O. The catalyst accelerates and promotes the feedstock's cellulose and lignin degradation during the HTC process for synthesizing qualified hydrochar.

9.5 Advantages and Disadvantages of Hydrothermal Carbonization

However, the hydrothermal carbonization process results in a well-qualified solid biofuel. The process also has certain advantages and disadvantages when compared to other thermo-chemical conversions (Fig. 9.1). The major and foremost advantage of HTC is the usage of wet waste as a feedstock material to transform into hydrochar (solid biofuel) with more water soluble-organics (Liu and Balasubramanian 2012). This reduces the energy-intensive pre-drying rather than incineration, pyrolysis, and torrefaction. The HTC also operates under lower temperatures thereby saves the higher energy costs (Lu et al. 2014; Libra et al. 2011). Apart from this the conversion process also limits to several hours, unlike the natural coalification process. The evolution of hydrolysis step through solvent media enhances the chemical quality of hydrochar (Kruse and Dahmen 2015). The generated process water contains rich organic acids, HMF, furfural, and TOC that are potential to create environmental footprint if disposed of unscientifically. However, that can be recycled to extract the value-added acids and the same also can reuse during the HTC process to improve the energy value of hydrochar. Likewise, many studies have recirculated HTC process water itself back and also integrated with another treatment technologies to reduce water-related footprint and have shown positive result on hydrochar mass yield, energy yield, energy recovery (Parthasarathy and Narayanan 2014; Kambo et al. 2018). Many studies concluded that the hydrothermal carbonization process is environmentally-benign as all the gases emitted during HTC is negligible compared to other thermo-chemical technologies. Even though various studies concluded HTC as an eco-friendly process, however, Life cycle assessment studies reported HTC process water (a by-product of HTC treatment) and CO_2 emissions can impose significant footprint (Berge et al. 2015; Saqib et al. 2019) at the large scale application.

The energy usage while the HTC process depends upon the operating conditions of the process. Therefore, in-detailed intensification of process and its integration is necessary to enhance the environmental/economic advantages of the HTC process (Sharma et al. 2020b). The major disadvantages of HTC process are unavailability of reaction kinetic information which includes mass transfer and reaction pathway for the design of kinetic reaction and process optimization via HTC technology (Munir et al. 2018). After the process completion, the synthesized hydrochar should be separated from process water. To separate the slurry, vacuum filtration is utilized that will be an additional expense to the HTC technology. However, filtration (nanomembrane) technology can come up as another possibility for hydrochar segregation to decrease the economy of the technology (Munir et al. 2018). Likewise, the use of wastewater source like landfill leachate and sewage sludge slurry could help in reducing the dependence on freshwater during HTC (Venna et al. 2021).

9.6 Potential Applications of Hydrochar

Due to the advantageous physicochemical characteristics, the applications of hydrochar have emerged in environmental and energy sectors. The various hydrochar applications include energy recovery and storage, catalyst, composting, soil remediation, excellent low-cost adsorbent, sequestration for decreasing GHGs, pathogen removal, wastewater treatment (removing heavy metals, organic pollutants, phosphates), and medicinal applications. Hydrochar fuel pellets are another excellent alternative solution to coal (Sharma and Dubey 2020a, b). Some studies of all the applications are reported in the current study. The synthesized hydrochar was found to reduce nitrogen dioxide (N_2O) emissions, increase CH_4, NH_3, and CO_2 compared to unamended soil (Malghani et al. 2013). The blending of hydrochar with soil reduces the soil bulk density and improves total pore volume, permitting a higher amount of water to be retained by soil (Abel et al. 2013). It is also studied that the hydrochars produced from plant biomass (organic fraction) are considered as an effective solid fuel due to the presence of low inorganic minerals and ash content in hydrochars. The hydrochar addition with nickel made nickel a good graphitization catalyst that enhanced the energy storage capacity of hydrochar by 149% (Ding et al. 2012). The antioxidant activity of hydrothermal extracts from different parts of watermelons was evaluated (Kim et al. 2014). Hydrothermal treatment of watermelons significantly increased the phenolic content and antioxidant activity of the flesh, green rind, and white rind parts of the fruit. Several new phenolic compounds such as catechol and its derivatives were identified in the hydrothermal extracts. Therefore, hydrothermal treatment may release phenolic compounds from watermelons, thereby increasing its antioxidant activity and levels of nutraceutical compounds in watermelon extracts.

9.7 Conclusion

In brief, hydrothermal carbonization (HTC) has proved as a solely futuristic thermochemical process for treating organic waste compared to other thermal processes to produce a value-added product as hydrolysis is the primary mechanism. The supremacy of HTC over other thermal and bio-based energy recovery techniques includes eco-friendly, simple and fast process, low activation energy and power consumption requirement, low operating temperatures, and no pre-drying requirement. However, the desirable hydrochars quality and yield completely rely upon the HTC process parameters. In addition, to determine the proper feedstock flow there should undergo a detailed study on mass and water balance of the chosen precursor during the HTC process. However, the utilization of hydrochar increased in a wide variety of applications, the in-depth research should be performed on reaction kinetic information which includes mass transfer and reaction pathway that are salient criteria's for the design of kinetic reaction and process optimization via HTC technology.

References

Abel S, Peters A, Trinks S, Schonsky H, Facklam M, Wessolek G (2013) Impact of biochar and hydrochar addition on water retention and water repellency of sandy soil. Geoderma 202–203:183–191. https://doi.org/10.1016/j.geoderma.2013.03.003

Akhtar J, Aishah N, Amin S (2011) A review on process conditions for optimum bio-oil yield in hydrothermal liquefaction of biomass. Renew Sustain Energy Rev 15(3):1615–1624. https://doi.org/10.1016/j.rser.2010.11.054

Akhtar J, Saidina N, Wood P (2012) A review on operating parameters for optimum liquid oil yield in biomass pyrolysis. Renew Sustain Energy Rev 16(7):5101–5109. https://doi.org/10.1016/j.rser.2012.05.033

Berge ND, Kammann C, Ro K, Libra J (2013) Technology: biochar production, carbon sequestration, and waste conversion. Wiley, New York

Berge ND, Li L, Flora JRV, Ro KS (2015) Assessing the environmental impact of energy production from hydrochar generated via hydrothermal carbonization of food wastes. Waste Manag 43:203–217. https://doi.org/10.1016/j.wasman.2015.04.029

Ding L, Wang Z, Li Y, Du Y, Liu H, Guo Y (2012) A novel hydrochar and nickel composite for the electrochemical supercapacitor electrode material. Mater Lett 74:111–114. https://doi.org/10.1016/j.matlet.2012.01.070

Jain A, Balasubramanian R, Srinivasan MP (2016) Hydrothermal conversion of biomass waste to activated carbon with high porosity: a review. Chem Eng J 283:789–805. https://doi.org/10.1016/j.cej.2015.08.014

Kambo HS, Minaret J, Dutta A (2018) Process water from the hydrothermal carbonization of biomass: a waste or a valuable product? Waste Biomass Valoriz 9(7):1181–1189. https://doi.org/10.1007/s12649-017-9914-0

Karago S, Bhaskar T, Muto A, Sakata Y (2006) Hydrothermal upgrading of biomass : effect of K 2 CO 3 concentration and biomass/water ratio on products distribution. Bioresour Technol 97:90–98. https://doi.org/10.1016/j.biortech.2005.02.051

Kent Hoekman S, Leland A, Felix L (2018) Hydrothermal carbonization (HTC) of biomass for energy applications. In: Biomass preprocessing and pretreatments for production of biofuels. CRC Press, Boca Raton, pp 196–254. https://doi.org/10.1201/9781315153735-8

Kersten SRA, Potic B, Prins W, Van Swaaij WPM (2006) Gasification of model compounds and wood in hot compressed water. Industr Eng Chem Res 45(12):4169–4177. https://doi.org/10.1021/ie0509490

Kim S, Matsushita Y, Fukushima K, Aoki D, Yagami S, Yuk H, Lee S (2014) Antioxidant activity of a hydrothermal extract from watermelons. LWT-Food Sci Technol 59(1):361–368. https://doi.org/10.1016/j.lwt.2014.04.041

Kruse A, Dahmen N (2015) Water – a magic solvent for biomass conversion. J Supercrit Fluids 96:36–45. https://doi.org/10.1016/j.supflu.2014.09.038

Kruse A, Funke A, Titirici MM (2013) Hydrothermal conversion of biomass to fuels and energetic materials. Curr Opin Chem Biol 17(3):515–521

Lachance RP, Fro M, Antal MJ, Peterson AA, Tester JW (2008) Thermochemical biofuel production in hydrothermal media : a review of sub- and supercritical water technologies. Energ Environ Sci 1:32–65. https://doi.org/10.1039/b810100k

Libra JA, Ro KS, Kammann C, Funke A, Berge ND, Neubauer Y et al (2011) Hydrothermal carbonization of biomass residuals: a comparative review of the chemistry, processes and applications of wet and dry pyrolysis. Biofuels 2(1):71–106. https://doi.org/10.4155/bfs.10.81

Liu H, Xie X, Li M, Sun R (2012) Hydrothermal liquefaction of cypress : effects of reaction conditions on 5-lump distribution and composition. J Anal Appl Pyrolysis 94:177–183. https://doi.org/10.1016/j.jaap.2011.12.007

Liu Z, Balasubramanian R (2012) Hydrothermal carbonization of waste biomass for energy generation. Procedia Environ Sci 16:159–166. https://doi.org/10.1016/j.proenv.2012.10.022

Lu X, Flora JRV, Berge ND (2014) Influence of process water quality on hydrothermal carbonization of cellulose. Bioresour Technol 154:229–239. https://doi.org/10.1016/j.biortech.2013.11.069

Malav LC, Yadav KK, Gupta N, Kumar S, Sharma GS, Krishnan S, Rezania S, Kamyab H, Pham QB, Yadav S, Bhattacharyya S, Yadav VK, Quang-Vu B (2020) A review on municipal solid waste as a renewable source for waste-to-energy project in India: current practices, challenges, and future opportunities. J Clean Prod 277:123227

Malghani S, Gleixner G, Trumbore SE (2013) Chars produced by slow pyrolysis and hydrothermal carbonization vary in carbon sequestration potential and greenhouse gases emissions. Soil Biol Biochem 62:137–146. https://doi.org/10.1016/j.soilbio.2013.03.013

Munir MT, Mardon I, Al-Zuhair S, Shawabkeh A, Saqib NU (2019) Plasma gasification of municipal solid waste for waste-to-value processing. Renew Sustain Energy Rev 116:109461

Munir MT, Soheil S, Udugama IA, Baroutian S, Gernaey KV, Young BR (2018) Resource recovery from organic solid waste using hydrothermal processing: opportunities and challenges. Renew Sustain Energy Rev 96(July):64–75. https://doi.org/10.1016/j.rser.2018.07.039

Parthasarathy P, Narayanan SK (2014) Effect of hydrothermal carbonization reaction parameters on the properties of hydrochar and pellets. Environ Prog Sustain Energy 33(3):676–680

Patel M, Zhang X, Kumar A (2016) Techno-economic and life cycle assessment on lignocellulosic biomass thermochemical conversion technologies: a review. Renew Sustain Energy Rev 53:1486–1499

Ramke H-G, Blöhse D, Lehmann H-J, Fettig J, Höxter ST (2009) Hydrothermal carbonization of organic waste. In: Twelfth international waste management and landfill symposium, Sardinia

Saqib NU, Sharma HB, Baroutian S, Dubey B, Sarmah AK (2019) Valorisation of food waste via hydrothermal carbonisation and techno-economic feasibility assessment. Sci Total Environ 690:261–276

Sharma HB, Dubey BK (2020a) Co-hydrothermal carbonization of food waste with yard waste for solid biofuel production: hydrochar characterization and its pelletization. Waste Manag 118:521–533

Sharma HB, Dubey BK (2020b) Binderless fuel pellets from hydrothermal carbonization of municipal yard waste: effect of severity factor on the hydrochar pellets properties. J Clean Prod 277:124295

Sharma HB, Sarmah AK, Dubey B (2020a) Hydrothermal carbonization of renewable waste biomass for solid biofuel production: a discussion on process mechanism, the influence of process parameters, environmental performance and fuel properties of hydrochar. Renew Sustain Energy Rev 123:109761

Sharma HB, Panigrahi S, Dubey BK, Narra S (2020b) Resource recovery from organic fraction of municipal solid waste using anaerobic digestion and hydrothermal carbonization. In: Bioresource utilization bioprocess. Springer, Singapore, pp 13–25

Sun P, Heng M, Sun S, Chen J (2010) Direct liquefaction of paulownia in hot compressed water: influence of catalysts. Energy 35(12):5421–5429. https://doi.org/10.1016/j.energy.2010.07.005

Titirici MM, Antonietti M (2010) Chemistry and materials options of sustainable carbon materials made by hydrothermal carbonization. Chem Soc Rev 39(1):103–116. https://doi.org/10.1039/b819318p

Venna S, Sharma HB, Reddy PHP, Chowdhury S, Dubey BK (2021) Landfill leachate as an alternative moisture source for hydrothermal carbonization of municipal solid wastes to solid biofuels. Bioresour Technol 320:124410

Wang G, Zhang J, Lee J, Mao X, Ye L, Xu W (2020) Hydrothermal carbonization of maize straw for hydrochar production and its injection for blast furnace. Appl Energy 266:114818. https://doi.org/10.1016/j.apenergy.2020.114818

Wang T, Zhai Y, Zhu Y, Li C, Zeng G (2018) A review of the hydrothermal carbonization of biomass waste for hydrochar formation: process conditions, fundamentals, and physicochemical properties. Renew Sustain Energy Rev 90:223–247. https://doi.org/10.1016/j.rser.2018.03.071

Wyczalkowska AK, Sengers JV (2002) Thermodynamic properties of sulfurhexafluoride in the critical region. J Chem Phys 111(4):1551–1560. https://doi.org/10.1063/1.479414
Xu Q, Qian Q, Quek A, Ai N, Zeng G, Wang J (2013) Hydrothermal carbonization of macroalgae and the effects of experimental parameters on the properties of hydrochars. ACS Sustain Chem Eng 1:1092–1101. https://doi.org/10.1021/sc400118f
Yan Y, Xu J, Li T, Ren Z (2021) Liquefaction of sawdust for liquid fuel. Fuel Energy 42:1–2
Yang H (2007) Characteristics of hemicellulose, cellulose and lignin pyrolysis. Fuel 86:1781–1788. https://doi.org/10.1016/j.fuel.2006.12.013
Zhang B, Keitz M, Von, Valentas K (2009) Thermochemical liquefaction of high-diversity grassland perennials. J Anal Appl Pyrolysis 84:18–24. https://doi.org/10.1016/j.jaap.2008.09.005
Zhao P, Shen Y, Ge S, Yoshikawa K (2014) Energy recycling from sewage sludge by producing solid biofuel with hydrothermal carbonization. Energ Conver Manage 78:815–821. https://doi.org/10.1016/j.enconman.2013.11.026

Pollutants Characterization and Toxicity Assessment of Pulp and Paper Industry Sludge for Safe Environmental Disposal

10

Pooja Sharma and Surendra Pratap Singh

10.1 Introduction

The pulp and paper industry processed wastewater contains a high amount of organic contaminants and heavy metals that pose environmental risks. Owing to the indiscriminate discharge of hazardous chemicals such as pesticides and solvents from different sources and manufacturing activities, heavy metal and metalloid pollution is a global environmental concern (Stefanowicz et al. 2016; Ngo et al. 2020). Recent research has shown that even after secondary treatment, pulp–paper mill sludge is a source of contamination that directly and indirectly toxicities phytoplankton and zooplankton by increasing parameters such as total suspended solids (TSS), chemical oxygen demand (COD), total dissolved solids (TDS), biological oxygen demand (BOD), and metals (Fe, Cu, Ni, Cr, Pb, Cd, Hg, As, Mg, and Mn) (Haq et al. 2016a, b; 2017, 2018; Selvi, et al. 2019; Haq et al. 2020a, b; Sharma et al. 2020b). Plant uptake of heavy metals from the contaminated industrial waste area can reduce the germination of seeds and impede the growth of shoots and roots, lead to chlorosis and wilting, decrease the content of chlorophyll and protein, and decrease the effectiveness of photosynthetic processes and various enzyme activities (Fahr et al. 2015; Kalaivanan and Ganeshamurthy 2016). The accumulation and translocation of heavy metals in a plant are regulated by various molecules such as metal ion transporters and complicating factors. Several advanced transporters, including channel proteins and H+, coupled transmembrane protein, located in the root cell plasma membrane, are important for metal ion uptake. Via the cell membranes, they

P. Sharma
Department of Environmental Microbiology, School for Environmental Sciences, Babasaheb Bhimrao Ambedkar Central University, Lucknow, Uttar Pradesh, India

S. P. Singh (✉)
Plant Molecular Biology Laboratory, Department of Botany, Dayanand Anglo-Vedic (PG) College, Chhatrapati Shahu Ji Maharaj University, Kanpur, India

I. Haq, A. S. Kalamdhad (eds.), *Emerging Treatment Technologies for Waste Management*, https://doi.org/10.1007/978-981-16-2015-7_10

could carry different metals and mediate the flow through roots to metal translocation shoots (Dal Corso et al. 2019).

More than 850 units of pulp–paper, including small paper industries, are estimated in India. The industry has a 400 TDP (tonne per day) paper production potential, which is not only a source of soil and aquatic contamination but also affects the food chain. Cost-effective and eco-friendly remediation technology is required to safeguard human health and the environment. This study, therefore, focused on the assessment of potential native plants growing on the sludge disposed site along with the drain of effluent discharge where the dissolved effluent contaminants communicate directly with these plants as well. These industries discharge pollutant-containing sludge, such as suspended solids, artificial colors, foam, surfactants, fabrics, biocides, plasticizers, sodium carbonate, bicarbonate, chlorides, sulfates, mercaptans, inorganic sulfides, cellulose, hemicellulose, tannins, fatty acids, chlorinated lignin, resin acids, hydrocarbons, and phenols, at different stages of the papermaking method (Chandra et al. 2018; Orrego et al. 2010; Chandra and Singh 2012). Recent studies have shown that the secondary wastewater treatment process does not adequately remove organic contaminants released into wastewater from pulp and paper mills (Chandra and Singh 2012). In species, many of the generated organic contaminants bioaccumulate and are graded as highly toxic, mutagenic, and estrogenic (Chandra et al. 2018; Orrego et al. 2010; Chandra et al. 2017; Ali and Sreekrishnan 2001). Today, hazardous organic pollutants, endocrine-disrupting chemicals (EDCs), and heavy metals emitted from the pulp and paper industries contaminate vast areas of agricultural and aquatic systems (Haq and Raj 2018; Chandra et al. 2018).

Lead (Pb), cadmium (Cd), copper (Cu), chromium (Cr), arsenic (As), mercury (Hg), and nickel (Ni) are the most common heavy metals in the world, according to the Environmental Protection Agency (EPA) (Selvi et al. 2019; O'connor et al. 2019, Jaskulak et al. 2019; Sharma et al. 2020c: Sharma et al. 2020d). Soil irrigation using wastewater results in an unnecessary accumulation of heavy metals in agricultural soils, as well as increased absorption of heavy metals by crops and vegetables, which can impact the quality and protection of food for human consumption (Muchuweti et al. 2006; Khan et al. 2018). Different parts of vegetable plants absorb more heavy metals from polluted soils, and the ingestion of these crops can lead to serious health problems (Khan et al. 2018). The development of an unattractive color of sludge is the responsibility of lignin released from the pulp and paper industries and prevents the growth of phototrophic species by reducing the amount of sunlight transferred to sludge (Karrasch et al. 2006). In *Selenastrum capricornutum* and *Chlorella vulgaris* green algae, *Eisenia fetida earthworm, Daphnia magna crustacean,* and *Oryzias latipes* fish, as well as several other freshwater species, toxic effects of chlorophenols and phenols have been recorded (Shigeoka et al. 1988; Miyazaki et al. 2002). Wastewater in the pulp and paper industry can accumulate chlorinated organic halides and fatty acid compounds in fish tissues and cause clastogenic, carcinogenic, endocrine, and mutagenic effects, which also pose a threat to human health (Savant et al. 2006).

In the treated pulp and paper industry, several other mutagenic and cytotoxic compounds, such as β-sitosterol trimethylsilyl ether, 2-methoxy phenol, and hexadecanoic acid, have also been confirmed to be present (Chandra et al. 2018). In pulp and paper industry wastewater, more than 700 organic and inorganic compounds were also found, such as adsorbable organic halogens (AOX), color, phenolic compounds (Karrasch et al. 2006; Buyukkamaci and Koken 2010). Metals that absorb through the gills, digestive tract, and body surface in aquatic organisms, including fish, can accumulate in the tissues and cause oxidative damges (Tao et al. 2001; Kamunde et al. 2002; Luoma and Rainbow 2008). The information gained from this chapter will encourage the scientific community to develop effective knowledge of it and evaluate the toxicity of discharged sludge in various plants and aquatic animals.

10.2 Process of Pulp and Paper Making

Paper plays a vital role in our everyday lives, and the papers were used for several coming years. Pulp and paper are manufactured from cellulose fibers and other plant materials (Haq and Raj 2019). Certain synthetic materials may be used to impart unique properties to the final product. Paper is made of wood fibers, but rags, flax, cotton lanterns, and bagasse (sugarcane residues) are often used in certain papers. Used paper is often recycled, and after purifying and often deinking, it is mostly mixed with new fibers and re-formed into the paper. Items such as cellulose acetate, radius, cellulose esters made from cellulose are used for packaging films, explosives. The goal of the pulping process is to extract lignin without sacrificing fiber power, thereby freeing the fibers and eliminating impurities that cause discoloration and potential future paper disintegration. Hemicellulose plays an important part in the bonding of fiber to fiber in paper production. This is similar in structure and function to cellulose. Many other extractives, like waxes, oleoresins, are found in wood but do not add to its strength properties; these two are extracted during the pulping process. For paper, the fiber derived from any plant can be used. However, the weight and consistency of the fiber and other considerations complicate the process of pulping. Wood that has been received from a pulp mill may be in various shapes. It depends on the pulping process and the origin of the raw material.

In wood digestion, where the wood chips are converted into masses of fiber by reacting it with sodium hydroxide (NaOH) and sodium sulfite (Na_2SO_4) at high temperature and pressure. The chemical reaction with wood fibers dissolves all the depository materials and these derivatives of depository materials are washed away from the fiber during the washing and dewatering process. During the process of pulp washing, the depositary material such as cellulose, lignin, phenols, fatty acid, tannins, and resins are released out from the pulp and form an alkaline viscus, dark black colored waste, called black liquor. The alkaline black liquor released from the digested pulp by pulp and paper industry wastewater constitutes only 10–15% of the total wastewater but it contributes approximately 90–95% of the total pollution load of pulp and paper industry wastewater (Grover et al. 1999). The method of producing pulp–paper consists of four fundamental steps: preparation of raw materials, pulping, like mechanical and chemical pulping, bleaching, and papermaking. Some of the toxic pollutants discharged from the pulp and paper industry are primarily manufactured at the various stages of the process of papermaking and wastewater production (Chandra et al. 2018). The bark is extracted at the first stage by tumbling in large steel drums where washing water can be applied, holding the logs in rotary drums (Smook 1992). Consequently, wood logs are processed at the pulping stage into wood chips that are digested at high temperature and pressure (160–170 °C) using a mixture of sodium hydroxide (NaOH) and sodium sulfide (Na_2S) for the kraft process, and wood chips are cooked in a mixture of sulfuric acid (H_2SO_3) and bisulfite ions (HSO_3-) for the sulfite pulping process (Saltman 1978). The third stage is pulp bleaching after the pulping process, in which raw pulp still contains lignin and decoloration, it must be bleached to produce white paper. The fibers are further "dignified" by solubilizing extra lignin from the cellulose by bleaching by chlorination and oxidation. Sodium bisulfite, sodium or zinc hydrosulfite, calcium or sodium hypochlorite, hydrogen, or sodium peroxide, and sulfur dioxide-borol dioxide are the chemicals used to bleach pulp.

10.3 Generation of Total Wastewater from Pulp and Industry

Concerning the natural world, the pulp and paper industry assumes a demanding role. The pulp–paper industry ranks sixth of the most polluting industry in the world that manufacture toxic effluent on a large scale after paper processing (Ugurlu et al. 2008). Chemical pulp production in recent times is 180 million tonnes per year, while eucalyptus pulp production has increased dramatically, particularly in the southern hemisphere (Cabrera 2017). The pulp and paper sector has historically been considered a significant user of natural resources, i.e. water and plants, and one of the main contributors of pollution to the ecosystem, such as air, water, and soil. Before the 1970s, increasing awareness of the impact of pulp and paper pollutants on the environment has led to a decrease in water and energy consumption and the discharge of hazardous compounds. One of the most important improvements undertaken in this respect was made inside the mill, where chlorine was replaced with chlorine dioxide as a bleaching chemical agent. The development of pulp and paper produces a significant amount of contaminants distinguished by the high concentration of suspended solids (SS), chemical oxygen demand (COD), toxicity, and the demand for biochemical oxygen (BOD) (Pokhrel and Viraraghavan 2004). Different combinations of unit processes involved in processing, the type of raw materials and chemicals consumed, the types of paper products, and the degree of water recovery depend on the volume and pollution load of the wastewater produced (Zarkovic et al. 2011). Aquatic and land habitats are threatened by the pollutants discharged from the paper industry. Several research studies have documented toxic effects on different fish species due to the exposure to industrial wastewater from pulp and paper, including liver damage, oxygenase activity of mixed functions, physiological changes, respiratory stress, toxicity, and lethal effects on fish exposed to this wastewater.

For each metric tonne of paper produced, large volumes (up to 70 m^3) of wastewater are created, depending on the nature of the raw material, the type of finished product, and the extent of the reuse of water (Latorre et al. 2007). Around 70% of its large intake of water is used by the pulp and paper industry as process water. Reducing water usage saves money and reduces the use of limited environmental resources by increasing internal water recirculation after the introduction of internal cleaning processes. The industry has decreased water usage by almost half over the past 20 years and by an incredible 95% per tonne of paper over the past 30 years (Blanco et al. 2004). Because the main pulping and papermaking unit operations are carried out in aqueous media, the availability of different chemical additives can significantly alter the properties of the effluents produced, making them harmful to the receiving environment. Complete water circuit closures are restricted by the accumulation of pollutants in the process water, which can cause rust, deposits, and odors and alter the paper machine's running output and the quality of the finished product. If further closure is needed, it is important to treat the effluent thoroughly so that it can be used again in the process.

10.4 Characterization to Total Pollutants from Sludge

Hydrophobic, non-volatile, diterpenic carboxylic acids commonly found in tree bark and softwood wood are resin acids. Compared to hardwood trees, resin acid concentrations can be up to several hundred parts per million and twice as high in softwoods (Ali and Sreekrishnan 2001). During the digestion and bleaching process of pulp in the pulp and paper industry, the resin acid is dissolved and released from the wood (Ali and Sreekrishnan 2001). Resin acids, i.e. abietane, pimarane, and isopimarane, are synthesized from three basic groups. Abietane derivatives such as abietic and dehydroabietic acids have an isopropyl substituent in their skeleton, but pimarane and isopimaric acid derivatives in the C13 position group have a vinyl and methyl group. The aromatic ring in the structure of dehydroabietic acid makes it very stable. Abietic acid, neoabietic acid, dehydroabietic acid, pimaric acid, isopimaric acid, palustric acid, levopimaric acid, and sandaracopimaric acid are the most used resin acids in the pulp and paper industry (Ali and Sreekrishnan 2001). The resin acids are referred to as weekly hydrophobic acids and are reported to be highly toxic to fish (200–800μg/l). It has also been stated that resin acid is responsible for 60–90 percent of the toxicity in the chemical mechanical pulping process. The solubility and toxicity of resin acid are heavily regulated by the pH of the wastewater (Ali and Sreekrishnan 2001). LC50 for resin acid of 0.4–1.7 mg L^{-1} was reported for rainbow trout at 96-hour exposure. The absorption and distribution properties of resin acids, dehydroabietic acids, and eight wood rosin acids were analyzed by calculating the quantities accumulated in various rainbow trout organs after 4 and 2 days of semi-static exposure at average concentrations of 1.2 and 1.4 mg L^{-1} of water, respectively. The liver and posterior kidneys were among the organs showing the highest uptake in both experiments. Dehydroabietic acid has been detected in blood plasma in contrast to these. Besides, in one or both experiments, the anterior (head) kidney, skin, heart, and brain all ranked high.

During pulp cleansing, molecular chlorine is often used as a bleaching agent. Many dioxins and furans, a well-known toxic compound in the environment, were identified from the pulp and paper effluent (Dioxin Homepage, EJnet.org website, Accessed July 16, 2010). The pulp industry has therefore substituted chlorine with chlorine dioxide, resulting in a pulp of high brightness with suitable wastewater properties. AOX, the key emission parameter used by regulatory agencies to assess the discharge of the chlorinated substance, was substantially reduced by this process. In the pulp–paper wastewater sediment and its discharge site from the paper industry, dioxins and furans have been found. Dioxins can be produced in the presence of a chlorine source at a temperature range of 200–600 °C through the degradation of carbon species. They can, however, be formed by precursor compounds, like pentachlorophenol, chlorinated diphenyl ethers, and polychlorinated biphenyls or chlorobenzene, which are used as a fungicidal and bactericidal agent in the pulp and paper industry for wood treatment. Polychlorinated dibenzofurans, polychlorinated dibenzodioxins, and dioxin-like polychlorinated biphenyls are structurally and chemically associated with halogenated aromatic hydrocarbons that are the most common type of dioxin compound detected in the pulp and paper industry. Various volatile sulfur compounds (VSCs) are formed during the process of kraft and sulfite pulping, producing mainly reduced sulfur compounds including methyl mercaptan, dimethyl disulfide, and hydrogen sulfide (Higgins et al. 2006) (Fig. 10.1).

10.5 Toxicity Assessment

A few research on the aquatic toxicity of wastewater in fish have been performed, but the genotoxic effects of wastewater from pulp and paper mills on plants were poorly known (Singh and Chandra 2019). Phytotoxic assessments have shown that pulp and paper contaminants impede plant germination and induce chromosomal aberrations of *Phaseolus mungo, Triticum aestivum, Aestivum sativa,* and *Allium cepa* (Chandra et al. 2018; Haq et al. 2017). The reproductivity of aquatic organisms can be impaired by discharged effluents with a toxic substance (Costigan et al. 2012; Waye et al. 2014). Long-chain fatty acids, developed during pulping operations, are one of the key sources of toxicity arising from pulping operations, according to Magnus et al. (2000). The toxic and sometimes carcinogenic effects of chlorinated organic compounds and dioxin have been noted (Wiegand et al. 2006). The primary inorganic contaminants Iron, copper, zinc, nickel, and magnesium, recorded for neurotoxicity, and are toxic to juvenile channel catfish (*Ictalurus punctatus*). Their accumulation is in order of gill > liver > ovary > muscle. Many of these molecules identified as chemicals are endocrine-disrupting chemicals (EDCs). Therefore, direct wastewater toxicity to the reproductive system is recorded in aquatic flora and fauna.

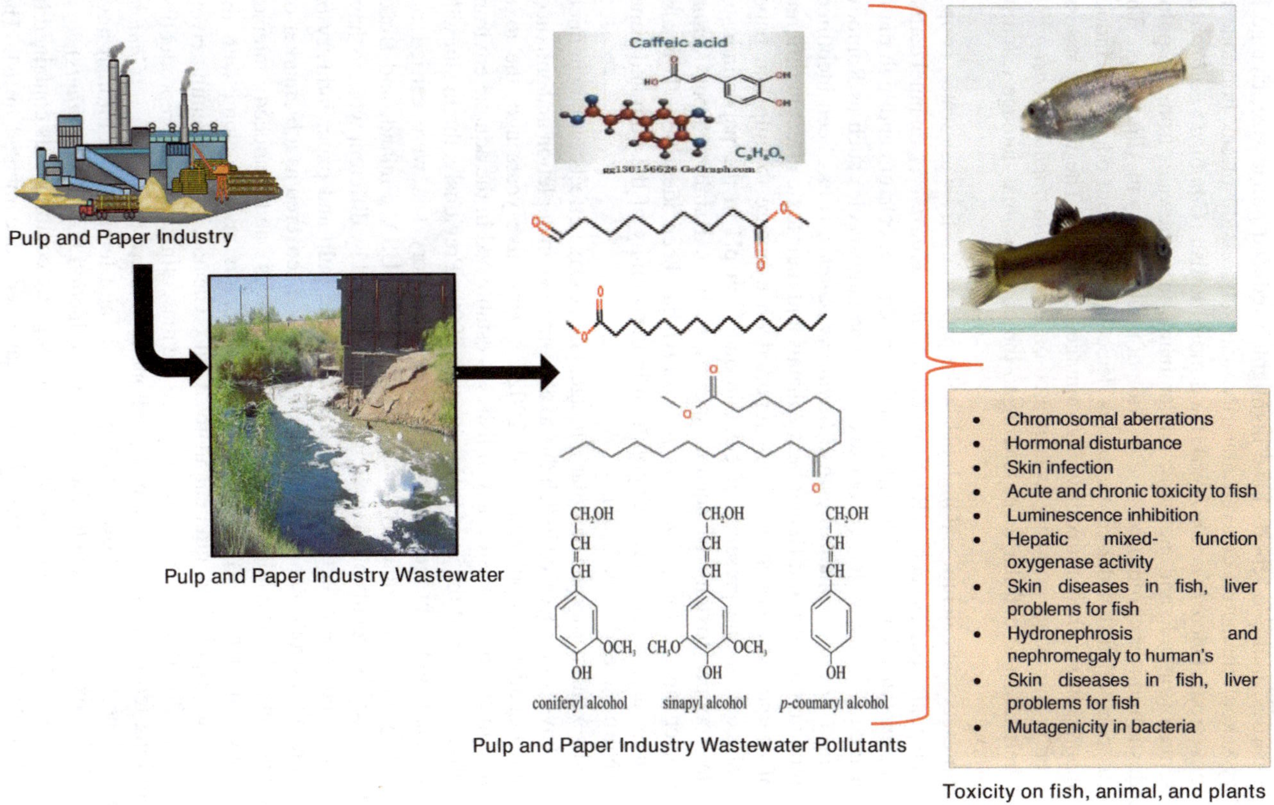

Fig. 10.1 Pulp and paper industry pollutants and their toxicity

In addition to in vitro studies of organic compounds on fish, *Salmonella typhimurium, Vibrio fischeri*, and *Saccharomyces* inhibition of growth and luminescence properties, multiple reports have highlighted reduced gonad size, change in secondary sexual character, delayed maturity, and suppression of sex hormone in fish rainbow trout (*Oncorhynchus mykiss*) and mosquitofish (*Gambusia holbrooki*). The presence of organic and inorganic contaminants in wastewater from the pulp–paper industry induces chromosomal aberration of phytotoxicity in *Allium cepa*. The report thus concluded that the contaminants detected create foul odors and induce fish hermaphroditism, hepatotoxicity, and mutagenic results. The toxicity assessment of pulp and paper industry wastewater on *Trigonella foenum-graecum* L showed their effect at cellular level leading to the chromosomal damage (Sharma et al. 2020a).

In the presence of calcium and high pH in the paper industry wastewater, copper toxicity to aquatic species such as Rainbow trout (*Salmo Gairdneri*), channel catfish (*Ictalurus punctatus*) has been identified. In the presence of calcium and high pH in the paper industry wastewater, copper toxicity to aquatic species such as Rainbow trout (*Salmo Gairdneri*), channel catfish (*Ictalurus punctatus*) has been identified. Bioaccumulation pattern of different heavy metals and their toxic effect in different organs of *Anabas testudineus* has been reported on exposure to pulp and paper industry effluent. In the presence of calcium and high pH in the paper industry wastewater, copper toxicity to aquatic species such as Rainbow trout (*Salmo Gairdneri*), channel catfish (*Ictalurus punctatus*) has been identified. This also suggests the source of groundwater pollution in the vicinity of the pulp and paper industry (Nurmesniemi et al. 2005).

Delayed sexual maturation, cellular damage, and adverse effects on biochemical parameters have been observed in fish due to effluent oxygen depletion and anoxia of the aquatic organism (Sepulveda et al. 2002). Twenty-five years ago, the toxic impact of pulp and paper wastewater in fish was established in the aquatic environment in different parts of the world (Hewitt et al. 2008). Previously, the in situ wild fish experiment and subsequently the in vivo laboratory test were carried out predominantly in New Zealand, Canada, Sweden, the USA, Finland, and Brazil (Hewitt et al. 2008). Details of the toxicity of fish reported by different workers have been summarized. The effect of a discharge from the pulp and paper industry on caged fish rainbow trout (*Oncorhynchus mykiss*) was discovered as a biomarker on the production of secondary sexual characteristics and induced intersex characteristics in juvenile rainbow trout (*Oncorhynchus mykiss*) along with the effluent discharge pollution gradient in the Chile, Canada, and Argentina rivers (Gustavo et al. 2015). In 2018, China reported the masculinization and reproductive effects of long-term exposure to androstenedione in western mosquitofish (*Gambusia affinis*). The study also showed that both male and female mosquitofish have adverse effects on growth, development, and reproduction (*Gambusia affinis*).

Furthermore, the individual effects of various organic contaminants contained in the effluent from the pulp and paper industry have been documented in fish. The mutagenicity, mortality, growth inhibition, skin diseases, and liver problems of pulp and paper wastewater in fish have been identified in several studies of organic pollutants. Besides, the toxic effect of organic contaminants found in the pulp and paper industry has also been documented in different species in humans (Field and Reed 1996; Lee et al. 1996). In microorganisms, the acute toxicity of pulp and paper industry effluents has also been observed via the microtox test by measuring the bioluminescence inhibition of *Vibrio fischeri*. Due to the presence of resin acid, fatty acid, and biocides that are found even after secondary treatment, the primary cause of toxicity of effluent has been identified by different staff. The toxicity of various heavy metals differs due to the presence of other non-metallic inorganic compounds and pulp and paper industry effluent hardness along with other metals (Commonwealth of Australia 2015). Toxicity of copper to aquatic organisms, i.e. rainbow trout (*Salmo Gairdneri*), channel catfish (*Ictalurus punctatus*) have been in the wastewater of pulp and paper industry. In certain organisms, arsenic is highly toxic to marine organisms and bio-accumulates. Arsenic is persistent and can hinder plant growth in the environment (ANZECC and ARMCANZ 2000) (Tables 10.1 and 10.2).

10.6 Management of Sludge after Secondary Treatment

Almost 3.5% of world industrial production and 2% of world trade was accounted by the paper industry worldwide. With an average annual growth in paper demand of about 3.7%, demand for paper and boards worldwide will hit 470 million tonnes in 2010. The market for paper and board is expected to hit 640 million tonnes in 2020. The world's timber consumption is estimated to grow from 861 million tonnes in 1995 to 1777 million tonnes in 2020. With only 17 units generating 0.11 million tonnes of paper in 1950, we currently have about 380 mills with a total installed capacity of approximately 4.2 million tonnes. There are 28 big mills and small paper mills are the remaining ones. In India, paper consumption per capita is a mere 3.2 kg compared to Asia, and the world average is 18 kg and 47.7 kg, respectively. Bamboo, wood, bagasse, waste paper, and agricultural residue such as wheat straw, rice straw, jute sticks, etc. are major raw materials used by the paper industry. In addition, the paper industry uses vast amounts of chemicals, such as caustic soda, sodium sulfide, sodium carbonate, chlorine, hypochlorite, mineral acid, tar, talcum powder, etc. Kraft pulping, mechanical pulping, semi-chemical pulping are process technologies used. In most mills, the bleaching processes used are CEH and CEHH, although very few mills use chlorine dioxide bleaching and oxygen delignification. The most significant factor limiting the growth of the Indian paper industry and its development into a globally competitive industry have been acute shortages and high costs of wood-based raw materials. The Kraft pulping process using bamboo, wood, and wastepaper is the basis of all large integrated paper mills in India. Small paper mills accounting for approximately 50% of India's total output are focused on soda pulping and wastepaper pulping and use as raw material agricultural residues, wheat straw, rice straw, bagasse, etc., and wastepaper. Kraft pulping and various mechanical processes using bamboo, wood, bagasse, etc. are the basis of newsprint mills. At various stages of pulp and production and paper application stages, solid waste occurs. Depending on the raw material consumed, in-plant control measures,

Table 10.1 Toxicity assessment of pulp and paper industry wastewater after secondary treatment

Organic compounds	Tested organism	Toxicity	References
Abietic acid, methyl dehydroabietic, ethyl dehydroabietic, ethyl abietate, dehydroabietic acid	*Saccharomyces cerevisiae*	Antiestrogenic activity	Terasaki et al. (2009)
Chlorophenols, chloroguaiacols, chlorosyringols, and chlorocatechols	Fish marine fish bile	Carcinogen	Soderstrom et al. (1994)
Trichlorotrihydroxybenzenes, bromomethylpropanylbenzene	*Salmonella typhimurium*	Mutagenicity	Bjørseth et al. (1979), Carlberg et al. (1980)
Trichlorohydroxyfuranone	*Salmonella typhimurium*	Mutagenicity	Holmbom et al. (1984)
Octadecenoic acid amide, linoleic acid	*Vibrio fischeri*	Luminescence inhibition	Svenson et al. (1996)
Chlorovanillins, chlorinated dehydroabietic acids	Fish	Acute and chronic toxicity to fish	Tavendale et al. (1995)
Nitrilotriacetic acid, EDTA	Human	Hydronephrosis and nephromegaly to human's	Lee et al. (1996)
Terpenes, methanol, phenol, alkylated phenols, decalone, dehydroabietic acid	Fish	Carcinogen	Botalova and Schwarzbauer (2011)
Nonylphenolethoxycarboxylates	Human	Kidneys and hormone levels	Field and Reed 1996
Sitosterol, wood sterols, O-cresol	Fish	Skin diseases in fish, liver problems for fish	Meriläinen and Oikari (2008)
Chlorinated dehydroabietic acids	Fish	Acute and chronic toxicity to fish	Tavendale et al. (1995)
Chlorinated pterostilbene	*Oncorhynchus mykiss*	Hepatic mixed-function oxygenase activity	Burnison et al. (1999)
Trichlorohydroxyfuranone	*Salmonella typhimurium*	Mutagenicity	Holmbom et al. (1984)
Dichloromethylpropanylbenzene	*Salmonella typhimurium*	Mutagenicity	Carlberg et al. (1980)
Abietic acid, dehydroabietic acid	Fish	Carcinogen	Botalova and Schwarzbauer (2011)
Linoleic acid isomer, benzoic acid	*Vibrio fischeri*	Luminescence inhibition	Svenson et al. (1996)

Table 10.2 Metals and their toxicity of kraft and pulp–paper industry wastewater

Source	Metals	Toxicity	References
Kraft pulping	Potassium	Cardiac, kidney, and lung diseases	Smyth et al. (2014)
Kraft pulping	Copper	Lipid peroxidation and DNA damage in the European eel *Anguilla anguilla*	Gravato et al. (2006)
Kraft pulping	Lead	Toadfish *Halobatrachus didactylus* and its effects on aminolevulinic acid dehydratase (ALA-D) activity, MT levels, and LPO in the liver, kidney, and blood were investigated	Center for Disease Control and Prevention (2005), Campana et al. (2003)
Kraft pulping	Nickel, cobalt	Caused oxidative stress to fish	Stohs and Bagchi (1995)
Kraft	Sulfur	Chronic exposure	U.S. Environmental Protection Agency (1991)
During uses of dye in the paper, digestion of wood	Arsenic	Carcinogenic, internal cancers, neurological problems, pulmonary disease, peripheral vascular disease, hypertension, and cardiovascular disease	Chowdhury et al. (2000), Smith et al. (2000)
Pulping	Phosphorus	Chronic kidney disease, hyperphosphatemia	Al-Azem and Khan (2012)
Sulfite pulping	Barium	Cardiac irregularities, tremors, and this toxicity may be caused by Ba2+ blocking potassium ion channels, which are critical to the proper function of the nervous system	Patnaik (2003)
Kraft pulping	Cadmium	Toxic to the kidney of the sea bass *Dicentrarchus labrax*	Romeo et al. (2000)
Chemical pulping process	Zinc	Accumulation gill > liver > ovary > muscle. The neurological and nervous system damage	Nurmesniemi et al. (2005)
Kraft pulping	Iron	Effect on medaka Oryzias latipes embryos. The deleterious effects of iron include DNA damage, lipid peroxidation (LPO), and oxidation of proteins	Li et al. (2009), Valko et al. (2005)

external control measures, housekeeping, waste utilization, disposal, and recycling methods, the forms and amounts of solid waste vary from plant to plant. Due to off-plant operation, in-plant operation, and application created, solid waste generation can be treated separately as solid waste generated. Different solid waste produced by off-plant activities in leaf processing, bark, unused branches of trees, during forest activity, solid waste generation during coal mining, limestone used in large quantities in the paper industry.

Solid waste is produced during plant operations at various stages of pulp and paper making. Solid waste generation is inevitable during different stages of the pulp and paper industry during processing, some solid waste such as chipper house dust, screen, and centric leaner rejects can be significantly reduced by in-plant control steps, loss of pulp due to overflow, leakages, waste of paper. Paper and board can be efficiently collected, and paper recycling can be strengthened by proper design of the municipal waste collection system, resulting in forest raw material conservation. Some of the areas that require urgent attention and consideration to solve the problems of solid waste disposal are the re-burning of lime sludge after desilication, the use of fly ash in the production of cement, the exploration of the possibility of using lime sludge in cement plants and for masonry cement, fly ash and effluent sludge as a soil conditioner by educating farmers, the better use of lime sludge. Since pulp and paper industry sludge contains large amounts of wastepaper, increased recycling will direct this valuable material from landfills through the proper collection and can provide an incentive for reuse that will minimize the effect on forests and reduce energy usage.

- Mud, sand, and other bamboo-washing particulates.
- Wood dust and bamboo in a chipper house.
- The pith from the pulping plant for bagasse.
- Rejects Knotter, screen, and centric leaner in the pulp mill.
- Lime dust from lime and slacker handling.
- Grit, dregs, in the recovery section, lime sludge.
- Rejections from paper machines by Centric Leaner.
- Contaminated wastepaper plant products.
- Effluent treatment plant sludge.
- Raw water clarifier sludge.
- Boiler bottom ash from the power generation plant and fly ash.

The quantity of sludge produced during pulp and paper production due to different activities. Production of sludge in different pulp and paper mills that are using agro-based products.

Solid waste	Quantity	Utilization
Sludge from the effluent treatment plant	50 kg per ton of paper	Can be used as manure after composting
Boilerhouse gas	220 kg per ton of paper	Used as absorbent
Straw, dust, fines	6.5% on the raw material basis	Used as landfill and fuel
Bagasse pith	12.15% on the raw material basis	Can be used as fuel

10.7 Future Prospective

The pulp and paper industry produces a large amount of waste material from sludge after secondary treatment. As stated, various biological, mechanical, and chemical processes carry out toxic metals, physico-chemical pollution parameters, and complete discharge of wastewater and sludge from this waste. Therefore, these findings demonstrate that the integration of biological treatment with advanced processes of oxidation is a sustainable way forward. Using the various techniques of clinker preparation, sorbent processing, biochar, and biofuel production, the residue after the treatment phase can be valorized. This will lead to the pulp and paper industry's zero waste development policy and make it environmentally friendly.

10.8 Conclusion

Different studies have shown that wastewater in the pulp and paper industry is one of the most polluting industries due to the emission of complex toxic gaseous and aquatic contaminants that have not yet been well recognized. Chlorophenols, chloroguaiacols, chlorosyringols, chlorocatechols, octacosane, pentane, silane, β-sitosterol, octadecenoic amide, octylphenol, chlorovanillins, terpenes, nonylphenol, ether trimethylsilyl, and estradiol, etc. are the key organic pollutants listed. Several of these compounds are screened by under the EDCs list. During bleaching and papermaking procedures, the mixture of various metallic and non-metallic contaminants such as nickel, iron, chromium, lead, sulfur, phosphorus, zinc, and arsenic is discharged. Female fish masculinization decreased gonad size and changes in other secondary sexual characteristics are reported by different researchers from different countries. Also, after secondary treatment of sludge in the pulp and paper industry, the presence of mutagenic and androgenic compounds as residual contaminants drains attention to tertiary treatment for pollution control in the aquatic environment.

Acknowledgments We are grateful to the Science and Engineering Research Board (SERB), a statutory body under the Department of Science and Technology (DST), Government of India for the Start-up Research Grant (SRG/2019/001940).

References

Al-Azem H, Khan AA (2012) Hypoparathyroidism. Best Pract Res Clin Endocrinol Metab 26 (4):517–522

Ali M, Sreekrishnan TR (2001) Aquatic toxicity from pulp and paper mill effluents: a review. Adv Environ Res 5(2):175–196

ANZECC and ARMCANZ (2000) National Water Quality Management Strategy: Australia and New Zealand water quality guidelines for fresh and marine water quality. Australian and New Zealand Conservation Council, Agriculture and Resource Management Council of Australia and New Zealand

Bjorseth A, Carlberg GE, Moller M (1979) Determination of halogenated organic compounds and mutagenicity testing of spent bleach liquors. Sci Total Environ 11(2):197–211

Blanco A, Negro C, Monte C, Fuente E, Tijero J (2004) Peer reviewed: the challenges of sustainable papermaking. Environ Sci Technol 38(21):414A–420A

Botalova O, Schwarzbauer J (2011) Geochemical characterization of organic pollutants in effluents discharged from various industrial sources to riverine systems. Water Air Soil Pollut 221 (1–4):77–98

Burnison BK, Comba ME, Carey JH, Parrott J, Sherry JP (1999) Isolation and tentative identification of compound in bleached-Kraft mill effluent capable of causing mixed-function oxygenase induction in fish. Environ Toxicol Chem Int J 18(12):2882–2887

Buyukkamaci N, Koken E (2010) Economic evaluation of alternative wastewater treatment plant options for pulp and paper industry. Sci Total Environ 408:6070–6078

Cabrera MN (2017) Pulp mill wastewater: characteristics and treatment. In: Biological wastewater treatment and resource recovery. Intechopen, New York, pp 119–139

Campana O, Sarasquete C, Blasco J (2003) Effect of lead on ALA-D activity, metallothionein levels, and lipid peroxidation in blood, kidney, and liver of the toadfish Halobatrachus didactylus. Ecotoxicol Environ Saf 55(1):116–125

Carlberg GE, Gjøs N, Møller M, Gustavsen KO, Tveten G, Renberg L (1980) Chemical characterization and mutagenicity testing of chlorinated trihydroxybenzenes identified in spent bleach liquors from a sulphite plant. Sci Total Environ 15(1):3–15

Centers for Disease Control and Prevention (CDC) (2005) Lead poisoning associated with use of litargirio--Rhode Island, 2003. MMWR Morb Mortal Wkly Rep 54(9):227

Chandra R, Sharma P, Yadav S, Tripathi S (2018) Biodegradation of endocrine-disrupting chemicals and residual organic pollutants of pulp and paper mill effluent by biostimulation. Front Microbiol 9:960

Chandra R, Singh R (2012) Decolourisation and detoxification of rayon grade pulp paper mill effluent by mixed bacterial culture isolated from pulp paper mill effluent polluted site. Biochem Eng J 61:49–58

Chandra R, Yadav S, Yadav S (2017) Phytoextraction potential of heavy metals by native wetland plants growing on chlorolignin containing sludge of pulp and paper industry. Ecol Eng 98:134–145

Chowdhury UK, Biswas BK, Chowdhury TR, Samanta G, Mandal BK, Basu GC, Roy S (2000) Groundwater arsenic contamination in Bangladesh and West Bengal, India. Environ Health Perspect 108(5):393–397

Commonwealth of Australia (2015) Department of Environment. National Pollutant Inventory: Substance Fact Sheet. http://www.npi.gov.au/substances/fact-sheets

Costigan SL, Werner J, Ouellet JD, Hill LG, Law RD (2012) Expression profiling and gene ontology analysis in fathead minnow (Pimephales promelas) liver following exposure to pulp and paper mill effluents. Aquat Toxicol 122:44–55

Dal Corso G, Fasani E, Manara A, Visioli G, Furini A (2019) Heavy metal pollutions: state of the art and innovation in phytoremediation. Int J Mol Sci 20(14):3412

Fahr M, Laplaze L, El Mzibri M, Doumas P, Bendaou N, Hocher V et al (2015) Assessment of lead tolerance and accumulation in metallicolous and non-metallicolous populations of Hirschfeldiaincana. Environ Exp Bot 109:186–192

Field JA, Reed RL (1996) Nonylphenol polyethoxy carboxylate metabolites of nonionic surfactants in US paper mill effluents, municipal sewage treatment plant effluents, and river waters. Environ Sci Technol 30(12):3544–3550

Gravato C, Teles M, Oliveira M, Santos MA (2006) Oxidative stress, liver biotransformation and genotoxic effects induced by copper in Anguilla anguilla L.–the influence of pre-exposure to β-naphthoflavone. Chemosphere 65(10):1821–1830

Grover R, Marwaha SS, Kennedy JF (1999) Studies on the use of an anaerobic baffled reactor for the continuous anaerobic digestion of pulp and paper mill black liquors. Process Biochem 34 (6–7):653–657

Gustavo C, Ricardo B, Mauricio D-J, Meyling R, Paulina B, Kelly RM (2015) Estrogenicity and intersex in juvenile rainbow trout (Oncorhynchus mykiss) exposed to pine/Eucalyptus pulp and paper production effluent in Chile. Aquat Toxicol 164(2015):126–134

Haq I, Raj A (2018) Endocrine-disrupting pollutants in industrial wastewater and their degradation and detoxification approaches. In: Bharagava RN, Chowdhary P (eds) Emerging and eco-friendly approaches for waste management. Springer Nature Singapore Pte Ltd, pp 121–142

Haq I, Raj A (2019) Pulp and paper mill wastewater: ecotoxicological effect and bioremediation approaches for environmental safety. In: Bharagava RN, Saxena G (eds) Biological agents and methods for industrial waste management. Springer Nature Singapore Pte Ltd, pp 333–356

Haq I, Kumar S, Kumari V, Singh SK, Raj A (2016a) Evaluation of bioremediation potentiality of ligninolytic Serratia liquefaciens for detoxification of pulp and paper mill effluent. J Hazard Mater 305:190–199

Haq I, Kumari V, Kumar S, Raj A, Lohani M, Bhargava RN (2016b) Evaluation of the phytotoxic and genotoxic potential of pulp and paper mill effluent using Vigna radiata and Allium cepa. Adv Biol 2016:1–10

Haq I, Kumar S, Raj A, Mohtashim L, Satyanarayana GNV (2017) Genotoxicity assessment of pulp and paper mill effluent before and after bacterial degradation using Allium cepa test. Chemosphere 169:642–650

Haq I, Raj A, Markandeya (2018) Biodegradation of Azure-B dye by Serratia liquefaciens and its validation by phytotoxicity, genotoxicity and cytotoxicity studies. Chemosphere 196:58–68

Haq I, Mazumder P, Kalamdhad AS (2020a) Recent advances in removal of lignin from paper industry wastewater and its industrial applications – a review. Bioresour Technol 312:123636

Haq I, Singh A, Kalamdhad AS (2020b) Arsenic: environmental contamination, health hazards, and bioremediation approaches for detoxification. In: Bioremediation for environmental sustainability toxicity, mechanisms of contaminants degradation, detoxification, and challenges. Elsevier, pp 121–142

Hewitt LM, Kovacs TG, Dub MG, Mac Latchy DL, Martel PH (2008) Altered reproduction of fish exposed to pulp and paper mill effluents: roles of individual compounds and mill operating procedures. Environ Toxicol Chem 27(5):1226–1226

Higgins MJ, Chen YC, Yarosz DP, Murthy SN, Maas NA, Glindemann D, Novak JT (2006) Cycling of volatile organic sulfur compounds in anaerobically digested biosolids and its implications for odors. Water Environ Res 78(3):243–252

Holmbom B, Voss RH, Mortimer RD, Wong A (1984) Fractionation, isolation and characterization of Ames-mutagenic compounds in Kraft chlorination effluents. Environ Sci Technol 18 (5):333–337

Jaskulak M, Grobelak A, Grosser A, Vandenbulcke F (2019) Gene expression, DNA damage and other stress markers in Sinapis alba L exposed to heavy metals with special reference to sewage sludge application on contaminated sites. Ecotoxicol Environ Saf 181:508–517

Kalaivanan D, Ganeshamurthy AN (2016) Mechanisms of heavy metal toxicity in plants. In: The abiotic stress physiology of horticultural crops. Springer, New Delhi, pp 85–10

Kamunde C, Clayton C, Wood CM (2002) Waterborne vs. dietary copper uptake in rainbow trout and the effects of previous waterborne copper exposure. Am J Phys Regul Integr Comp Phys 283(1):R69–R78

Karrasch B, Parra O, Cid H, Mehrens M, Pacheco P, Urrutia R, Valdovinos C, Zaror C (2006) Effects of pulp and paper mill effluents on the microplankton and microbial self-purification capabilities of the Biobìo River. Chile Sci Total Environ 359:194–208

Khan N, Bano A, Zandi P (2018) Effects of exogenously applied plant growth regulators in combination with PGPR on the physiology and root growth of chickpea (Cicer arietinum) and their role in drought tolerance. J Plant Interact 13(1):239–247

Latorre A, Malmqvist A, Lacorte S, Welander T, Barceló D (2007) Evaluation of the treatment efficiencies of paper mill whitewaters in terms of organic composition and toxicity. Environ Pollut 147(3):648–655. https://doi.org/10.1016/j.envpol.2006.09.015

Lee HB, Peart TE, Kaiser KL (1996) Determination of nitrolotriacetic, ethylenediaminetetraacetic and diethylenetriaminepentaacetic acids in sewage treatment plant and paper mill effluents. J Chromatogr A 738(1):91–99

Li H, Zhou Q, Wu Y, Fu J, Wang T, Jiang G (2009) Effects of waterborne nano-iron on medaka (Oryzias latipes): antioxidant enzymatic activity, lipid peroxidation and histopathology. Ecotoxicol Environ Saf 72(3):684–692

Luoma SN, Rainbow PS (2008) Sources and cycles of trace metals. In: Metal contamination in aquatic environments: science and lateral management. Cambridge University Press, Cambridge, UK, pp 47–66

Magnus E, Hoel H, Carlberg GE (2000) Treatment of an NSSC effluent in a biological high-efficiency compact reactor. TAPPI J 83(1):149–155

Meriläinen P, Oikari A (2008) Uptake of organic xenobiotics by benthic invertebrates from sediment contaminated by the pulp and paper industry. Water Res 42(6–7):1715–1725

Miyazaki T, Wada M, Kawahara H, Sato M, Baba H, Shimada S (2002) Dynamic load at baseline can predict radiographic disease progression in medial compartment knee osteoarthritis. Ann Rheum Dis 61(7):617–622

Muchuweti M, Birkett JW, Chinyanga E, Zvauya R, Scrimshaw MD, Lester JN (2006) Heavy metal content of vegetables irrigated with mixtures of wastewater and sewage sludge in Zimbabwe: implications for human health. Agric Ecosyst Environ 112(1):41–48

Ngo HH, Bui XT, Nghiem LD, Guo W (2020) Green technologies for sustainable water. Bioresour Technol 317:123978

Nurmesniemi H, Pöykiö R, Perämäki P, Kuokkanen T (2005) The use of a sequential leaching procedure for heavy metal fractionation in green liquor dregs from a causticizing process at a pulp mill. Chemosphere 61(10):1475–1484

O'Connor D, Zheng X, Hou D, Shen Z, Li G, Miao G et al (2019) Phytoremediation: climate change resilience and sustainability assessment at a coastal brownfield redevelopment. Environ Int 130:104945

Orrego R, Guchardi J, Krause R, Holdway D (2010) Estrogenic and anti-estrogenic effects of wood extractives present in pulp and paper mill effluents on rainbow trout. Aquat Toxicol 99 (2):160–167

Patnaik P (2003) Handbook of inorganic chemicals, vol 529. McGraw-Hill, New York

Pokhrel D, Viraraghavan T (2004) Treatment of pulp and paper mill wastewater-a review. Sci Total Environ 333(1–3):37–58

Romeo M, Bennani N, Gnassia-Barelli M, Lafaurie M, Girard JP (2000) Cadmium and copper display different responses towards oxidative stress in the kidney of the sea bass Dicentrarchus labrax. Aquat Toxicol 48(2–3):185–194

Saltman D (1978) Paper basics: forestry, manufacture, selection, purchasing, mathematics, and metrics, recycling. Van Nostrand Reinhold Company, San Francisco, CA

Savant DV, Abdul-Rahman R, Ranade DR (2006) Anaerobic degradation of adsorbable organic halides (AOX) from pulp and paper industry wastewater. Bioresour Technol 97:1092–1104

Selvi A, Rajasekar A, Theerthagiri J, Ananthaselvam A, Sathishkumar K, Madhavan J, Rahman PK (2019) Integrated remediation processes toward heavy metal removal/recovery from various environments-a review. Front Environ Sci 7:66

Sepulveda MS, Johnson WE, Higman JC, Denslow ND, Schoeb TR, Gross TS (2002) An evaluation of biomarkers of reproductive function and potential contaminant effects in Florida largemouth bass (Micropterus salmoides floridanus) sampled from the St Johns River. Sci Total Environ 289(1–3):133–144

Sharma P, Tripathi S, Chandra R (2020c) Highly efficient phytoremediation potential of metal and metalloids from the pulp paper industry waste employing *Eclipta alba* (L) and *Alternanthera philoxeroide* (L): biosorption and pollution reduction. Bioresour Technol 319:124147

Sharma P, Tripathi S, Chandra R (2020d) Phytoremediation potential of heavy metal accumulator plants for waste management in the pulp and paper industry. Heliyon 6(7):e04559

Sharma P, Tripathi S, Chaturvedi P, Chaurasia D, Chandra R (2020b) Newly isolated *Bacillus* sp. PS-6 assisted phytoremediation of heavy metals using Phragmites communis: potential application in wastewater treatment. Bioresour Technol 320:124353

Sharma P, Tripathi S, Vadakedath N, Chandra R (2020a) In-situ toxicity assessment of pulp and paper industry wastewater on *Trigonella foenum-graecum* L: potential source of cytotoxicity and chromosomal damage. Environ Technol Innovation 21:101251

Shigeoka T, Sato Y, Takeda Y, Yoshida K, Yamauchi F (1988) Acute toxicity of chlorophenols to green algae, Selenastrum capricornutum and *Chlorella vulgaris*, and quantitative structure-activity relationships. Environ Toxicol Chem Int J 7(10):847–854

Singh AK, Chandra R (2019) Pollutants released from the pulp paper industry: aquatic toxicity and their health hazards. Aquat Toxicol 211:202–216

Smith AH, Lingas EO, Rahman M (2000) Contamination of drinking-water by arsenic in Bangladesh: a public health emergency. Bull World Health Organ 78:1093–1103

Smook GA (1992) Handbook for pulp and paper technologist. Angus Wilde Publications, Vancouver, BC

Smyth A, Dunkler D, Gao P, Teo KK, Yusuf S, O'donnell MJ, Clase CM (2014) The relationship between estimated sodium and potassium excretion and subsequent renal outcomes. Kidney Int 86(6):1205–1212

Soderstrom M, Wachtmeister CA, Forlin L (1994) Analysis of chlorophenolics from bleach Kraft mill effluents (BKME) in bile of perch (Percafluviatilis) from the Baltic Sea and development of an analytical procedure also measuring chlorocatechols. Chemosphere 28(9):1701–1719

Stefanowicz AM, Stanek M, Woch MW, Kapusta P (2016) The accumulation of elements in plants growing spontaneously on small heaps left by the historical Zn-Pb ore mining. Environ Sci Pollut Res 23(7):6524–6534

Stohs SJ, Bagchi D (1995) Oxidative mechanisms in the toxicity of metal ions. Free Radic Biol Med 18(2):321–336

Svenson A, Norin H, Hynning PA (1996) Toxicity-directed fractionation of effluents using the bioluminescence of Vibrio fischeri and gas chromatography/mass spectroscopy identification of the toxic components. Environ Toxicol Water Qual Int J 11(3):277–284

Tao S, Wen Y, Long A, Dawson R, Cao J, Xu F (2001) Simulation of acid–base condition and copper speciation in the fish gill microenvironment. Comput Chem 25(3):215–222

Tavendale MH, Wilkins AL, Langdon AG, Mackie KL, Stuthridge TR, McFarlane PN (1995) Analytical methodology for the determination of freely available bleached Kraft mill effluent-derived organic constituents in recipient sediments. Environ Sci Technol 29(5):1407–1414

Terasaki M, Shiraishi F, Fukazawa H, Makino M (2009) Development and validation of chemical and biological analyses to determine the antiestrogenic potency of resin acids in paper mill effluents. Environ Sci Technol 43(24):9300–9305

U.S. Environmental Protection Agency (1991) Reregistration eligibility document facts: sulfur. US EPA Pesticides and Toxic Substances, Washington, DC

Ugurlu M, Gurses A, Dogar C, Yalçın M (2008) The removal of lignin and phenol from paper mill effluents by electrocoagulation. J Environ Manag 87(3):420–428

Valko MMHCM, Morris H, Cronin MTD (2005) Metals, toxicity, and oxidative stress. Curr Med Chem 12(10):1161–1208

Waye A, Annal M, Tang A, Picard G, Harnois F, Guerrero-Analco JA, Trudeau VL (2014) Canadian boreal pulp and paper feedstocks contain neuroactive substances that interact in vitro with GABA and dopaminergic systems in the brain. Sci Total Environ 468:315–325

Wiegand P, Unwin J, Cook D, Ebert E (2006) A review of contemporary topics on dioxin and pulp and paper industry effluents and wastewater residuals. NCASI Special Rep 06–08:1–79

Zarkovic DB, Todorović ZN, Rajakovic LV (2011) Simple and cost-effective measures for the improvement of paper mill effluent treatment-a case study. J Clean Prod 19(6–7):764–774

Use of Flue Gas as a Carbon Source for Algal Cultivation

11

Pandian Prabakaran, Pradeepa Virumandi, Sundaram Ravikumar, Nagasundaram Rashiya, Nagarajan Padmini, and Gopal Selvakumar

11.1 Introduction

The increase in world population density, the burning of fossil fuels, and deforestation on a large scale (Benis and Ferrão 2017) has contributed to the increasing of greenhouse gases concentration in atmosphere, resulting in global warming (Wang et al. 2008; Kondili and Kaldellis 2007). Global warming was not only be felt many decades from now it is already happening and its impacts are clearly visible. The chemistry of the atmosphere is varying because of emission of carbon monoxide, nitrogen oxides, and volatile organic compounds, among other species, in addition to the changes in the greenhouse gases. This temperature rise on the planet caused by anthropic activities has become one of the most worrying environmental problems in the world (Houghton 2005).

Continued use of fossil fuels becomes unsustainable as this feature is limited and combustion for power generation results in excessive emission of greenhouse gases into the atmosphere. There are the gases emitted in these processes, carbon dioxide (CO_2), nitrogen oxide (NO_x), and sulfur (SO_x) are the main ones. Among these ones, CO_2 is considered the biggest enhancer of greenhouse effect. The increase in the concentration of atmospheric carbon dioxide is considered to be one of the main causes of global warming. Microalgae and cyanobacteria can contribute to the

P. Prabakaran · S. Ravikumar
Department of Biomedical Sciences, Department of Microbiology, Science Campus, Alagappa University, Karaikudi, Tamil Nadu, India

P. Virumandi
Department of Zoology, Seethalakshmi Achi College for Women, Sivaganga District, Tamil Nadu, India

N. Rashiya · N. Padmini · G. Selvakumar (✉)
Department of Microbiology, Science Campus, Alagappa University, Karaikudi, Tamil Nadu, India
e-mail: skumarg@alagappauniversity.ac.in

I. Haq, A. S. Kalamdhad (eds.), *Emerging Treatment Technologies for Waste Management*, https://doi.org/10.1007/978-981-16-2015-7_11

reduction of atmospheric carbon dioxide by using this gas as carbon source. This process is known as biofixation of carbon dioxide, as the fixation is done by biological and not synthetic means. Strategies have been checked by reducing these gases released into the atmosphere, the use of microalgae as an alternative to CO_2 mitigation. Several studies employ microalgae in fixing flue gases emitted by municipal waste incinerators (Douskova et al. 2009) and thermal power plants (Costa et al. 2015; Vaz et al. 2016; Duarte et al. 2016).

The biological fixation of carbon dioxide has been considered a promising approach compared to other methods such as geological and chemical (Kumar et al. 2011). Microalgae have received attention in recent years due to high cell growth rates and higher photosynthetic efficiency compared to terrestrial plants (Anjos et al. 2013). These microorganisms have capability to convert CO_2 into oxygen and biomass, accumulating high-value biocompounds, with applicability in the most different knowledge areas (Tang et al. 2011). CO_2 fixation by microalgae has emerged as a promising option for CO_2 mitigation. Intensive research work has been carried out to develop a feasible system for removing CO_2 from industrial exhaust gases. In this way, the chapter aimed to approach the implementation of flue gas as a carbon source in microalgal cultivation, as well as the metabolism of CO_2 biofixation and involved environmental issues that justify and make it important to study this issue.

11.2 Greenhouse Gases

Global warming and the excessive emission of greenhouse gases (GHG) into the atmosphere have caused great concern worldwide. The three most powerful long lived greenhouse gases in the atmosphere are carbon dioxide, methane, and nitrous oxide. In this way, solutions to sustainable development have become the focus of discussions and international actions. Deforestation is the main activity that causes the loss of ecosystems, followed by mining and hydroelectric power generation. Deforestation, increase in industrial production, and intensive use of fossil fuels have contributed to the raise of CO_2 concentration in the atmosphere, the main greenhouse gas (I. P. O. C. CHANGE 2007). According to IPCC (*Intergovernmental Panel on Climate Change*) (I. P. O. C. CHANGE 2014), the annual emission of GHG increased 10 Gt_{CO2eq} between 2000 and 2010 derived directly of supply energy sectors (47%), industries (30%), transport (11%), and constructions (3%). The global energy matrix is still based on non-renewable sources, such as oil, coal, and natural gas. These sources represent about 80% of energy consumption in the world (Mota and Monteiro 2013). Fossil fuel includes coal, natural gas, petroleum, shale oil, and bitumen. During fossil fuels burning like oil, natural gas and coal released flue gases that may contain up to 142 compounds. Among these are oxygen (O_2), water steam (H_2O), nitrogen (N_2), carbon dioxide (CO_2), nitrogen oxides (NO_x), sulfur oxides (SO_x), carbon monoxide (CO), hydrocarbons from incomplete combustion (CxHy), acids of halogens and particulates (Van Den Hende et al. 2012).

Environmental pollution, flora and fauna degradation, and climate changes that are occurring in the world are associated with oil companies, which produce comfort and social well-being in twentieth century. Modern society still enjoys the facilities and benefits provided by oil refineries as the use of high energy fuels, plastics and chemicals production. However, the society charges the negative environmental cost caused in the last century by excessive CO_2 emissions, making it a challenge for twenty-first century (Mota and Monteiro 2013). Human activities directly or indirectly affect the environment adversely. A stone crusher adds a lot of suspended particulate matter and noise into the atmosphere. Automobiles emit from their tail pipes oxides of nitrogen, sulfur dioxide, carbon dioxide, carbon monoxide, and a complex mixture of unburnt hydrocarbons and black soot which pollute the atmosphere. Domestic sewage and run off from agricultural fields, laden with pesticides and fertilizers, pollute water bodies. Effluents from tanneries contain many harmful chemicals and emit foul smell. These are only a few examples which show how human activities pollute the environment.

All major climate changes, including natural ones, are disruptive. Past climate changes led to extinction of many species, population migrations, and pronounced changes in the land surface and ocean circulation. The quickness of the current climate change is faster than most of the past events, making it more difficult for human societies and the natural world to adapt. Climate changes and the warming of the atmosphere and oceans, melting of Antarctica and Greenland glaciers, seas level rising, and more frequent and long heat waves are threatening the planet (Moreira and Pires 2016; Dawson et al. 2011; Meinshausen et al. 2009; Rockström et al. 2009). These climatic changes can result in harmful effects to human life by increasing the number of diseases and heat related deaths and the spread of disease by variations in precipitations and temperature (Moreira and Pires 2016).

11.3 Greenhouse Gases Reduction

Stabilizing atmospheric greenhouse gas concentrations requires that discharges be equally balanced by the processes that remove greenhouse gases from the atmosphere. For CO_2, this means that anthropogenic emissions must in the long run decline toward zero as the ocean and atmosphere come into equilibrium. In contrast, CO_2 emissions are rising today and, absent actions considered to alter this situation, are projected to continue to rise for many decades into the future. Meeting the objective of stabilizing greenhouse gas concentrations will, therefore, require vital changes in the way the world produces and uses energy, as well as in many other greenhouse gas-emitting activities within the industrial, agricultural, and land-use areas of the global economy. It is widely acknowledged that new and improved technologies could substantially reduce the economic burden of such changes. Recent industrial economies are dependent on fossil fuels with their associated greenhouse gas emissions. Not unpredictably, many governments view measures to foster technological change as integral to their policies toward climate change.

During the last two decades, several strategies of CO_2 mitigation that have been investigated by research had included approaches in relation to chemical reactions, biological process, renewable energy sources use, and injection of CO_2 at oceans and geological formations (Singh and Ahluwalia 2013; Zheng et al. 2011). The chemical fixation to remove CO_2 from industrial combustion has been used due to advantages of operation and high absorption efficiency (Peng et al. 2012). This process is based on a chemical reagent (methyl diethanolamine or potassium carbonate) or a system consisting of more than one absorbent. Usually skill of a chemical absorbent is described as the loading capacity of CO_2 or H_2S (acid gases) (Rufford et al. 2012). A disadvantage of this process of absorption using monoethanolamine is the high energy consumption for regeneration of the absorbent and CO_2 separation after absorption (Meldon and Morales-cabrera 2011). The geological storage of carbon dioxide aims to store the CO_2 safely in oceans and geological formations (I. P. O. C. CHANGE 2007). This process consists in a separation and capture of CO_2 from industrial processes to an isolated storage location for a long time. The CO_2 can be stored in oil reservoirs, coal seams, deep saline aquifers, salt caverns or may be injected directly into the oceans (I. P. O. C. CHANGE 2007; Kharaka et al. 2009).

Generally, photosynthetic system provides critical oxygen renewal along with the recycling of carbon into potentially beneficial biomass. The efficiency of such system depends on the type of the organism used. Microalgae are the most promising production facilities. They are capable of fixing several-fold more CO_2 per unit area than trees or crops. Such CO_2 fixation by photoautotrophic algal cultures has the potential to diminish the release of CO_2 into the atmosphere, helping alleviate the trend toward global warming. To realize effective biological CO_2 fixation systems, selection of optimal microalgae species is vital. The choice of optimal microalgae species depends on specific strategies employed for CO_2 sequestration.

Biological fixation from microalgae has been highlighted besides producing high-value products from biomass. The microalgae application is widely accepted in practice, one of the best strategies outlined by bioengineering (Tsai et al. 2016). The use of microalgae in CO_2 capture has the advantages of high growth rate compared to terrestrial plants (Minowa et al. 1995; Demirbas 2009a) and low impact on food demand (Schenk et al. 2008). These microorganisms have the ability and specificity in CO_2 mitigation (Lee and Lee 2003) and biomass with high added value for application in nutrition, pharmaceutical products (Morais 2015), fertilizers (Wuang et al. 2016), and biofuels (Milano et al. 2016). In particular, the ability of these cells to absorb CO_2 suggests microalgae cultivation as an attractive alternative for CO_2 sequestration that can be applied to fossil fuel power plant gas effluents to facilitate the reduction of greenhouse gas emissions.

Among the studies in CO_2 mitigation area by microalgae highlighted by employing flue gas as a nutrient source. Radmann et al. (2011) and Duarte et al. (2016) use simulated flue gas of coal in the cultivation of different species of microalgae. Costa et al. (2015) in studies with real flue gas from coal burn found that this promotes an increase in *Spirulina* biomass productivity. Another study by Rosa et al. (2015) proposes the use of chemical and biological fixation integrated.

The authors described that the chemical absorbent monoethanolamine (MEA) use in cultivation with CO_2 injection can increase the concentration of carbon dissolved in the liquid medium contributing to *Spirulina* sp. LEB 18 growth and to macromolecules production in biomass. Different gases can also be used in microalgae cultivation such as SO_2, NO, mainly from coal power plants. The addition of residues (CO_2, SO_2, NO, and ashes) *Chlorella fusca* LEB 111 culture was reported by Duarte et al. (2016) and Vaz et al. (2016) to evaluate the growth parameters during cultivation, CO_2 biofixation, and characterization of microalgal biomass. Duarte et al. (2016) have found more efficient CO_2 fixing (50 ± 0.8%, w w^{-1}) using 10% of CO_2, 200 ppm of SO_2 and NO, and 40 ppm of ashes. The microalgal biomass composition was similar in all cultures, about 19.7% (w w^{-1}) of carbohydrates, 15.5% (w w^{-1}) of lipids, and 50.2% (w w^{-1}) of protein. Combining strategies through microalgae cultivation, CO_2 mitigation, and bioproducts production make the process economically viable and environmentally sustainable. One of the possible strategies may be enhanced by CO_2 fixation of flue gases from industries and power plants (Costa et al. 2015).

11.4 Microalgae Production

In a cyclic microalgal production process, the system returns to the same starting point after a series of transformations, where the intermediates are microalgal cultivation, biomass production, one or more bioproducts obtaining, in the effluent generation and use the same as a source of nutrients for cultivation (Fig. 11.1). Microalgal biomass production has great potential to contribute to world energy supplies and to control CO_2 emissions as the demand for energy increases. This technology makes productive use of arid and semi-arid lands and highly saline water, resources that are not suitable for agriculture and other biomass technologies (Duarte et al. 2016). This process aims to make it more sustainable and profitable microalgal cultivation for obtaining bioproducts, reducing effluents emissions and cultivation costs using these wastes as a nutritional source. The major components of industrial wastewaters are also those from microalgal culture medium, carbon and nitrogen, which, together, represent more than 50% of the compounds used in cultivations.

Microalgal cells are sunlight-driven cell factories that can convert carbon dioxide (CO_2) into raw materials for producing biofuels (e.g., biohydrogen, biodiesel, and bioethanol), animal food chemical feedstocks, and high-value bioactive compounds (e.g., Docosahexaenoic acid (DHA)) (Razzak et al. 2013).

The most studied products obtained by microalgae biofuels are biodiesel, bioethanol, bio-kerosene, bio-oil and biomethane. The obtaining and application of microalgal biofuels are limited because of the production costs involved in the process (Pandey et al. 2014). Thus, the implementation of microalgae cyclic production systems beyond biorefinery, where more than one bioproduct is obtained from biomass, is alternative to make biofuels production more feasible (da Silva 2016). Although the emission of gases such as CO_2 by biofuels is reduced when compared

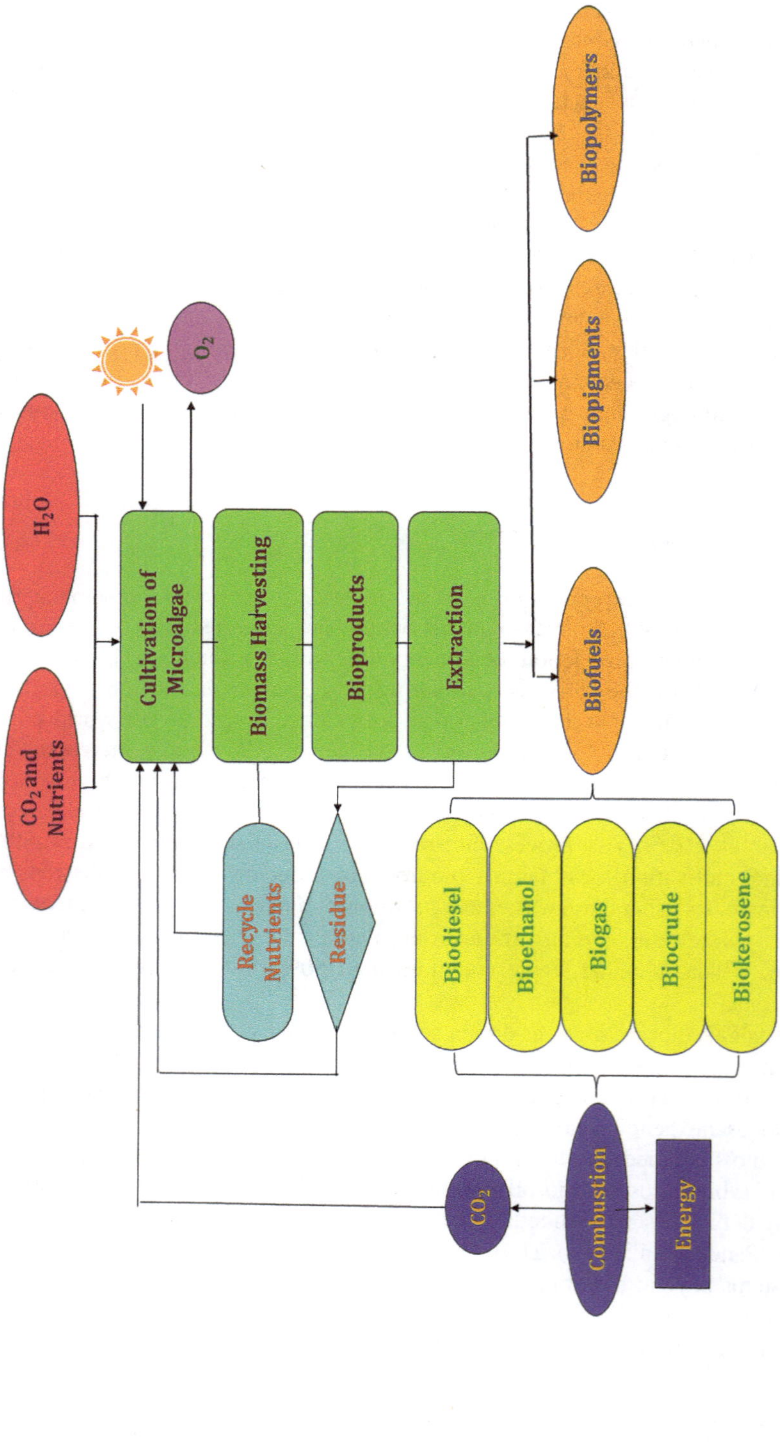

Fig. 11.1 Diagram of microalgae production

Table 11.1 Bioproducts of microalgae cultivation, effluent generated, potential use of effluent as a nutritional source in cyclical process

Bioproducts	Effluent	Nutritional source	Reference
Biofuels	CO_2	Inorganic carbon	Morais and Costa (2007a)
Biodiesel	Glycerol	Inorganic carbon	Morais (2015)
Biomethane	Anaerobic digestion waste	Organic carbon and nitrogen	Borges et al. (2013)
Biopolymer	PHB extraction waste	Nitrogen	da Silva (2016)

to fossil fuels, this amount emitted can return to microalgal cultivation for use as inorganic carbon (Morais and Costa 2007a). In addition, the portion emitted in use, which is generated during such obtaining biofuels may also be used, such as biodiesel and biomethane effluent (Morais 2015; Borges et al. 2013). While CO_2 is still released when fuels derived from algal biomass are burned, integration of microalgal farms for flue gas capture approximately doubles the amount of energy produced per unit of CO_2 released. Resources derived from microalgal biomass also can be used for other long-term uses, serving to sequester CO_2. Flue gas has the potential to provide sufficient quantities of CO_2 for such large-scale microalgae farms (Kumar et al. 2011).

Borges et al. (2013) have applied the effluent from *Spirulina* sp. LEB 18 biomass anaerobic digestion as a substrate for this strain cultivation. The effluent composed by carbon, nitrogen, and phosphorus added at a concentration, that represents 20% of the culture medium, did not alter the productivity of microalgae or maximum specific growth rate when added to the cultures. Morais (2015) has studied the cultivation with glycerin for different microalgae for biodiesel production. The authors have found efficiency up to 50% of biodiesel in batch of *Spirulina* sp. LEB 18 using 0.01 mol L^{-1} of the effluent (Table 11.1).

The polyhydroxybutyrate (PHB) extraction by cyanobacteria generates a waste that is not very explored. da Silva (2016) has studied the application of the extraction PHB waste from *Spirulina* sp. LEB 18 biomass that contains sodium hypochlorite and acetone for the cultivation of the same strain. This study verifies that there is a increase in polymer concentration when the effluent was employed with a reduction in the carbon source of Zarrouk medium (sodium bicarbonate).

11.5 Flue Gases in Microalgae Cultivation

The emissions of the flue gases from power plants account for more than 7% of total CO_2 emissions in the world (Kadam 1997). Among the flue gases, it stands out the CO_2 produced from the carbon combustion of fuel in concentrations which can vary between 5% and 6% of CO_2 for natural gas and 10% to 15% for coal (US DOE 2010). In order to reduce the high CO_2 concentrations released to the atmosphere, several processes have been studied and applied. In this context, the CO_2 biofixation by microalgae has been considered a promising method for post-combustion

Table 11.2 Examples of high CO_2– tolerant species of microalgae

Species	Known maximum CO_2 (%)
Cyanidium caldarium	100
Scenedesmus sp.	80
Chlorococcum littorale	60
Synechococcus elongatus	60
Euglena gracilis	45
Chlorella sp.	40
Eudorina spp	20
Dunaliella tertiolecta	15
Nannochloris sp.	15
Chlamydomonas sp.	15
Tetraselmis sp.	14

processes, capture and CO_2 storage. Microalgae have a high capacity to fix and store carbon through photosynthesis, converting CO_2 and water into organic compounds, without consuming additional energy and without providing secondary pollution (Zhao and Su 2014). As a distinct advantage, biomass from microalgae for energy consumption is provided after CO_2 capture. The performance of CO_2 fixation via microalgae and biomass production depends on the cultivation conditions (e.g., temperature, light, pH, and nutrient availability), species of microalgae, CO_2 concentration, and toxic pollutants in the flue gas (Zhao and Su 2014). CO_2 fixation via microalgae has many benefits, such as a high photosynthesis rate (e.g., 6.9 × 104 cells/mL/h (Suali and Sarbatly 2012) a rapid growth rate (0.7–3.2 day–1) (Ryu et al. 2009), good environmental adaptability, and low cost of operation.

The high CO_2 and nutrient requirements of microalgae can be supplied by flue gases and wastewater from other industrial processes, providing ecological advantages while lowering the cost of biomass production. Microalgae can be used as bioremediation agents to remove inorganic nutrients from wastewaters cost of wastewater treatment about 0.15–6.0 USD/m3 (Fu et al. 2013; Kumar et al. 2010) and to improve water quality due to their high capacity for nutrient uptake (Razzak et al. 2013). Microalgae that grow in the bioremediation of waste are further processed into a wide spectrum of fuel products, including (1) hydrogen (H2) via direct and indirect biophotolysis, (2) biodiesel through transesterification, (3) biomethane via anaerobic digestion, (4) bioethanol and biobutanol via fermentation, (5) bio-oil via thermochemical conversion, and (6) green diesel and gasoline through direct catalytic hydrothermal liquefaction (Demirbas 2009b, c; Damartzis and Zabaniotou 2011; Nigam and Singh 2011; Huang and Tan 2014).

Among nutritional factors for the cultivation of microalgae, carbon is considered one of the essential elements for growth, account for about 50% of the cellular matter (Rosa et al. 2015; Borges et al. 2013; Richmond 2004). The carbon can be provided from various sources, such as atmospheric CO_2 from flue gases and chemically fixed in the form of soluble carbonates (Kumar et al. 2010; Becker 1994). Since the atmospheric CO_2 concentration is low (0.03–0.06%), this may limit the growth of microalgae, necessitating the addition of carbon as soluble carbonates or compressed

CO_2 which results in increased production costs. Direct utilization of power plant flue gas has been considered for CO_2 sequestration systems (Tang et al. 2011).

The benefit of utilizing flue gas directly is the reduction of the cost of separating CO_2 gas. Since power plant flue gas contains a higher concentration of CO_2 (I. P. O. C. CHANGE 2007, 2014) identifying high CO_2 tolerant species is important. Although CO_2 concentrations vary depending on the flue gas source (Anjos et al. 2013) 15%–20% v/v is typically assumed. Several species have been tested under CO_2 concentrations of over 15%, as given in Table 11.2 (Douskova et al. 2009).

According to Molina Grima et al. (2003) the compressed CO_2 supply can represent up to 41% of the total costs of production of microalgal biomass. Flue gases contain 6–15% CO_2 and are considered a promising source of carbon for the microalgae cultivation and a potentially more efficient route for CO_2 fixation. In addition, the use of flue gas from industrial units (power plants) contributes to the mitigation of this GHG emissions to the atmosphere (Mata et al. 2010). According to Van Den Hende et al. (2012) the direct use of flue gases becomes an economically and environmentally better option compared to the compressed CO_2, especially when the industrial facilities are near the cultivation site of the microalgae. One of the high CO_2 tolerant species is *Euglena gracilis*. Growth of this species was improved under 5%–45% concentration of CO_2. The finest growth was observed with 5% CO_2 concentration. Nevertheless, the species did not grow under greater than 45% CO_2 (Meinshausen et al. 2009). It is also reported that *Chlorella* sp. can be grown under 40% CO_2 conditions (Van Den Hende et al. 2012). Furthermore, Maeda et al. (1995) found a strain of *Chlorella* sp. T-1 which could grow under 100% CO_2 although the maximum growth rate occurred under a 10% concentration, while *Scenedesmus* sp. might grow under 80% CO_2 conditions but the maximum cell mass was observed in 10%–20% CO_2 concentrations.

Cyanidium caldarium (Mota and Monteiro 2013) and some other species of *Cyanidium* can grow in pure CO_2 (Tsai et al. 2016). Normally, phototrophic microalgal growth requires a supply of carbon dioxide as a carbon source. CO_2 supply contributes to control the pH of the culture (Minowa et al. 1995). Chemical analysis has shown that algal biomass consists of 40% to 50% carbon, which suggests that about 1.5 to 2.0 kg of CO_2 is required to produce 1.0 kg of biomass (Demirbas 2009a). According to previous studies, the supply of carbon to microalgal mass culture systems is one of the principal difficulties and limitations that must be solved (Tang et al. 2011; Schenk et al. 2008; Lee and Lee 2003). Several studies have been conducted to evaluate the use of flue gas as a source of nutrients for the cultivation of microalgae, confirming promising results regarding the growth and sequestration of CO_2 (Duarte et al. 2016; Radmann et al. 2011; Chiu et al. 2011). Application of microalgal cultivation in industrial grade using flue gas is still little known. Among the examples, there is the *Seambiotic* Ltd, an Israeli company that patented the method of cultivation of photosynthetic organisms with flue gases from fossil fuels power plant (Weiss 2008). *Cyanotech Corporation*, located in Hawaii, patented a system that uses a microalgae production plant integrated with the generation of electricity. The CO_2 produced by burning fossil fuels for power generation is recovered and used as a nutrient for cultivation (Jensen et al. 1997). Microalgae response to varying CO_2 concentrations has been widely investigated. *Chlorella vulgaris* was cultivated under various light intensities in a gas recycling photobioreactor. The light intensity affected the algal growth and the CO_2 concentration in the exit gas. In the linear growth phase, CO_2 concentration in the exit gas ranged between 4.6% and 6.0% (v/v) when 20% (v/v) CO_2 balanced with 80 % (v/v) N_2 was introduced into the photobioreactor (Zhao and Su 2014). In a different study (Van Den Hende et al. 2012) *Chlorella* species showed much higher log phase growth rates, while *Scenedesmus* species was better able to tolerate very high CO_2 concentrations than *Chlorella*. However, both algae had about the same growth rate when the CO_2 concentration was in the range 10%–30%. Scenedesmus was completely inhibited by 100% CO_2. This inhibition was reversible since growth was resumed when CO_2 concentration was returned to 20%. Other microalgae species, *Chlorella minutissima* grown under extreme carbon dioxide concentrations (0.036%–100%), strongly increase the microalgal biomass through photochemical and non-photochemical changes in the photosynthetic apparatus (Mata et al. 2010).

Yun and Moon Park (1997) cultivated *Chlorella vulgaris* in wastewater liquidated from a steel-making plant with the aim of developing an economically feasible system to remove ammonia from wastewater and CO_2 from flue gas simultaneously (since no phosphorus compounds existed in wastewater, external phosphate (15.3–46.0 g^{-3}) was added to the wastewater). After adaptation to 5% (v/v) CO_2, the growth of *C. vulgaris* was expressively improved at a typical concentration of CO_2 in flue gas of 15% (v/v). Growth of *C. vulgaris* in raw wastewater was better than that in wastewater buffered with HEPES at 15% (v/v) CO_2. CO_2 fixation and ammonia removal rates were estimated as 260 g CO_2 $m^{-3}h^{-1}$ and 0.92 g NH_3 $m^{-3}h^{-1}$, respectively, when the alga was cultivated in wastewater supplemented with 460 g $PO_4{}^{-3}m^{-3}$ without pH control at 15% (v/v) CO_2.

Microalgae can be cultured in transparent polyethylene bags, called plastic bag photobioreactors. These bags are hung or placed in a cage under sunlight irradiation, while the microalgae are mixed with air in the bottom of the bags (Razzak et al. 2013). Transparent polyethylene sleeves that are sealed at the bottom in a conical shape are used to prevent cell settling. With the project, several studies have been developed both in the laboratory scale and pilot scale in order that to facilitate and maximize the CO_2 fixation, as well as contribute to the reduction of this GHG. Research has contemplated strains isolation (Radmann et al. 2011; Morais and Costa 2007b), CO_2 fixation in different photobioreactors and CO_2 concentrations (de Morais et al. 2015; de Morais and Costa 2007), mass transfer study with different configurations of diffuser (Moraes et al. 2016), use of real and simulated flue gas (Duarte et al. 2016), integration of chemical and biological fixation (Rosa et al. 2016), nutrients recycle (de Morais et al. 2015), and evaluation of biomass composition (Moraes et al. 2016; Rosa et al. 2016; Radmann and Costa 2008; de Morais and Costa 2008).

11.6 Factors Influencing CO_2 Fixation from Flue Gas by Microalgae

Several studies have been performed not only to facilitate the use of microalgae in fixing CO_2 from flue gases but also to check the high concentration of CO_2, the presence of other gases and particulate matter (Pires et al. 2012). CO_2 fixation by microalgae biomass and productivity depend on the cultivation conditions applied. In addition to the species of microalgae, the factors that influence CO_2 fixation usually include physicochemical (*e.g.*, CO_2 concentration, pH, toxic pollutants in flue gas, temperature and light) and hydrodynamic parameters (*e.g.*, flow rate, mixing and mass transfer) (Zhao and Su 2014).

11.6.1 Microalgae Strains

Different algae show different responses to high oxygen concentrations and high light intensities. Generally, inhibition of photosynthesis (CO_2 fixation and growth) increases with increasing oxygen concentration and with increasing light intensity (at light intensities greater than saturation) (Moraes et al. 2016). This is very much in favor of using these biological systems for the production of biofuel in various climates. The selection of suitable strains for CO_2 fixation has a significant effect on the competitiveness of costs and effectiveness of the process. In order to achieve high CO_2 fixation rates, desirable features include high cell growth rate, CO_2 utilization efficiency, tolerance to flue gases constituents, skill to produce bioproducts, ease of harvest, tolerance to high temperatures of the liquid medium, and capability of being cultivated with wastewater (Brennan and Owende 2010).

According to Ono and Cuello (2007), the use of species that can tolerate high concentrations of CO_2, contamination of flue gases, and high temperatures can minimize pre-gas treatment costs. Morais and Costa (2007b) confirm that one of the ways to select strains that are more tolerant to adverse conditions offered by the cultures with flue gases is with the isolation of native species in places near the emission sources such as thermal power plants. Morais and Costa (2007b) verify that the strains *Chlorella kessleri* LEB 15 and *Scenedesmus obliquus* LEB 22 isolated from lakes near the UTPM thermal power plant showed cell growth when exposed to CO_2 concentrations (6%, 12% and 18% v/v), indicating the potential of native strains in the fixation of CO_2 emitted in the combustion of coal in power plants. Growing *Chlorella fusca* LEB 111 isolated from wastewater stabilization ponds near the UTPM thermal power plant, Vaz et al. (2016) found that the strain showed tolerance to the flue gas from coal and ashes, besides high CO_2 fixation rate.

11.6.2 CO_2 Concentration in Flue Gas

Several research have focused on CO_2 fixation by microalgae evaluate the tolerance of the microalgae against high concentrations of CO_2. Studies have shown the positive effect of using pure CO_2, real and simulated flue gases as a carbon source (Morais and Costa 2007b). Sung et al. (1999a) have evaluated the effect of CO_2 (10%, 30%, 50%, and 70%), pH, temperature, and light intensity in cultivations of *Chlorella* sp. KR-1. The authors have found that the highest growth rate was found with the injection of 10% of CO_2. The microalgae have also proved tolerant to pH and temperature changes, allowing the authors to conclude that the KR-1 strain has the potential for large-scale cultivation to fix CO_2 from flue gas.

From Costa et al. (2015), the use of gas produced from the combustion of coal promoted a 35% increase in productivity of *Spirulina* sp. LEB 18 biomass and reduction of 24% CO_2 from the flue gas. Another study applying flue gases as carbon source was performed by Chiu et al. (2011). These authors have found that the growth rate of *Chlorella* sp. MTF-7 was approximately 1.5, 2.5, and 2.0 times higher than those achieved by *Chlorella* sp. WT with 2%, 10%, and 25% CO_2, respectively. Extreme CO_2 concentrations about 1000 times higher than the ambient one can be easily metabolized from the unicellular green alga to biomass and can be used, on a local scale at least, for the future development of microalgal photobioreactors for the mitigation of the point source-produced carbon dioxide. Similar conclusion was derived by Prof. Shiraiwa's Laboratory that some microalgae could grow very rapidly at a CO_2 concentration higher than 40%, those cells being referred to as extremely high CO_2 cells.

11.6.3 pH

The effect of pH on growth varies with different strains of microalgae. In general a pH of 7 for freshwater algae and 8 for marine microalgae is the optimum for growth, while at pH below 4 most microalgae cease to grow (Huntley and Redalje 2007). The presence of CO_2, SO_x, NO_x all affect the pH of the media, the most significant effect on pH being associated with high concentrations of SO_x in flue gases. In general CO_2 is the main mechanism which dictates the pH of the cultivation media (Tsai et al. 2012) The pH of the medium is considered an important factor which significantly affects the growth of microalgae (Sankar et al. 2011), solubility, and bioavailability of nutrients (Stumn and Morgan 1981). The transport of substances through the cytoplasmic membrane, activity of intracellular and extracellular enzymes, and electron transport in photosynthesis and osmotic potential of the cytoplasm can be influenced by modifying the pH (Walsby 1982; Poza-Carrión et al. 2001).

During microalgal cultivation, a complex relationship between pH and CO_2 concentration occurs due to chemical equilibrium between the carbon species in the liquid medium ($CO_{2(aq)} \leftrightarrow H_2CO_3 \leftrightarrow HCO_3^- \leftrightarrow CO_3^{2-}$) (Richmond and Grobbelaar 1986). The addition of flue gases in the cultivation of microalgae can reduce the pH of medium, not only due to CO_2 but also due to dissolution of NO_x and SO_x. Moreover, the injection of these gases in the cultivation can cause growth inhibition (Pires et al. 2012). Maeda et al. (1995) have observed a reduction in the pH in the cultivation of *Chlorella* sp. T-1 with the addition of high concentrations of SO_x and NO_x. The authors have found a decrease to levels below pH 3 when were added 15% CO_2, 50–80 ppm of SO_x, and 150–240 ppm NO_x. However, the T-1 strain showed tolerance to reduced pH and observed cell growth when high concentration of CO_2 was added (50%–100%). Tang et al. (2011), in cultivation of *Chlorella pyrenoidosa* SJTU-2, reported a reduction from pH 8.7 to 5.2 with increasing CO_2 concentration (0.03%–50% v/v).

11.6.4 NO_x, SO_x, and Particulate Materials

The direct utilization of the flue gas reduces the cost of pre-treatment, but requires extreme cultivation conditions for microalgae, such as high concentrations of CO_2 (10–15%) and the presence of growth inhibiting compounds, such as NO_x, SO_x (Lee et al. 2002; Chen et al. 2013) particulate materials, among others (Vaz et al. 2016; Duarte et al. 2016). In flue gas, the NO_x emission level varies from several hundred to several thousand ppm with more than 90–95% NO and 5–10% NO_2. After the flue gas de-NO_x process, NO is still at the level of 50–200 ppm. It is difficult for NO to directly impact the growth of microalgae via pH in the cultivation medium. The NO concentration usually has a two-sided influence on the growth of microalgae. An extremely low concentration of NO may even be absorbed by the cultivation medium and transformed into $NO_2{}_-$ as the source of nitrogen nutrition for microalgae when using inorganic forms (Zhao and Su 2014).

Flue gases exhibit different species of NO_x, with nitric oxide (NO) and nitrogen dioxide (NO_2) being the main ones (Skalska et al. 2010). The addition of NO_x in the cultivation of microalgae can have toxic effects (Van Den Hende et al. 2012). However, some research show that NO_x present in low concentrations can be used by some species of microalgae as a source of nitrogen (Nagase et al. 2001). The tolerance of the microalgae to NO_x depends on factors such as cell density, concentration and flow of the gas, the cultivation, agitation, bioreactor, and microalgal species (Zhao and Su 2014). Unlike the NO_x, sulfur oxides, including sulfur dioxide (SO_2) stand out for having high solubility in aqueous medium and are found in higher concentrations in the flue gas (Van Den Hende et al. 2012). The main effect of SO_2 verified in the cultivation of microalgae is the reduction in pH due to the release of H^+ ions by the hydrolysis of the SO_2, which can result in decreased CO_2 fixation efficiency by microalgae (Cheah et al. 2015). Several authors report the toxic effects of NO_x and SO_x in the cultivation of microalgae. Hirata et al. (1996) have proved the toxic effect suffered by the microalgae *Chlorella* when exposed to high NO_x (>150 ppm) and SO_x (>50 ppm) concentrations. Lee et al. (2000) have checked that the alkalization of the medium with calcium carbonate can reduce the toxicity of SO_2 while maintaining the pH above 6 to prevent the formation of sulfites.

Some microalgae species may show tolerance to high levels of the SO_x and NO_x present in flue gases. Duarte et al. (2016) found that the addition of 60 ppm SO_2, 100 ppm NO, and 40 ppm of ash in the cultivation of *Chlorella fusca* LEB 111 promoted results similar to those observed in the control culture, without the addition of industrial waste. Furthermore, the authors observed that the microalgae were able to tolerate SO_2 and NO concentrations of up to 400 ppm. According to

Kao et al. (2012a), the microalgae *Chlorella* sp. MTF-15 when they are cultivated with power plant flue gas (24% CO_2, 25–30 ppm of NO_x (NO and NO_2) and 15–20 ppm SO_2) showed removal efficiency of 80% for NO_x and 93% for SO_2. Particulate materials from combustion processes can vary with emission sources and are sorted in aerosols, fly ash, and soot (Van Den Hende et al. 2012; Merker et al. 2006). The effect of flue gas ashes in the cultivation of *Spirulina* sp. LEB 18 and *Chlorella fusca* LEB 111 was evaluated by Vaz et al. (2016). The authors have checked that addition of 40 ppm of ash from the burning coal of UTPM (Candiota, Brazil) in the culture medium has not caused inhibition of growth of microalgae; likewise, the use of this particulate material may act as a source of trace elements for the metabolism of microalgae. Similarly, Duarte et al. (2016) have observed that the addition of simulated flue gas and ash in the same concentration (40 ppm) in the cultivation of *Chlorella fusca* LEB 111 did not affect cell growth and CO_2 fixation efficiency by microalgae.

Typically, SO_x and NO_x in flue gas can be treated separately by flue gas desulfurization (FGD) and selective catalytic reduction (SCR) processes, respectively, or simultaneously by the combined treatment systems before the treated gas stream enters a microalgae reactor. The FGD commonly uses $CaCO_3$ to absorb SO_2 to form $CaSO_4$. SCR uses ammonia or urea and a catalyst such as titanium oxide, silico-alumino-phosphate, zeolite, Al_2O_3, etc. to decompose NO_x into N_2 and H_2O (Van Den Hende et al. 2012). The examples of combined treatment systems are the DeSoNox or SNOX processes, where a catalytic reduction of NO_x is combined with a catalytic oxidation of SO_2 (Trozzi et al. 2010).

11.6.5 Temperature and Light

The temperature of cultivation is a determinant factor in CO_2 fixation by microalgae and in biomass production that is why photosynthesis is a process composed of a number of physicochemical dependent temperature reactions (Cheah et al. 2015). At low temperatures, the reduction of ribulose bisphosphate carboxylase-oxygenase (Rubisco) enzymatic activity can occur, leading to reduced photosynthetic rate. However, at high temperatures, inhibition of the metabolic rate of microalgae is verified along with the reduction of CO_2 solubility in the liquid medium (Zhao and Su 2014).

Since flue gases from point sources such as power plants have high temperatures (around 120°C), the use of algae tolerant of high temperatures would achieve significant reductions in cooling costs of the gases. Most microalgal species considered for carbon mitigation are mesophilic microalgae (optimum growth in temperatures of 34.5°C) with high tolerance to CO_2 while there has also been some studies carried out in the production of thermophilic cyanobacteria (temperatures of 42–75 °C) to address the problem with cooling costs; however, high productivity cannot be achieved with cyanobacteria (Ono and Cuello 2003). Another problem with cultivation of microorganisms at these high temperatures is loss of water through evaporation. Within the mesophilic range, an increase in temperature will lead to an increase in productivity up to a critical temperature limit at which growth is hindered for the specific strain of microalgae (Sung et al. 1999b).

Flue gases of power plants can reach high temperatures (120 °C), making the use of thermophilic microalgae a form of enabling the CO_2 sequestration from flue gases (Kumar et al. 2011). Besides being used for CO_2 sequestration, microalgae can contribute to reducing cooling costs of the exhaust gases and produce biomass rich in secondary metabolites (Singh and Ahluwalia 2013). Hsueh et al. (2009) in studies with *Thermosynechococcus* sp. TCL-1, at 50 °C and CO_2 injection of 10%, found the specific growth rate of 2.7 d^{-1} and a biomass with a high concentration of carbohydrates (60.6%). Because of the thermophilic characteristic, the TCL-1 strain demonstrated potential for application in cultivation using the flue gases. Light is considered one of the most important factors in the CO_2 fixation process and biomass production. In microalgal cultivation using the CO_2 from flue gases, the light source is generally divided into solar, which is applied both in open and closed systems, and artificial, that is mainly applied in closed cultivation. Generally, the effect of the light in microalgal cultivation is evaluated for intensity and dark/light cycles (Zhao and Su 2014). Ho et al. (2012) found the highest specific growth rates (1.3 d^{-1}) and CO_2 fixation rate (853.7 mg L^{-1} d^{-1}) for *Scenedesmus obliquus* CNW-N when it is exposed to a light intensity of 180 $\mu mol_{photons}$ $m^{-2}s^{-1}$ and 2.5% CO_2. In their studies with four strains of microalgae, Gonçalves et al. (2014) observed that the application of higher light intensity (180 μE m^{-2} s^{-1}) and light cycle (24:0) resulted in higher rates of growth and CO_2 fixation.

da Rosa et al. (2011) have evaluated the growth of microalga *Spirulina* sp. LEB 18 and *Chlorella kessleri* under controlled and uncontrolled light and temperature conditions. The authors have observed that in uncontrolled conditions *Spirulina* microalgae had a maximum cell concentration (X_{max}) (4.95 g L^{-1}) when cultivated with 6% CO_2, while for the *Chlorella* cultivation, the X_{max} values were obtained under concentrations of 12% and 18% CO_2. Among all the light sources available, that from the sun is clearly the most abundant, as its radiation provides the highest energy flow of ca. 5.7 1024 J/year, which is about 10,000 times more than the total energy consumed by human beings every year. At mid-day, the sunlight provides the highest light intensity, at 1100 W/m^2 (Miyake et al. 1999), which exceeds the intensity required for efficient production of microalgae. Most of the commercial cultivation of microalgae is carried out in open pond systems, with solar light energy being directly utilized (Pulz 2001).

11.6.6 Mass Transfer in Bioreactors

There is a complex mass transfer process in a Microalgal-PBR. It involves a process of three-phase mass transfer containing mass transfer between gas (CO_2) liquid (medium), gas (CO_2) solid (microalgae), and liquid (medium) solid (microalgae). In the gas aerating method, mass transfer performance and biochemical reaction rate depend on bubble size, gas hold-up, gas-liquid contact area and CO_2 concentration and gas/liquid ratio, etc. Experiments showed that a PBR could reach to higher CO_2

removal efficiency if the high-performance microalgal species were used under optimized operating conditions, e.g., for microalgae *Chlorella* sp. at CO_2 concentrations of 2%, 5%, 10%, and 15% (v/v), the CO_2 removal efficiency was 58%, 27%, 20%, and 16%, respectively (Chiu et al. 2008). Other microalgal species, *S. obliquus* WUST4 can adapt to the acural flue gas and reached to the CO2 removal efficiency of 67% under 12,000–13,000 lx light intensity, 12% CO2 concentration, and 0.1 vvm aeration rate (Li et al. 2011).

The carbon dioxide mass transfer to the liquid medium and the mixing conditions are critical parameters in the process of CO_2 fixation by microalgae in photobioreactors (Chai and Zhao 2012). The CO_2 mass transfer process for microalgae cultivation consists of three stages: the gas phase (CO_2)—liquid phase (culture medium), gas phase (CO_2)—solid phase (microalgae), and the liquid phase (culture medium)—solid phase (microalgae) (Zhao and Su 2014). Factors such as flow rate, bubbles size, gas retention time in the liquid medium, gas-liquid contact area, carbon dioxide concentration may influence the CO_2 mass transfer into microalgae cultivation (Ryu et al. 2009; Fan et al. 2007, 2008).

The mass transfer coefficient (k_La) is a parameter commonly used to evaluate the photobioreactor performance. The volumetric mass transfer coefficient depends on factors such as stirring rate, diffuser configuration, surfactants/antifoams agents, and temperature (Ugwu et al. 2008). The k_La parameter characterizes the mass transfer capacity of CO_2 in the bioreactor, and thus determines whether a particular reactor will be able to sustain a given cell growth rate. As regards the transfer of CO_2, this parameter is critical in the design stages, in the scale-up, and in the operation of a cultivation system for the production of biomass (Talbot et al. 1991). Moraes et al. (2016) have examined the effect of different diffusers configurations and CO_2 flow rates in vertical tubular photobioreactors on mass transfer and CO_2 fixation by *Spirulina* sp. LEB 18. The authors have found that the use of a smaller flow injection and porous diffusers can promote lower CO_2k_La values for the liquid medium, however, under these conditions higher CO_2 fixation efficiencies are observed for the microalgae.

11.6.7 Bioreactor Application in CO_2 Fixation by Microalgae

Bioreactors can be classified into closed (photobioreactors) or open systems (Pires et al. 2012). The main advantage of open ones is the relative ease of construction and operation compared to most closed ones (Ugwu et al. 2008). However, these systems usually have a disadvantage as to controlled parameters, such as the availability of light, agitation, temperature, pH, and nutrient content, but it does away with lighting costs (Brennan and Owende 2010). When used in CO_2 fixation, these systems can result in losses due to exhaust gas to atmosphere and low gas residence time in the liquid medium due to lower liquid columns offered by these ones (Carvalho et al. 2006).

Biomass yield is higher in photobioreactors as compared to open raceway ponds and this is due to adequate nutritional control mechanisms and monospecific culture growth conditions in photobioreactors (Singh and Gu 2010). In addition, lipid productivity is higher in photobioreactors as compared to open raceway ponds since all the growth parameters are amenable for optimization of higher lipid productivity in photobioreactors. There is higher light utilization efficiency in photobioreactors as compared to open raceway ponds and this is mainly due to a large surface area in the open system (Campbell et al. 2011). Biomass yields of 0.5–1 g/L are accepted as standard for raceway ponds. Photobioreactors are generally limited to 4 g/L for photobioreactors before the shading effect greatly limits further growth (Davis et al. 2011). In order to enable the use of open bioreactors in the CO_2 fixation process, research is being carried out in laboratory and pilot scale. Duarte-Santos et al. (Duarte-Santos et al. 2016) have evaluated the CO_2 molar fraction and gas flow rate for control of pH in microalgae cultivation in raceway equipped with sump in pilot scale. At this study, it was found that the use of low CO_2 molar fraction (2–6%) or gas flow rate (75–100 L min^{-1}) resulted in carbon efficiency in the sump was higher than 95% and 85% of the injected carbon being converted into biomass.

Several studies have reported that photobioreactors (closed systems) have better results when used in CO_2 fixation process by microalgae (Ho et al. 2011). The photobioreactors can have several configurations such as vertical and horizontal tubular, bubble columns, air-lift, flat plates, helical type, stirred tank, hybrid type (Kumar et al. 2011), and membrane photobioreactor (Lam et al. 2012). This bioreactors' configuration has advantages over open systems, such as high photosynthetic efficiency, high CO_2 fixation efficiency by microalgae, high biomass productivity, better parameters control of cultivation, and larger CO_2 mass transfer efficiency (Kumar et al. 2010; Cheah et al. 2015; Ho et al. 2011). Morais and Costa (2007a) have evaluated different bioreactors configurations in fixing CO_2 in the cultivation of four strains of microalgae. The maximum CO_2 fixation was observed in the *Spirulina* sp. LEB 18 cultivation in vertical tubular photobioreactors with volume of 2 L (9.3%) and 4 L (9.2%) with injection of 6% CO_2. Chai and Zhao (2012) used a bubble tank-type stirrer to enhance the CO_2 removal efficiency and improve the photobioreactor mixing conditions in the cultivation of *Chlorococcum* sp. The authors found that the use of this system promoted 94% removal of CO_2 from the gas stream enriched with 10% CO_2.

Radmann et al. (2011; Radmann and Costa 2008) have found the greatest results of maximum daily CO_2 fixation (14.9% and 13.4%) of *Spirulina* sp. LEB 18 and *Chlorella vulgaris* LEB 106, respectively, in the cultivations with vertical tubular photobioreactors in series with simulated coal flue gas injection (12% of CO_2, 100 ppm of NO, and 60 ppm of SO_2). The authors concluded that these microalgae have higher CO_2 fixation capacity in medium containing SO_2 and NO, and can be cultivated in thermal power plants with coal flue gas, contributing to the reduction of global warming. Flat plate photobioreactors are very effective for the biomass cultivation of microalgae. These photobioreactors provide a high surface area to volume ratio for illumination and have a convenient modular design for scale-up (Barbosa et al. 2005). Airlift photobioreactors are simple and cost effective reactors for the mass cultivation of various types of microalgae. Acrylic glass is used as a material to construct airlift photobioreactors because it is inexpensive and available. There are two zones in airlift photobioreactors, including dark (called rinser) and irradiated zones. The airlift photobioreactors are believed to meet the desired criteria for the new generation photobioreactors of high light penetration and biomass production, ease of maintenance, and minimal contamination (Barbosa 2003).

From studies using a photobioreactor with a polytetrafluoroethylene membrane system, Chai et al. (2012) have reported high concentrations of CO_2 dissolved in the liquid medium, resulting in high transfer rates of CO_2 in the cultivation of microalgae *Chlorococcum* sp. At high concentrations of dissolved carbon and longer retention time in this medium, higher fixation rates and lower losses of CO_2 due to exhaustion are observed.

11.6.8 CO_2 Biofixation Metabolism

Biotechnological techniques for CO_2 fixation are conducted to reduce CO_2 emission. These processes are based on the use of reactors to generate photosynthetic reactions in which microalgae are used as biocatalysts in a series of biochemical reactions that are responsible for the conversion of CO_2 into photosynthetic metabolic products (Jacob-Lopes et al. 2010). Microalgal biomass contains approximately 50% carbon by dry weight (Mirón et al. 2003). This carbon is typically obtained from carbon dioxide; roughly, 100 tons of microalgal biomass fixes 183 tons of CO_2 (Bonura et al. 2011).

The biological CO_2 fixation can be attributed to fixation made by the higher plants and the microalgae through photosynthesis. Although they are very similar mechanisms, the fixing by microalgae is considered more efficient, because these microorganisms develop rapidly and CO_2 fixing efficiency is 10–50 times higher than plants (Demirbas 2011; Ho et al. 2013). In microalgae biofixation process, CO_2 can be captured directly from the emission source, while higher plants capture this gas from atmosphere. Furthermore, the organic material produced through photosynthesis can be processed into several products such as biofuels, food additives, and health and beauty products. This represents additional benefits of CO_2 reduction process by microalgae (de Morais and Costa 2008). The CO_2 biofixation mechanism by microalgae is based on the efficiency to perform photosynthesis. In photosynthesis, plants and photosynthetic microorganisms convert solar energy into chemical energy through the following reaction: $6\ CO_2 + 6H_2O + \text{energy solar} \rightarrow C_6H_{12}O_6 + 6O_2$. In this reaction, the light intensity and the quality of radiation carry vital role in cell growth rate (Ravelonandro et al. 2008).

A favorable pathway to capture CO_2 in the form of microalgal biomass is provided by biological processes via photosynthesis. In microalgae, photosynthesis releases oxygen and can be called "oxygenic photosynthesis." CO_2 is converted into lipids and other hydrocarbons in this process, explaining the designation of "CO_2 fixation process." In oxygenic photosynthesis, water is the electron donor, and oxygen is released after hydrolysis. The equation for photosynthesis can be written as follows:

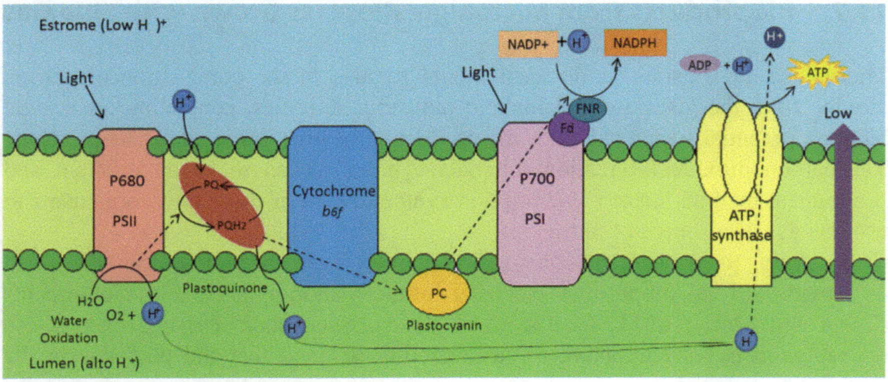

Fig. 11.2 Schematic view of the photosynthetic electron transport system in the thylakoid membrane

$$H_2O + CO_2 + \text{Photons (light)} \rightarrow [CH_2O]n + O_2$$

In this reaction, the standard free energy for the synthesis of glucose is 2870 kJ/mol (Zhao and Su 2014). Photosynthesis occurs in two stages: light phase, which happens only when the cells are illuminated, and carbon fixation reactions, also known as dark phase, which occurs both in presence and absence of light. In the center of the photosynthetic apparatus there are two main photoactive complexes, photosystem I (PSI) and photosystem II (PSII), that absorb sunlight to the electron transport chain through an excited chlorophyll dimer (Fig. 11.2) (Iverson 2006).

Photosynthesis starts in PSII complex and then transfers electrons through the excitation of P680 chlorophyll dimer to the PSI. Thus, PSI transfers the received electrons from PSII via chlorophyll dimer P700, which is oxidized by the antenna complex reducing NADPH (Červený et al. 2009) and ferredoxin, which is subsequently used in CO_2 fixation reactions through the Calvin cycle (dark reaction) (Iverson 2006; Calvin 1989). In Calvin cycle, three molecules of ATP and two molecules of NADPH are consumed for every CO_2molecule converted to carbohydrate according to the reaction:

$$3CO_2 + 9ATP + 6NADPH + \text{water} \rightarrow \text{glyceraldehyde3-fosfato} + 8Pi + 6NADP^+$$

The enzymes D-ribulose 1,5-bisphosphate carboxylase-oxygenase(Rubisco) and phosphorylase kinase are essential to Calvin cycle, they are responsible for carbon dioxide sequestration and catalyze the regeneration of the CO_2 acceptor. In the central reaction of carbon fixation, an inorganic carbon is converted to organic carbon. The available CO_2 combines with ribulose 1,5-bisphosphate and water to yield two molecules of 3-phosphoglycerate compound. The fixation reaction is catalyzed by the stromal chloroplast ribulose bisphosphate carboxylase (Giordano et al. 2005).

11.7 Bioproducts from Microalgal Biomass Grown with Flue Gas

Microalgae are existing as an ideal model for use of the combustion of carbon dioxide from thermal power plants and industrial activities, generating microalgal biomass potentially applicable for the production of high value-added bioproducts. The carbon dioxide biofixation microalgae have proven to be an efficient and economical method, mainly due to photosynthetic skill of these microorganisms to use this gas as a source of nutrient.

According to the chemical composition of microalgae, the biomass can be applied to provide several bioproducts such as biofuels, biopolymers, and biopigments (Costa and Morais 2011; Coelho et al. 2015). Such characteristics make these microorganisms an alternative to sustainability demand that the planet is in the current times.

11.7.1 Biofuels

The present patterns of energy generation and consumption are based on fossil fuels, which causes emissions of local pollutants, greenhouse gases and put at risk the long-term energy supply on the planet. Microalgae are presented as an alternative substrate for the production of biofuels such as biodiesel, bioethanol, biogas, and biohydrogen (Lam and Lee 2013; Cerón-García et al. 2013; Guo et al. 2013; Zamalloa et al. 2012; Rashid et al. 2011). Depending on cultivation conditions, microalgae can produce compounds easily converted into biofuels (Costa and Morais 2011).

The extensive drying of microalgae biomass is required for biofuel production, as the presence of water will inhibit several downstream processes, such as lipid extraction and transesterification. However, the drying step is energy intensive, which adds to the cost complexity of the overall production process. The microalgal slurry moisture content must be reduced to at least 10% by drying and dehydration. There are many types of drying technologies, such as drum drying, oven drying, freeze drying, and spray drying, that have been used with microalgae. The selection of the best drying method depends on the required operation scale and desired product value. Sun drying is an old and cheap drying method that can be performed easily by exposure to solar radiation. However, solar drying is not feasible, taking a long drying time and requiring a large drying area due to limited sunlight and uncertain sunlight time (Taher et al. 2011; Lam and Lee 2012).

Among the different biofuel obtainable derived from microalgal biomass, biodiesel is highlighted due to their chemical characteristics similar to diesel oil (Pandey et al. 2014). For biodiesel production, it is desired high concentrations of saturated fatty acids in microalgal biomass. Depending on the genus and species, microalgae produce different types of lipids (Guschina and Harwood 2006) and its fatty acid profile can be influenced by the carbon dioxide concentration injected in the cultures (de Morais and Costa 2008). After extraction of lipids for biodiesel production, the residual biomass obtained can be used as a substrate to produce bioethanol and biogas (Harun et al. 2011). Ethanol is a biofuel obtained in the biochemical process, by fermentation of rich in carbohydrates biomass and after hydrolysis of starch and cellulose (Cardoso et al. 2011). Several carbohydrates are present in microalgae, such as cellulose, xylose, galactose, arabinose, glucose, and mannose, and they can be used as substrates for bioethanol production (Silva and Bertucco 2016; Möllers et al. 2014). Depending on the cultivation conditions, the microalgae biomass can provide high concentrations of carbon compounds, and these compounds can be used directly in the fermentation or after pre-treatment and used for ethanol production (Pandey et al. 2014). The microalgae cultivation has lower environmental impact, note competing for space with food crops, unlike biofuels produced from other raw materials (Chisti 2007; Harun et al. 2010; Zhu et al. 2014).

Biogas is the product of anaerobic digestion of organic matter mainly consisting of methane and carbon dioxide in the presence of methanogenic bacteria (Harun et al. 2011). Substrates normally used for anaerobic digestion involve domestic and industrial effluents (Kim et al. 2011; Rani et al. 2012), swine and bovine manure (Alvarez and Lide 2009; Amon et al. 2007), and algal biomass (Sialve et al. 2009; Costa et al. 2012). Microalgae present proteins, carbohydrates, and lipids in their composition that can be digested to produce methane, which by burning, can generate carbon free electricity (Harun et al. 2011). The biogas production when compared to other biofuels is operationally simple and eliminates drying and chemical pre-treatment of the raw material. Moreover, the CO_2 produced during the anaerobic process can be consumed as a nutrient in the microalgal cultivation, thus increasing the biogas purity (Kao et al. 2012b) and residual biomass from anaerobic digester can pass through a reprocessing for use as a fertilizer and/or animal feed (Sialve et al. 2009).

Another auspicious biofuel that can be produced by using microalgal cultivation for industrial flue gases is biohydrogen (Costa and Morais 2011). Hydrogen is an alternative fuel compared to conventional fossil; it releases energy in combustion without pollutants (Pandey et al. 2014). Bioprocesses for biological hydrogen production can be water biophotolysis by algae; hydrogen production in the absence of light during the acidogenic phase in anaerobic digestion of organic matter; photo-fermentative hydrogen production (Kapdan and Kargi 2006; Romagnoli et al. 2011). Im et al. (2014) proposed the method for the in situ transesterification of wet *Nannochloropsis oceanica* (65% moisture content) using chloroform, methanol, and sulfuric acid in a one-pot reaction. The addition of chloroform can improve the performance of transesterification with methanol by reducing the diffusion limit of the reactant to the liquid reaction phase. This study provided a yield of greater than 90% without the addition of extra energy inputs for drying microalgae prior to in situ transesterification. Thus, this method can reduce the production cost of biodiesel from microalgae.

Microalgae are highlighted for this proceeding, because it can be cultivated in closed systems, allowing the capture of the hydrogen. The photobiological hydrogen production can be increased according to the concentration of carbon in biomass (Benemann 1997). The biomass obtained from hydrogen production process can be burned for energy generation due to the calorific value of these microorganisms is greater than some coals (Pandey et al. 2014). Microalgae can be used to produce biomethanol as a renewable fuel. *Spirulina* sp. is converted to methanol by gasification (Hirano et al. 1998). The *Spirulina* biomass was partially oxidized at temperatures of up to 1000 °C to produce H_2, CO, CO_2, and hydrocarbon. The maximum yield was 0.64 g of methanol per gram of microalgal biomass at 1000 °C. The carbon dioxide produced by the reaction was removed by absorption using alkanolamine.

11.7.2 Biopigments

The biopigments can be found in several microorganisms, such as microalgae and cyanobacteria. Microalgae are presented as excellent biological resource for obtaining a range of pigments such as chlorophyll, carotenoids, and phycobiliproteins (Benavente-Valdés et al. 2016). These pigments have health beneficial properties such as vitamin precursors, antioxidants, strengtheners of the immune system, and anti-inflammatory agents (Pangestuti and Kim 2011). The pigment content in microalgae is a specific feature of each species. Its evaluation is essential as an indirect measure of cell growth, as well as a parameter to check the trophic level of waters. Components of algae are frequently used in cosmetics as thickening agents, water-binding agents, and antioxidants. Some microalgal species are established in the skin care market, the main ones being *Arthrospira* and *Chlorella* (Stolz and Obermayer 2005). Microalgae extracts can be mainly found in face and skin care products (e.g., anti-aging cream, refreshing or regenerant care products, emollient, and as an anti-irritant in peelers).

Chlorophylls are green pigments, hydrophobic, soluble in lipids found in algae, higher plants, and cyanobacteria (Rasmussen and Morrissey 2007). The green color present in algae is due to the presence of chlorophylls *a*, *b*, c_1, c_2, and *d*, which absorb wavelength corresponding to its absorption spectrum (450–675 nm) and reflect the wavelengths of its color. Its structure is composed of a chain of phytol and a porphyrin ring and appears to be dissolved in the lipid membrane of chloroplasts, while at the center of the porphyrin ring structure there is a magnesium atom carrying the electrons to the photosynthetic reactions occur (Carvalho et al. 2011). Several factors can influence chlorophyll content in cyanobacteria biomass as culture medium composition, cultivation time, and light intensity (Rangel-Yagui et al. 2004). Nitrogen (N) added in the cultivation favoring the accumulation of chlorophylls (*a* and *b*) in microalgal biomass, due to chlorophyll is N rich, and this is easily accessible, while high luminosity can reduce the chlorophyll content in biomass (Ördög et al. 2012; Seyfabadi et al. 2011). The electromagnetic field is another promising technique that can stimulate chlorophyll accumulation in microalgae cultures (Benavente-Valdés et al. 2016). Chlorophyll can be used as a natural colorant and antioxidant in foods and as antimutagenic agent.

Microalgal pigment has marketable uses as a natural food coloring and cosmetic ingredient. Some microalgae contain substantial amounts of carotene (besides beta carotene). Other types of coloring appear in microalgae as well. Beta carotene is used as a food coloring (with a major application in providing the yellow color to margarine), as a food additive to enhance the color of the flesh of fish and the yolk of eggs, and to improve the health and fertility of grain-fed cattle (see survey by Borowitzka and Borowitzka 1987). Carotenoids are widely nature distributed pigments found in plants, algae, and some fungi and bacteria species. Chemically, carotenoid compounds have difference in a few characteristics, which give rise to different structures, the class of hydrocarbons (carotenes) and their oxygenated derivatives (xanthophylls) (Esteban et al. 2015). There are over than 600 types of carotenoids that perform several functions in various organisms, including microalgae. The biopigments present in algae and cyanobacteria are β-carotene, astaxanthin, lutein, canthaxanthin, violaxanthin, neoxanthin, zeaxanthin, and myxoxanthophyll. The green algae are present themselves as a source of lutein, whereas cyanobacteria can be exploited for the production of zeaxanthin and myxoxanthophyll and β-carotene is found in all classes and orders of microalgae (Paliwal et al. 2016).

The phycobiliproteins are water-soluble proteins, photosynthetic light collector complex antenna component, bright color, highly fluorescent, present in cyanobacteria, red algae, and cryptophytes. These proteins are divided into two major groups based on their colors, phycoerythrin (red) and phycocyanin (blue). The phycocyanins include C-phycocyanin, R-phycocyanin, and allophycocyanin (Kuddus et al. 2013), being phycocyanin the main pigment found in microalgae cells, while allophycocyanin is present in less quantity (Moraes et al. 2010). C-phycocyanin is strongly fluorescent, and a blue color, together with the chlorophyll, which has green color, are responsible for the blue-green staining of cyanobacteria (Gantar et al. 2012). It can be used as biopigment in food and cosmetics, as well as present as promising source for the treatment of cancer and support for tissue engineering (Figueira et al. 2016). Among the existing cyanobacteria, *Spirulina* receives emphasis, since the C-phycocyanin may be present up to 20% dry weight of total proteins of the microorganism (Vonshak 2002). Additionally, *Spirulina* is presented as excellent CO_2 fixator, using this gas as a carbon source in cultivation for its growth, making simultaneous production of C-phycocyanin a valuable product with excellent benefits (Chen et al. 2013).

11.7.3 Biopolymers

Biopolymers or organic plastics are a form of plastics derived from renewable biomass sources such as vegetable oil, corn starch, pea starch unlike fossil fuel plastics derived from petroleum. Biopolymers provide the twin advantages of conservation of fossil resources and reduction in CO_2 emissions, which make them an important innovation of sustainable development. Polyhydroxybutyrate (PHB) is a biodegradable plastic with similar properties to conventional plastics, oil based, synthesized by a variety of organisms through intracellular material storage. Polyhydroxybutyrate can be produced by prokaryotic microorganisms

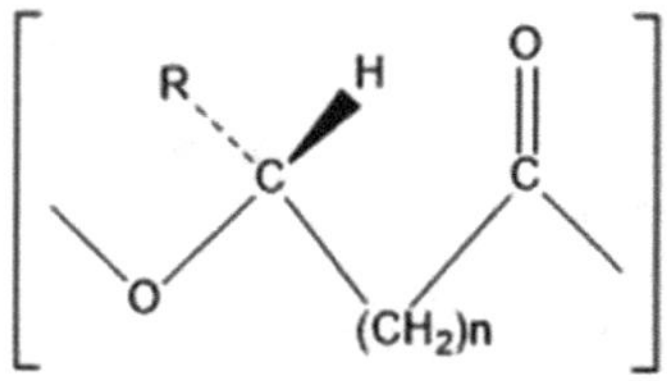

(100 - 30000)

n=1	R=H	Poly(3-hydroxypropienate)	P(3HP)
	R=CH_3	Poly(3-hydroxybutyrate)	P(3HB)
	R=C_2H_5	Poly(3-hydroxyvalerate)	P(3HV)
	R=C_3H_7	Poly(3-hydroxyhexanoate)	P(3HX)
n=2	R=H	Poly(4-hydroxybutyrate)	P(4HB)
n=3	R=H	Poly(5-hydroxyvalerate)	P(5HV)

Fig. 11.3 Structure of polyhydroxyalkanoates

such as *Spirulina*, where it functions as carbon and energy reserve (Richmond and Grobbelaar 1986). PHB belongs to polyhydroxyalkanoates (PHA) group, which are hydroxyalkanoic acid polyesters which can be obtained from renewable carbon sources and organic waste being produced by various bacteria and cyanobacteria as a reserve material as intracellular granules (Laycock et al. 2013; Wang et al. 2013). These microorganisms can be considered an alternative system for biopolymer production, due to its photoautotrophic nature (Balaji et al. 2013). PHBs are a class of polymers formed by monomeric units which can have different numbers of carbon atoms, in which the most studied is the poly(3-hydroxybutyrate) (P(3HB)) and poly(3-hydroxybutyrate-co-3-hydroxyvalerate) (P(3HB-co-3HV)). The monomeric unit depends on the composition or the side chain of the radical R and the value of n (Fig. 11.3).

PHB biosynthesis in microalgae cell is directly related to the physicochemical conditions and nutrition of the microorganism. Cultivation submitted to limiting nutrients as phosphorus and nitrogen, carbon excess as CO_2, stress conditions, temperature, pH, metabolism, and flow control modes of cultivation may increase the amount of PHB in the cells (Mata et al. 2010; Campos et al. 2014; Shrivastav et al. 2010). Carbon sources such as glycerin obtained as by-product of biodiesel production, and carbon dioxide from industries and power plants, can be used as raw material in cell growth. Use of industrial wastes has become increasingly promising to become a sustainable alternative, in addition to closing the carbon cycle (Ashby et al. 2012; Drosg et al. 2015).

de Morais and Costa (2007) observed that the addition of *Spirulina* LEB 18 biomass to nanofibers results in a strong increase in conductivity. Nanofibers with *Spirulina* LEB 18 biomass free of beads were produced with diameters of 107 nm (Douskova et al. 2009). The general finding is that nanofibers composed of *Spirulina* LEB 18 PHB have surprisingly enhanced mechanical properties when compared with nanofibers composed of commercial PHB. These nanofibers had high porosity, and the cells filled the matrix structure, thus enabling the arrival of nutrients and growth factors and removal of metabolic products (data not published) (Fig. 11.2) (Hirata et al. 1996). Mallick et al. (2007) have studied the PHB-HV production in *Nostoc muscorum* using BG-11 medium with propionate addition, achieving 28.2% of biopolymers. Martins et al. (2014) obtained a PHB yield of 44.2% when *Spirulina* sp. LEB 18 was cultivation in Zarrouk medium modified containing 8.4 g L^{-1} of sodium bicarbonate and 0.25 g L^{-1} of sodium nitrate. Bhati e Mallick (2012) cultivated *Nostoc muscorum* Agardh cyanobacteria in different carbon source like glucose, acetate, fructose, propionate, and valerate to (P(3HB-co-3HV)) production, obtaining biopolymer yields of 28% e 26% using 0.4% of acetate and 0.4% of glucose, respectively. However, when adding 0.4% of valerate and 0.4% of acetate, with nitrogen restriction, the biopolymer yield reaches 60%. Kaewbai-ngam et al. (2016) evaluate the polydroxybutyrate accumulation in *Calothrix scytonemicola* TISTR 8095 using carbon dioxide and observe an PHB accumulation of 25.4% (w w^{-1}) de PHB in biomass, presenting PHB productivity equivalent to the CO_2 consumption of 729.2 mg L^{-1}.

In modern studies where nanofibers incorporating phycocyanin were developed, resistance to the thermal degradation of this biopigment increased when compared with the phycocyanin alone. This showed that the nanofibers produced via electrospinning may protect the added bioactive compounds (data not published) (Lee et al. 2002).

11.8 Conclusion

Concerning what was presented, the CO_2 biofixation process by microalgae from flue gas contributes to the reduction of greenhouse gases in atmosphere. The advantages of using these microorganisms for this purpose reduce the production costs of a biomass rich in high-value compounds with the viability of the operation of large industries. Furthermore, microalgae species and cultivation conditions should be used to increase the biofixation process efficiency and with this producing biomass with applicability in the obtention of several bioproducts such as biofuels, biopigments and biopolymers.

References

Alvarez R, Lide G (2009) Low temperature anaerobic digestion of mixtures of llama, cow and sheep manure for improved methane production. Biomass Bioenergy 33(3):527–533

Amon T, Amon B, Kryvoruchko V, Zollitsch W, Mayer K, Gruber L (2007) Biogas production from maize and dairy cattle manure – influence of biomass composition on the methane yield of biomass composition on the methane yield. Agric Ecosyst Environ 118:173–182

Anjos M, Fernandes BD, Vicente AA, Teixeira JA, Dragone G (2013) Optimization of CO_2 bio-mitigation by *Chlorella vulgaris*. Bioresour Technol 139:149–154

Ashby RD, Solaiman DKY, Strahan GD, Zhu C, Tappel RC, Nomura CT (2012) Glycerine and levulinic acid: renewable co-substrates for the fermentative synthesis of short-chain poly (hydroxyalkanoate) biopolymers. Bioresour Technol 118:272–280

Balaji S, Gopi K, Muthuvelan B (2013) A review on production of poly-β-hydroxybutyrates from cyanobacteria for the production of bioplastics. Algal Res 2(2):278–285

Barbosa MJGV (2003) Microalgal photobioreactors: scale-up and optimization. Wageningen University, The Netherlands, p 168. https://edepot.wur.nl/121438

Barbosa MJ, Zijffers JW, Nisworo A, Vaes W, van Schoonhoven J, Wijffels RH (2005) Optimization of biomass, vitamins, and carotenoid yield on light energy in a flat-panel reactor using the A-stat technique. Biotechnol Bioeng 89(2):233–242

Becker EW (1994) *Microalgae: Biotechnology and Microbiology*. Cambridge University Press

Benavente-Valdés JR, Aguilar C, Contreras-Esquivel JC, Méndez-Zavala A, Montañez J (2016) Strategies to enhance the production of photosynthetic pigments and lipids in chlorophycae species. Biotechnol Reports 10:117–125

Benemann JR (1997) Feasibility analysis of photobiological hydrogen production. Int J Hydrogen Energy 22:979–987

Benis K, Ferrão P (2017) Potential mitigation of the environmental impacts of food systems through urban and peri-urban agriculture (UPA) e a life cycle assessment approach. J Clean Prod:784–795

Bhati R, Mallick N (2012) Production and characterization of poly(3-hydroxybutyrate-co-3-hydroxyvalerate) co-polymer by a N_2-fixing cyanobacterium, *Nostoc muscorum* Agardh. J Chem Technol Biotechnol 87(4):505–512

Bonura G, Arena F, Mezzatesta G, Cannilla C, Spadaro L, Frusteri F (2011) Role of the ceria promoter and carrier on the functionality of Cu-based catalysts in the CO_2-to-methanol hydrogenation reaction. Catal Today 171:251–256

Borges JA, Rosa GM, Meza LHR, Henrard AA, Souza M d RAZ, Costa JAV (2013) *Spirulina* sp. LEB-18 culture using effluent from the anaerobic digestion. Brazilian J Chem Eng 30 (2):277–287

Borowitzka MA, Borowitzka LJ (1987) Vitamins and fine chemicals from micro-algae. In: Borowitzka MA, Borowitzka LJ (eds) Micro-algal biotechnology. Cambridge University Press, New York

Brennan L, Owende P (2010) Biofuels from microalgae – a review of technologies for production, processing, and extractions of biofuels and co-products. Renew Sustain Energy Rev 14 (2):557–577

Calvin M (1989) Forty years of photosynthesis and related activities. Photosynth Res 21:3–16

Campbell PK, Beer T, Batten D (2011) Life cycle assessment of biodiesel production from microalgae in ponds. Bioresour Technol 102(1):50–56

Campos MI, Figueiredo TVB, Sousa LS, Druzian JI (2014) The influence of crude glycerin and nitrogen concentrations on the production of PHA by *Cupriavidus necator* using a response surface methodology and its characterizations. Ind Crop Prod 52:338–346

Cardoso AS, Vieira GEG, Marques AK (2011) O uso de microalgas para a obtenção de biocombustíveis. Rev Bras Biociências 9(4):542–549. (in Portuguese)

Carvalho AP, Meireles LA, Malcata FX (2006) Microalgal reactors: a review of enclosed system designs and performances. Biotechnol Prog 22(6):1490–1506

Carvalho AP, Silva SO, Baptista JM, Malcata FX (2011) Light requirements in microalgal photobioreactors: an overview of biophotonic aspects. Appl Microbiol Biotechnol 89:1275–1288

Cerón-García MC, Macías-Sánchez MD, Sánchez-Mirón A, García-Camacho F, Molina-Grima E (2013) A process for biodiesel production involving the heterotrophic fermentation of *Chlorella protothecoides* with glycerol as the carbon source. Appl Energy 103:341–349

Červený J, Šetlík I, Trtílek M, Nedbal L (2009) Photobioreactor for cultivation and real-time, *in-situ* measurement of O_2 and CO_2 exchange rates, growth dynamics, and of chlorophyll fluorescence emission of photoautotrophic microorganisms. Eng Life Sci 9(3):247–253

Chai X, Zhao X (2012) Enhanced removal of carbon dioxide and alleviation of dissolved oxygen accumulation in photobioreactor with bubble tank. Bioresour Technol 116:360–365

Chai X, Zhao Z, Baoying W (2012) Biofixation of carbon dioxide by *Chlorococcum* sp. in a photobioreactor with polytetrafluoroethene membrane sparger. Afr J Biotechnol 11 (29):7445–7453

Cheah WY, Show PL, Chang J-S, Ling TC, Juan JC (2015) Biosequestration of atmospheric CO_2 and flue gas-containing CO_2 by microalgae. Bioresour Technol 184:190–201

Chen C, Kao P, Tsai C, Lee D, Chang J (2013) Engineering strategies for simultaneous enhancement of C-phycocyanin production and CO_2 fixation with *Spirulina platensis*. Bioresour Technol 145:307–312

Chisti Y (2007) Biodiesel from microalgae. Biotechnol Adv 25:294–306

Chiu SY, Kao CY, Chen CH, Kuan TC, Ong SC, Lin CS (2008) Reduction of CO_2 by a high-density culture of Chlorella sp. in a semicontinuous photobioreactor. Bioresour Technol 99 (9):3389–3396

Chiu SY, Kao CY, Huang TT, Lin CJ, Ong SC, Chen CD, Chang JS, Lin CS (2011) Microalgal biomass production and on-site bioremediation of carbon dioxide, nitrogen oxide and sulfur dioxide from flue gas using *Chlorella* sp. cultures. Bioresour Technol 102(19):9135–9142

Coelho VC, Silva C d SK, Terra AL, Costa JAV, de Morais MG (2015) Polyhydroxybutyrate production by *Spirulina* sp. LEB 18 grown under different nutrient concentrations. Afr J Microbiol Res 9(24):1586–1594

Costa JAV, Morais MG (2011) The role of biochemical engineering in the production of biofuels from microalgae. Bioresour Technol 102(1):2–9

Costa JC, Gonçalves PR, Nobre A, Alves MM (2012) Biomethanation potential of macroalgae *Ulva* spp. and *Gracilaria* spp. and in co-digestion with waste activated sludge. Bioresour Technol 114:320–326

Costa JAV, Morais MG, Radmann EM, Santana FB, Camerini FV, Souza M d RAZ, Henrard AA, da Rosa APC, Brusch LRF (2015) Biofixation of carbon dioxide from coal station flue gas using *Spirulina* sp. LEB 18 and *Scenedesmus obliquus* LEB 22. Afr J Microbiol Res 9(44):2202–2208

Damartzis T, Zabaniotou A (2011) Thermochemical conversion of biomass to second generation biofuels through integrated process design – a review. Renew Sustain Energy Rev 15:366–378

Davis R, Aden A, Pienkos PT (2011) Techno-economic analysis of autotrophic microalgae for fuel production. Appl Energy 88(10):3524–3531

Dawson TP, Jackson ST, House JI, Prentice IC, Mace GM (2011) Beyond predictions: biodiversity conservation in a changing climate. Science 332:53–58

Demirbas A (2009a) Production of biodiesel from algae oils. Energy Source 31:163–168

Demirbas MF (2009b) Biorefineries for biofuel upgrading: a critical review. Appl Energy 86:151–161

Demirbas A (2009c) Biofuels securing the planet's future energy needs. Energ Conver Manage 50:2239–2249

Demirbas A (2011) Biodiesel from oilgae, biofixation of carbon dioxide by microalgae: a solution to pollution problems. Appl Energy 88(2):3541–3547

Douskova I, Doucha J, Livansky K, Machat J, Novak P, Umysova D, Zachleder V, Vitova M (2009) Simultaneous flue gas bioremediation and reduction of microalgal biomass production costs. Appl Microbiol Biotechnol 82:179–185

Drosg B, Fritz I, Gattermayr F, Silvestrini L (2015) Photo-autotrophic production of poly (hydroxyalkanoates) in cyanobacteria. Chem Biochem Eng Q 29(2):145–156

Duarte JH, Fanka LS, Costa JAV (2016) Utilization of simulated flue gas containing CO_2, SO_2, NO and ash for *Chlorella fusca* cultivation. Bioresour Technol 214:159–165

Duarte-Santos T, Mendoza-Martín JL, Fernández FGA, Molina E, Costa JAV (2016) Optimization of carbon dioxide supply in raceway reactors: influence of carbon dioxide molar fraction and gas flow rate. Bioresour Technol 212:72–81

Esteban R, Moran JF, Becerril JM, García-Plazaola JI (2015) Versatility of carotenoids: an integrated view on diversity, evolution, functional roles and environmental interactions. Environ Exp Bot 119:63–75

Fan L, Zhang Y, Cheng L, Zhang L, Tang D, Chen H (2007) Optimization of carbon dioxide fixation by *Chlorella vulgaris* cultivated in a membrane-photobioreactor. Chem Eng Technol 30 (8):1094–1099

Fan LH, Zhang YT, Zhang L, Chen HL (2008) Evaluation of a membrane-sparged helical tubular photobioreactor for carbon dioxide biofixation by *Chlorella vulgaris*. J Membr Sci 325 (1):336–345

Figueira FS, Gettens JG, Costa JAV, de Morais MG, Moraes CC, Kalil SJ (2016) Production of nanofibers containing the bioactive compound C-phycocyanin. J Nanosci Nanotechnol 16:944–949

Fu X, Li D, Chen J, Zhang Y, Huang W, Zhu Y, Yang J, Zhang C (2013) A microalgae residue based carbon solid catalyst for biodiesel production. Bioresour Technol 146:767–770

Gantar M, Simović D, Djilas S, Gonzalez WW, Miksovska J (2012) Isolation, characterization and antioxidative activity of C-phycocyanin from *Limnothrix* sp. strain 37-2-1. J Biotechnol 159:21–26

Giordano M, Beardall J, Raven JA (2005) CO_2 concentrating mechanisms in algae: mechanisms, environmental modulation, and evolution. Annu Rev Plant Biol 56:99–131

Gonçalves AL, Simões M, Pires JCM (2014) The effect of light supply on microalgal growth, CO_2 uptake and nutrient removal from wastewater. Energ Conver Manage 85:530–536

Guo H, Daroch M, Liu L, Qiu G, Geng S, Wang G (2013) Biochemical features and bioethanol production of microalgae from coastal waters of Pearl River Delta. Bioresour Technol 127:422–428

Guschina IA, Harwood JL (2006) Lipids and lipid metabolism in eukaryotic algae. Prog Lipid Res 45:160–186

Harun R, Danquah K, Forde GM (2010) Microalgal biomass as a fermentation feedstock for bioethanol production. J Chem Technol Biotechnol 85:199–203

Harun R, Davidson M, Doyle M, Gopiraj R, Danquah M, Forde G (2011) Technoeconomic analysis of an integrated microalgae photobioreactor, biodiesel and biogas production facility. Biomass Bioenergy 35:741–747

Hirano A, Hon-Nami K, Kunito S, Hada M, Ogushi Y (1998) Temperature effect on continuous gasification of microalgal biomass: theoretical yield of methanol production and its energy balance. Catal Today 45:399–404

Hirata S, Taya M, Tone S (1996) Characterization of chlorella cell cultures in batch and continuous operations under a photoautotrophic condition. J Chem Eng Japan 29(6):953–959. https://doi.org/10.1252/jcej.29.953

Ho S, Chen C, Lee D, Chang J (2011) Perspectives on microalgal CO_2 -emission mitigation systems – a review. Biotechnol Adv 29(2):189–198

Ho SH, Chen CY, Chang JS (2012) Effect of light intensity and nitrogen starvation on CO_2 fixation and lipid/carbohydrate production of an indigenous microalga *Scenedesmus obliquus* CNW-N. Bioresour Technol 113:244–252

Ho S-H, Kondo A, Hasunuma T, Chang J-S (2013) Engineering strategies for improving the CO_2 fixation and carbohydrate productivity of *Scenedesmus obliquus* CNW-N used for bioethanol fermentation. Bioresour Technol 143:163–171

Houghton J (2005) Global warming. Rep Prog Phys 68:1343–1403

Hsueh HT, Li WJ, Chen HH, Chu H (2009) Carbon bio-fixation by photosynthesis of *Thermosynechococcus* sp. CL-1 and *Nannochloropsis oculta*. J Photochem Photobiol B Biol 95(1):33–39

Huang CH, Tan CS (2014) A review: CO_2 utilization. Aerosol Air Qual Res 14:480–499

Huntley ME, Redalje DG (2007) CO_2 mitigation and renewable oil from photosynthetic microbes: a new appraisal. Mitig Adapt Strat Glob Chang 12(4):573–608

I. P. O. C. CHANGE (2007) Report IPCC/ONU – new climate scenarios – initiative ecolatina 1 (in Portuguese)

I. P. O. C. CHANGE (2014) SPM: summary for policymakers contents. Mitig Clim Chang Summ Policymakers

Im H, Lee H, Park MS, Yang JW, Lee JW (2014) Concurrent extraction and reaction for the production of biodiesel from wet microalgae. Bioresour Technol 152:534–537

Iverson TM (2006) Evolution and unique bioenergetic mechanisms in oxygenic photosynthesis. Curr Opin Chem Biol 10:91–100

Jacob-Lopes E, Scoparo CHG, Queiroz MI, Franco TT (2010) Biotransformations of carbon dioxide in photobioreactors. Energ Conver Manage 51(5):894–900

Jensen G, Hi K-K, Eric HR, Pnnceton N (1997) Integrated microalgae production and electricity cogeneration. U.S. Patent 5,659,977, August 26

Kadam KL (1997) Power plant flue gas as a source of CO_2 for microalgae cultivation: economic impact of different process options. Energ Conver Manage 38:505–510

Kaewbai-ngam A, Incharoensakdi A, Monshupanee T (2016) Increased accumulation of polyhydroxybutyrate in divergent cyanobacteria under nutrient-deprived photoautotrophy: an efficient conversion of solar energy and carbon dioxide to polyhydroxybutyrate by *Calothrix scytonemicola* TISTR 8095. Bioresour Technol 212:342–347

Kao CY, Chiu SY, Huang TT, Dai L, Hsu LK, Lin CS (2012a) Ability of a mutant strain of the microalga *Chlorella* sp. to capture carbon dioxide for biogas upgrading. Appl Energy 93:176–183

Kao C, Chiu S, Huang T, Dai L, Wang G, Tseng C, Chen C, Lin C (2012b) A mutant strain of microalga *Chlorella* sp. for the carbon dioxide capture from biogas. Biomass Bioenergy 36:132–140

Kapdan IK, Kargi F (2006) Bio-hydrogen Production from Waste Materials. Enzyme Microb Technol 38:569–582

Kharaka YK, Thordsen JJ, Hovorka SD, Nance HS, Cole DR, Phelps TJ, Knauss KG (2009) Potential environmental issues of CO_2 storage in deep saline aquifers: geochemical results from the Frio-I Brine Pilot test, Texas, USA. Appl Geochem 24:1106–1112

Kim H, Nam J, Shin H (2011) A comparison study on the high-rate co-digestion of sewage sludge and food waste using a temperature-phased anaerobic sequencing batch reactor system. Bioresour Technol 102(15):7272–7279

Kondili EM, Kaldellis JK (2007) Biofuel implementation in East Europe: current status and future prospects. Renew Sustain Energy Rev 11:2137–2151

Kuddus M, Singh P, Thomas G, Al-hazimi A (2013) Recent developments in production and biotechnological applications of C-phycocyanin. Biomed Res Int

Kumar A, Ergas S, Yuan X, Sahu A, Zhang Q, Dewulf J, Malcata FX, van Langenhove H (2010) Enhanced CO_2 fixation and biofuel production via microalgae: recent developments and future directions. Trends Biotechnol 28(7):371–380

Kumar K, Nag C, Nayak B, Lindblad P, Das D (2011) Development of suitable photobioreactors for CO_2 sequestration addressing global warming using green algae and cyanobacteria. Bioresour Technol 102(8):4945–4953

Lam MK, Lee KT (2012) Microalgae biofuels: a critical review of issues, problems and the way forward. Biotechnol Adv 30:673–690

Lam MK, Lee KT (2013) Effect of carbon source towards the growth of *Chlorella vulgaris* for CO_2 bio-mitigation and biodiesel production. Int J Greenh Gas Control 14:169–176

Lam MK, Lee KT, Mohamed AR (2012) Current status and challenges on microalgae-based carbon capture. Int J Greenh Gas Control 10:456–469

Laycock B, Halley P, Pratt S, Werker A, Lant P (2013) The chemomechanical properties of microbial polyhydroxyalkanoates. Prog Polym Sci 38(3–4):536–583

Lee J-S, Lee J-P (2003) Review of advances in biological CO_2 mitigation technology. Biotechnol Bioprocess Eng 8:354–359

Lee JN, Lee JS, Shin CS, Park SC, Kim SW (2000) Methods to enhance tolerances of *Chlorella* KR-1 to toxic compounds in flue gas. Appl Biochem Biotechnol 84–86(x):329–342

Lee J, Kim D, Lee J, Park S, Koh J, Cho H, Kim S (2002) Effects of SO_2 and NO on growth of *Chlorella* sp. KR-1. Bioresour Technol 82:2–5

Li FF, Yang ZH, Zeng R, Yang G, Chang X, Yan JB, Hou YL (2011) Microalgae capture of CO_2 from actual flue gas discharged from a combustion chamber. Ind Eng Chem Res 50 (10):6496–6502

Maeda K, Owada M, Kimura N, Omata L, Karube I (1995) CO_2 fixation from the flue gas on coalfired thermal power plant by microalgae. Energ Conver Manage 36(6–9):717–720. https://doi.org/10.1016/0196-8904(95)00105-M

Mallick N, Sharma L, Singh AK (2007) Poly-β-hydroxybutyrate accumulation in *Nostoc muscorum*: effects of metabolic inhibitors. J Plant Physiol 164:312–317

Martins RG, Gonçalves IS, de Morais MG, Costa JAV (2014) Bioprocess engineering aspects of biopolymer production by the cyanobacterium *Spirulina* strain LEB 18. Int J Polym Sci 2014:1–7

Mata TM, Martins AA, Caetano NS (2010) Microalgae for biodiesel production and other applications: a review. Renew Sustain Energy Rev 14(1):217–232

Meinshausen M, Meinshausen N, Hare W, Raper SCB, Frieler K, Knutti R, Frame DJ, Allen MR (2009) Greenhouse-gas emission targets for limiting global warming to 2 °C. Nature 458 (30):1158–1162

Meldon JH, Morales-cabrera MA (2011) Analysis of carbon dioxide absorption in and stripping from aqueous monoethanolamine. Chem Eng J 171:753–759

Merker G, Schwarz C, Stiesch G, Otto F (2006) *Simulating combustion*, 1st edn. Springer-Verlag, Berlin Heidelberg

Milano J, Chyuan H, Masjuki HH, Chong WT, Kee M (2016) Microalgae biofuels as an alternative to fossil fuel for power generation. Renew Sustain Energy Rev 58:180–197

Minowa T, Yokoyama S, Kishimoto M, Okakurat T (1995) Oil production from algal cells of *Dunaliella tertiolecta* by direct thermochemical liquefaction. Butterworth Heinemann 74 (12):1735–1738

Mirón AS, Garcia MC, Gómez AC, Camacho FG, Grima EM, Chisti Y (2003) Shear stress tolerance and biochemical characterization of Phaeodactylum tricornutum in quasi steady-state continuous culture in outdoor photobioreactors. Biochem Eng J 16:287–297

Miyake J, Wakayama T, Schnackenberg J, Arai T, Asada Y (1999) Simulation of the daily sunlight illumination pattern for bacterial photo-hydrogen production. J Biosci Bioeng 88(6):659–663

Molina Grima E, Acién Fernández F, Robles Medina A, Chisti Y (2003) Recovery of microalgal biomass and metabolites: process options and economics. Biotechnol Adv 20:491–515

Möllers KB, Cannella D, Jørgensen H, Frigaard N-U (2014) Cyanobacterial biomass as carbohydrate and nutrient feedstock for bioethanol production by yeast fermentation. Biotechnol Biofuels 64(7)

Moraes CC, Burkert JFDM, Kalil SJ (2010) C-phycocyanin extraction process for large-scale use. J Food Biochem 34:133–148

Moraes L, Rosa GM, Cardias BB, Oliveira L, Costa JAV (2016) Microalgal biotechnology for greenhouse gas control: carbon dioxide fixation by *Spirulina* sp. at different diffusers. Ecol Eng 91:426–431

Morais EG (2015) Cultivo de microalgas utilizando efluentes da indústria bioenergética e avaliação da conversão em biodiesel. Msc dissertation (in Portuguese), Federal University of Rio Grande, Brazil

de Morais MG, Costa JAV (2007) Carbon dioxide fixation by *Chlorella kessleri*, *C. vulgaris*, *Scenedesmus obliquus* and *Spirulina* sp. cultivated in flasks and vertical tubular photobioreactors. Biotechnol Lett 29(9):1349–1352

Morais MG, Costa JAV (2007a) Biofixation of carbon dioxide by *Spirulina* sp. and *Scenedesmus obliquus* cultivated in a three-stage serial tubular photobioreactor. J Biotechnol 129(3):439–445

Morais MG, Costa JAV (2007b) Isolation and selection of microalgae from coal fired thermoelectric power plant for biofixation of carbon dioxide. Energ Conver Manage 48(7):2169–2173

de Morais MG, Costa JAV (2008) Fatty acids profile of microalgae cultivated with carbon dioxide. Sci Agrotechnol 32(4):1245–1251 de Morais MG, Silva CK, Henrard AA, Costa JAV (2015) Carbon dioxide mitigation by microalga in a vertical tubular reactor with recycling of the culture medium. Afr J Microbiol Res 9 (33):1935–1940

Moreira D, Pires JCM (2016) Atmospheric CO_2 capture by algae: negative carbon dioxide emission path. Bioresour Technol 215:371–379

Mota CJA, Monteiro RS (2013) Quim. Nova. Química e Sustentabilidade Novas Front em Biocombustíveis 36(10):1483–1490

Nagase H, Yoshihara K, Eguchi K, Okamoto Y, Murasaki S, Yamashita R, Hirata K, Miyamoto K (2001) Uptake pathway and continuous removal of nitric oxide from flue gas using microalgae. Biochem Eng J 7:241–246

Nigam PS, Singh A (2011) Production of liquid biofuels from renewable resources. Prog Energy Combust Sci 37:52–68

Ono E, Cuello JL (2003) Selection of optimal microalgae species for CO_2 sequestration. Proceedings 2nd Annual Conference on Carbon Sequestration, Alexandria, pp 1–7

Ono E, Cuello JL (2007) Carbon dioxide mitigation using thermophilic cyanobacteria. Biosyst Eng 96(1):129–134

Ördög V, Stirk WA, Bálint P, Van Staden J, Lovász C (2012) Changes in lipid, protein and pigment concentrations in nitrogen-stressed *Chlorella minutissima* cultures. J Appl Phycol 24:907–914

Paliwal C, Ghosh T, George B, Pancha I, Maurya R, Chokshi K, Ghosh A, Mishra S (2016) Microalgal carotenoids: potential nutraceutical compounds with chemotaxonomic importance. Algal Res 15:24–31

Pandey A, Lee D-J, Chisti Y, Soccol CR (2014) *Biofuels from Algae*. Elsevier

Pangestuti R, Kim S (2011) Biological activities and health benefit effects of natural pigments derived from marine algae. J Funct Foods 3(4):255–266

Peng Y, Zhao B, Li L (2012) Advance in post-combustion CO_2 capture with alkaline solution: a brief review. Energy Procedia 14:1515–1522

Pires JCM, Martins FG, Simões M (2012) Carbon dioxide capture from flue gases using microalgae: engineering aspects and biorefinery concept. Renew Sustain Energy Rev 16 (5):3043–3053

Poza-Carrión C, Fernández-Valiente E, Piñas FF, Leganés F (2001) Acclimation of photosynthetic pigments and photosynthesis of the cyanobacterium *Nostoc* sp. strain UAM206 to combined fluctuations of irradiance, pH, and inorganic carbon availability. J Plant Physiol 158:1455–1461

Pulz O (2001) Photobioreactors: production systems for phototrophic microorganisms. Appl Microbiol Biotechnol 57(3):287–293

Radmann EM, Costa JAV (2008) Lipid content and fatty acids composition variation of microalgae exposed to CO_2, SO_2 and NO. Quim Nova 31(7):1609–1612

Radmann EM, Camerini FV, Santos TD, Costa JAV (2011) Isolation and application of SO_x and NO_x resistant microalgae in biofixation of CO_2 from thermoelectricity plants. Energ Conver Manage 52:3132–3136

Rangel-Yagui CDO, Danesi EDG, de Carvalho JCM, Sato S (2004) Chlorophyll production from *Spirulina platensis*: cultivation with urea addition by fed-batch process. Bioresour Technol 92:133–141

Rani RU, Kumar SA, Kaliappan S, Yeom I, Banu JR (2012) Low temperature thermo-chemical pretreatment of dairy waste activated sludge for anaerobic digestion process. Bioresour Technol 103(1):415–424

Rashid N, Lee K, Mahmood Q (2011) Bio-hydrogen production by *Chlorella vulgaris* under diverse photoperiods. Bioresour Technol 102:2101–2104

Rasmussen RS, Morrissey MT (2007) Marine biotechnology for production of food ingredients. Adv Food Nutr Res 52

Ravelonandro PH, Ratianarivo DH, Joannis-Cassan C, Isambert A, Raherimandimby M (2008) Influence of light quality and intensity in the cultivation of *Spirulina platensis* from Toliara (Madagascar) in a closed system. J Chem Technol Biotechnol 83(2):842–848

Razzak SA, Hossain MM, Lucky RA, Bassi AS (2013) Integrated CO_2 capture, wastewater treatment and biofuel production by microalgae culturing – a review. Renewable Sustainable Energy Rev 27:622–653

Richmond A (2004) Handbook of microalgae culture: biotechnology and applied phycology. 296 (2)

Richmond A, Grobbelaar JU (1986) Factors affecting the output rate of *Spirulina platensis* with reference to mass cultivation. Biomass 10(4):253–264

Rockström J, Steffen W, Noone K, Persson Å, Chapin FS, Lambin EF, Lenton TM et al (2009) A safe operating space for humanity. Nature 461(24):472–475

Romagnoli F, Blumberga D, Pilicka I (2011) Life cycle assessment of biohydrogen production in photosynthetic processes. Int J Hydrogen Energy 36:7866–7871

da Rosa APC, Carvalho LF, Goldbeck L, Costa JAV (2011) Carbon dioxide fixation by microalgae cultivated in open bioreactors. Energ Conver Manage 52(8–9):3071–3073

Rosa GM, Moraes L, Cardias BB, Souza M d RAZ, Costa JAV (2015) Chemical absorption and CO_2 biofixation via the cultivation of *Spirulina* in semicontinuous mode with nutrient recycle. Bioresour Technol 192:321–327

Rosa GM, Moraes L, Souza M d RAZ, Costa JAV (2016) *Spirulina* cultivation with a CO_2 absorbent: influence on growth parameters and macromolecule production. Bioresour Technol 200:528–534

Rufford TE, Smart S, Watson GCY, Graham BF, Boxall J, Diniz JC, May EF (2012) The removal of CO_2 and N_2 from natural gas: a review of conventional and emerging process technologies. J Petrol Sci Eng 94–95:123–154

Ryu HJ, Oh KK, Kim YS (2009) Optimization of the influential factors for the improvement of CO_2 utilization efficiency and CO_2 mass transfer rate. J Ind Eng Chem 15:471–475

Sankar V, Daniel DK, Krastanov A (2011) Carbon dioxide fixation by *Chlorella minutissima* batch cultures in a stirred tank bioreactor. Biotechnol Biotechnol Equip 25(3):2468–2476

Schenk PM, Thomas-hall SR, Stephens E, Marx UC, Mussgnug JH, Posten C, Kruse O, Hankamer B (2008) Second generation biofuels: high-efficiency microalgae for biodiesel production. Bioenergy Res 1:20–43

Seyfabadi J, Ramezanpour Z, Khoeyi ZA (2011) Protein, fatty acid, and pigment content of *Chlorella vulgaris* under different light regimes. J Appl Phycol 23:721–726

Shrivastav A, Mishra SK, Mishra S (2010) International Journal of Biological Macromolecules Polyhydroxyalkanoate (PHA) synthesis by *Spirulina subsalsa* from Gujarat coast of India. Int J Biol Macromol 46:255–260

Sialve B, Bernet N, Bernard O (2009) Anaerobic digestion of microalgae as a necessary step to make microalgal biodiesel sustainable. Biotechnol Adv 27(4):409–416

da Silva CK (2016) Produção biotecnológica de biopolímeros aplicando processo cíclico de reutilização de resíduo da extração de poli-hidroxibutirato (PHB). Msc dissertation (in Portuguese), Federal University of Rio Grande, Brazil

C. E. de F. Silva and A. Bertucco, "Bioethanol from microalgae and cyanobacteria: a review and technological outlook technological outlook," Process Biochem., 51(11) 2016.

Singh UB, Ahluwalia AS (2013) Microalgae: a promising tool for carbon sequestration. Mitig Adapt Strat Glob Chang 18(1):73–95

Singh J, Gu S (2010) Commercialization potential of microalgae for biofuels production. Renew Sustain Energy Rev 14(9):2596–2610

Skalska K, Miller JS, Ledakowicz S (2010) Trends in NO_x abatement: a review. Sci Total Environ 408(19):3976–3989

Stolz P, Obermayer B (2005) Manufacturing microalgae for skin care. Cosmetics Toiletries 120:99–106

Stumn W, Morgan J (1981) *Aquatic chemistry*, 2nd edn. Wiley, New York

Suali E, Sarbatly R (2012) Conversion of microalgae to biofuel. Renew Sustain Energy Rev 16:4316–4342

Sung K, Lee J, Shin C, Park S, Choi M-J (1999a) CO_2 fixation by *Chlorella* sp. KR-1 and its cultural characteristics. Bioresour Technol 68:269–273

Sung KD, Lee JS, Shin CS, Park SC (1999b) Isolation of a new highly CO_2 tolerant fresh water Microalga Chlorella sp. KR-1. Renew Energy 16(1):1019–1022

Taher H, Al-Zuhair S, Al-Marzouqui AH, Haik Y, Farid MM (2011) A review of enzymatic transesterification of microalgal oil-based biodiesel using supercritical technology. Enzyme Res 46:82–92. https://doi.org/10.4061/2011/468292

Talbot P, Gortares MP, Lencki RW, de la Noiie J (1991) Absorption of CO_2 in algal mass culture systems: a different characterization approach. Biotechnol Bioeng 37:834–842

Tang D, Han W, Li P, Miao X, Zhong J (2011) CO_2 biofixation and fatty acid composition of *Scenedesmus obliquus* and *Chlorella pyrenoidosa* in response to different CO_2 levels. Bioresour Technol 102(3):3071–3076

Trozzi C, Rentz O, Oertel D, Woodfield M, Stewart R (2010) Energy industries. Combustion in energy and transformation industries. Air pollutant emission inventory guidebook. European Environment Agency

Tsai DDW, Ramaraj R, Chen PH (2012) Growth condition study of algae function in ecosystem for CO_2 bio-fixation. J Photochem Photobio B: Bio 107:27–34

Tsai DD-W, Ramaraj R, Chen PH (2016) Carbon dioxide bio-fixation by algae of high rate pond on natural water medium. Ecol Eng 92:106–110

Ugwu CU, Aoyagi H, Uchiyama H (2008) Photobioreactors for mass cultivation of algae. Bioresour Technol 99:4021–4028

US DOE (2010) National algal biofuels technology roadmap. May, U.S. Department of Energy, Office of Energy Efficiency and Renewable Energy

Van Den Hende S, Vervaeren H, Boon N (2012) Flue gas compounds and microalgae: (Bio-) chemical interactions leading to biotechnological opportunities. Biotechnol Adv 30:1405–1424

Vaz BS, Costa JAV, de Morais MG (2016) CO_2 biofixation by the cyanobacterium *Spirulina* sp. LEB 18 and the green alga *Chlorella fusca* LEB 111 grown using gas effluents and solid residues of thermoelectric origin. Appl Biochem Biotechnol 178:418–429

Vonshak A (2002) *Spirulina Platensis Arthrospira: physiology, cell-biology and biotechnology*. CRC Press

Walsby A (1982) Cell-water and cell-solute relations. In: Carr N, Whitton B (eds) *The biology of cyanobacteria*. Blackwell Science Publications, Oxford, pp 237–262

Wang B, Li Y, Wu N, Lan CQ (Jul. 2008) CO_2 bio-mitigation using microalgae. Appl Microbiol Biotechnol 79(5):707–718

Wang B, Pugh S, Nielsen DR, Zhang W, Meldrum DR (2013) Engineering cyanobacteria for photosynthetic production of 3-hydroxybutyrate directly from CO_2. Metab Eng 16:68–77

Weiss H (2008) Method for growing photosynthetic organisms – United States patent application US20080220486 A1. Google Patents, September 11

Wuang SC, Khin MC, Chua PQD, Luo YD (2016) Use of *Spirulina* biomass produced from treatment of aquaculture wastewater as agricultural fertilizers. Algal Res 15:59–64

Yun Y-S, Moon Park J (1997) Development of gas recycling photobioreactor system for microalgal carbon dioxide fixation. J Korean J Chem Eng 14(4):297–300. https://doi.org/10.1007/BF02706827

Zamalloa C, Boon N, Verstraete W (2012) Anaerobic digestibility of *Scenedesmus obliquus* and *Phaeodactylum tricornutum* under mesophilic and thermophilic conditions. Appl Energy 92:733–738

Zhao B, Su Y (2014) Process effect of microalgal-carbon dioxide fixation and biomass production: a review. Renew Sustain Energy Rev 31:121–132

Zheng Y, Chen Z, Lu H, Zhang W (2011) Optimization of carbon dioxide fixation and starch accumulation by *Tetraselmis subcordiformis* in a rectangular airlift photobioreactor. Afr J Biotechnol 10(10):1888–1901

Zhu LD, Hiltunen E, Antila E, Zhong JJ, Yuan ZH, Wang ZM (2014) Microalgal biofuels: flexible bioenergies for sustainable development. Renew Sustain Energy Rev 30:1035–1046

Kolibri 360 GmbH
Ferdinand-Jühlke-Straße 7
99095 Erfurt, Deutschland
produktsicherheit@kolibri360.de